Selected Titles in This Series

661 **J. P. C. Greenlees,** Rational S^1-equivariant stable homotopy theory, 1999

660 **Dale E. Alspach,** Tensor products and independent sums of $\mathcal{L}_p$-spaces $1 < p < \infty$, 1999

659 **R. D. Nussbaum and S. M. Verduyn Lunel,** Generalizations of the Perron-Frobenius theorem for nonlinear maps, 1999

658 **Hasna Riahi,** Study of the critical points at infinity arising from the failure of the Palais-Smale condition for n-body type problems, 1999

657 **Richard F. Bass and Krzysztof Burdzy,** Cutting Brownian paths, 1999

656 **W. G. Bade, H. G. Dales, and Z. A. Lykova,** Algebraic and strong splittings of extensions of Banach algebras, 1999

655 **Yuval Z. Flicker,** Matching of orbital integrals on $GL(4)$ and $GSp(2)$, 1999

654 **Wancheng Sheng and Tong Zhang,** The Riemann problem for the transportation equations in gas dynamics, 1999

653 **L. C. Evans and W. Gangbo,** Differential equations methods for the Monge-Kantorovich mass transfer problem, 1999

652 **Arne Meurman and Mirko Primc,** Annihilating fields of standard modules of $\mathfrak{sl}(2,\mathbb{C})^\sim$ and combinatorial identities, 1999

651 **Lindsay N. Childs, Cornelius Greither, David J. Moss, Jim Sauerberg, and Karl Zimmermann,** Hopf algebras, polynomial formal groups, and Raynaud orders, 1998

650 **Ian M. Musson and Michel Van den Bergh,** Invariants under Tori of rings of differential operators and related topics, 1998

649 **Bernd Stellmacher and Franz Georg Timmesfeld,** Rank 3 amalgams, 1998

648 **Raúl E. Curto and Lawrence A. Fialkow,** Flat extensions of positive moment matrices: Recursively generated relations, 1998

647 **Wenxian Shen and Yingfei Yi,** Almost automorphic and almost periodic dynamics in skew-product semiflows, 1998

646 **Russell Johnson and Mahesh Nerurkar,** Controllability, stabilization, and the regulator problem for random differential systems, 1998

645 **Peter W. Bates, Kening Lu, and Chongchun Zeng,** Existence and persistence of invariant manifolds for semiflows in Banach space, 1998

644 **Michael David Weiner,** Bosonic construction of vertex operator para-algebras from symplectic affine Kac-Moody algebras, 1998

643 **Józef Dodziuk and Jay Jorgenson,** Spectral asymptotics on degenerating hyperbolic 3-manifolds, 1998

642 **Chu Wenchang,** Basic almost-poised hypergeometric series, 1998

641 **W. Bulla, F. Gesztesy, H. Holden, and G. Teschl,** Algebro-geometric quasi-periodic finite-gap solutions of the Toda and Kac-van Moerbeke hierarchies, 1998

640 **Xingde Dai and David R. Larson,** Wandering vectors for unitary systems and orthogonal wavelets, 1998

639 **Joan C. Artés, Robert E. Kooij, and Jaume Llibre,** Structurally stable quadratic vector fields, 1998

638 **Gunnar Fløystad,** Higher initial ideals of homogeneous ideals, 1998

637 **Thomáš Gedeon,** Cyclic feedback systems, 1998

636 **Ching-Chau Yu,** Nonlinear eigenvalues and analytic-hypoellipticity, 1998

635 **Magdy Assem,** On stability and endoscopic transfer of unipotent orbital integrals on p-adic symplectic groups, 1998

634 **Darrin D. Frey,** Conjugacy of Alt_5 and $\mathrm{SL}(2,5)$ subgroups of $\mathrm{E}_8(\mathbb{C})$, 1998

633 **Dikran Dikranjan and Dmitri Shakhmatov,** Algebraic structure of pseudocompact groups, 1998

(See the AMS catalog for earlier titles)

Rational S^1-Equivariant Stable Homotopy Theory

MEMOIRS
of the
American Mathematical Society

Number 661

Rational S^1-Equivariant Stable Homotopy Theory

J. P. C. Greenlees

March 1999 • Volume 138 • Number 661 (end of volume) • ISSN 0065-9266

American Mathematical Society
Providence, Rhode Island

1991 *Mathematics Subject Classification.*
Primary 55N91, 55P41, 55P42, 55P62;
Secondary 18E30, 19D55, 19L47, 20C99, 55T15, 55T25.

Library of Congress Cataloging-in-Publication Data

Greenlees, J. P. C. (John Patrick Campbell), 1959–
Rational S^1-equivariant stable homotopy theory / J. P. C. Greenlees.
p. cm. — (Memoirs of the American Mathematical Society, ISSN 0065-9266 ; no. 661)
"Volume 138, number 661 (end of volume)."
Includes bibliographical references and indexes.
ISBN 0-8218-1001-4 (alk. paper)
1. Homotopy theory. I. Title. II. Series.
QA3.A57 no. 661
[QA612.7]
510 s—dc21
[514′.24]

98-53107
CIP

Memoirs of the American Mathematical Society

This journal is devoted entirely to research in pure and applied mathematics.

Subscription information. The 1999 subscription begins with volume 137 and consists of six mailings, each containing one or more numbers. Subscription prices for 1999 are $448 list, $358 institutional member. A late charge of 10% of the subscription price will be imposed on orders received from nonmembers after January 1 of the subscription year. Subscribers outside the United States and India must pay a postage surcharge of $30; subscribers in India must pay a postage surcharge of $43. Expedited delivery to destinations in North America $35; elsewhere $130. Each number may be ordered separately; *please specify number* when ordering an individual number. For prices and titles of recently released numbers, see the New Publications sections of the *Notices of the American Mathematical Society.*

Back number information. For back issues see the *AMS Catalog of Publications.*

Subscriptions and orders should be addressed to the American Mathematical Society, P. O. Box 5904, Boston, MA 02206-5904. *All orders must be accompanied by payment.* Other correspondence should be addressed to Box 6248, Providence, RI 02940-6248.

Memoirs of the American Mathematical Society is published bimonthly (each volume consisting usually of more than one number) by the American Mathematical Society at 201 Charles Street, Providence, RI 02904-2294. Periodicals postage paid at Providence, RI. Postmaster: Send address changes to Memoirs, American Mathematical Society, P. O. Box 6248, Providence, RI 02940-6248.

This publication is indexed in *Science Citation Index*®, *SciSearch*®, *Research Alert*®, *CompuMath Citation Index*®, *Current Contents*®*/Physical, Chemical & Earth Sciences.*
Printed in the United States of America.

∞ The paper used in this book is acid-free and falls within the guidelines established to ensure permanence and durability.
Visit the AMS home page at URL: `http://www.ams.org/`

10 9 8 7 6 5 4 3 2 1 04 03 02 01 00 99

Contents

ABSTRACT. We make a systematic study of rational S^1-equivariant cohomology theories, or rather of their representing objects, rational S^1-spectra.

In Part I we construct a complete algebraic model for the homotopy category of S^1-spectra, reminiscent of the localization theorem. The model is of homological dimension one, and simple enough to allow practical calculations; in particular we obtain a classification of rational S^1-equivariant cohomology theories.

In Part II we identify the algebraic counterparts of all the usual change of groups functors associated to S^1-spectra. This enables us in Part III to give a rational analysis of a number of interesting phenomena, such as ordinary cohomology, the Atiyah-Hirzebruch spectral sequence, the Segal conjecture, K-theory and topological cyclic homology.

Finally in Part IV we make a more thorough study of the algebraic models. This culminates in an identification of the algebraic models of the smash product and function spectrum.

The author is grateful to the Nuffield Foundation for its support.

CHAPTER 0

General Introduction.

0.1. Motivation.

Spaces with actions of the circle group $\mathbb{T}$ are of particular interest[1]. Loops occur in many constructions, and it is often appropriate to take into account the action of the circle by rotation; in particular the free loop space has been the object of much study. This in turn leads towards the use of the circle group in cyclic cohomology; the refinements of topological Hochschild homology and topological cyclic constructions are also important in algebraic K-theory. More prosaically the circle is simply the first infinite compact Lie group, and it plays a fundamental role in the understanding of all positive dimensional groups. For any one of these reasons it is important to understand equivariant cohomology theories for spaces with circle action.

To obtain a reasonably broad and simple picture, we consider the case of rational cohomology theories; these have been considered before for special classes of spaces (see for example [**5**]), but this appears to be the first attempt to obtain a complete algebraic picture. In any case, the understanding of the rational case is a necessary first step towards a general understanding of $\mathbb{T}$-equivariant cohomology theories. It is well known [**14**] that, for finite groups, all cohomology theories are products of ordinary cohomology theories, but this is false for the circle group. A test case of particular interest is rational topological K-theory. The example of J.-P.Haeberly [**17**] shows that, by contrast with the case of finite groups of equivariance, there is no Chern character isomorphism. It follows that $\mathbb{T}$-equivariantly some topological structure remains, even after rationalization. The author began the present work to understand the $\mathbb{T}$-equivariant Chern character, the $\mathbb{T}$-equivariant Segal conjecture, the Tate construction on $\mathbb{T}$-equivariant K-theory and several other $\mathbb{T}$-equivariant rational objects that had come to light. A list of examples treated here can be gleaned from the contents pages.

From now on we let $\mathbb{T}$ denote the circle group. We only consider closed subgroups, and the letters H, K and L will denote finite subgroups. The family of all finite subgroups will be denoted by $\mathcal{F}$. We work rationally throughout, without displaying this in the notation; for example S^n denotes the rationalized n-sphere.

0.2. Overview.

Equivariant cohomology theories are represented by equivariant spectra, and we shall conduct most of the investigation at the represented level. This gives more precise information both about individual theories and about natural transformations between them;

[1]Received by the editor March 25, 1996.

indeed, the only loss is any geometric interpretation of the cohomology theory concerned, which is inevitable in any general study. It is important to be explicit that we only consider cohomology theories which admit suspension isomorphisms for arbitrary representations; these are sometimes known as 'genuine' or '$RO(G)$-graded' cohomology theories. The corresponding representing objects are G-spectra. For these too there are adjectives to emphasize the type of spectra concerned: they are 'genuine' G-spectra or G-spectra 'indexed on a complete G-universe'. Since these cohomology theories and these G-spectra form the most natural classes to consider, we shall not use these adjectives unless required for emphasis. As made clear by the title, we consider the circle group $G = \mathbb{T}$.

Before summarizing our results, we begin by putting the circle group into context. In fact the circle stands at a watershed: for finite groups of equivariance rational cohomology theories may be analysed completely, and any group more complicated than the circle is substantially harder to understand.

The main problem in analysing spectra is to choose basic objects which are easy to work with and which give theorems of practical use. It is natural to be guided by one's favourite algebraic invariant, and this suggests analysis in terms of Moore spectra or Eilenberg-MacLane spectra. For finite groups of equivariance both approaches work well, and one may analyse rational spectra completely in terms of either one. There are two reasons for this: firstly the group has no topology, and secondly the classifying space has no rational cohomology. The first fact means the category of Mackey functors is very simple, and the second means that the classes of Eilenberg-MacLane spectra, of Moore spectra and of Brown-Comenetz spectra coincide, so that all their characteristic properties can be used at once. Both simplifying factors fail for infinite groups, and the three basic classes are distinct. This means that different methods must be used: in essence we base our analysis for the circle group on a slightly embellished version of equivariant homotopy with its primary operations. The reason such a simple invariant suffices is that the rank of the circle group is one. In general the injective dimension of the category of rational Mackey functors and the Krull dimension of the cohomology of its classifying space are both equal to the rank of the group. When the rank is one there is no room for extension problems, and some hope of a simple answer. However, even for the group $O(2)$ **[12]**, it is necessary to take into account a topology on the space of subgroups, and to work with sheaves: it is no longer possible to treat different conjugacy classes of subgroups entirely separately. This explains why it is worthwhile to treat the single case of the circle in such detail.

The work is broken into four parts. The foundation for the rest is Part I, in which we construct the algebraic models for various classes of $\mathbb{T}$-spectra. In the other parts we use the models of Part I to study natural questions: in Part II we identify the algebraic counterparts of various change of groups functors, in Part III we consider several classes of examples of particular interest, and finally in Part IV we examine the algebra of the models in some detail, and thus obtain models for the smash product and the function spectrum. Each part has a detailed introduction of its own, but we give a general outline here.

Part I begins by discussing K-theory. On the one hand, we give Haeberly's example showing that K-theory cannot be described simply using ordinary cohomology. On the other hand, we give a generalization of McClure's result that the K-theory Atiyah-Hirzebruch spectral sequence collapses for $\mathcal{F}$-free spaces. This suggests the necessity of the present work and that it is practical. We then turn to the main business of constructing a model: in this introduction we describe the model in an aesthetically satisfying way, but do not attempt to explain the proof that it is a model. The introduction to Part I gives a different approach to the model which does suggest the proof. We would prefer to achieve these two ideals simultaneously.

To motivate the form of the model, one should recall the classical localization theorem for semifree $\mathbb{T}$-spaces. This states that if X is a finite space which only has isotropy groups $\mathbb{T}$ and 1, then the inclusion of the fixed point space $X^{\mathbb{T}} \longrightarrow X$ induces an isomorphism in Borel cohomology once the Euler classes $\mathcal{E} = \{1, c_1, c_1^2, \dots\}$ are inverted:

$$\mathcal{E}^{-1}H^*(E\mathbb{T}_+ \wedge_{\mathbb{T}} X) \xrightarrow{\cong} \mathcal{E}^{-1}H^*(E\mathbb{T}_+ \wedge_{\mathbb{T}} X^{\mathbb{T}}) = \mathcal{E}^{-1}H^*(B\mathbb{T}_+) \otimes H^*(X^{\mathbb{T}}).$$

We conclude that $N = H^*(E\mathbb{T}_+ \wedge_{\mathbb{T}} X)$, regarded as a module over $\mathbb{Q}[c_1] = H^*(B\mathbb{T}_+)$ is very nearly enough to identify the homology of the fixed point space $X^{\mathbb{T}}$, but we need to pick out a vector subspace $V = H^*(X^{\mathbb{T}})$ of $\mathcal{E}^{-1}N$ which is a basis in the sense that $\mathcal{E}^{-1}N \cong \mathcal{E}^{-1}H^*(B\mathbb{T}_+) \otimes V$. In particular, if X is free then N is $\mathcal{E}$-torsion.

Now $\mathbb{T}$-equivariant cohomology theories are represented by $\mathbb{T}$-spectra, and the localization theorem suggests a model which turns out to be a complete invariant. To describe it, we first note that there is a natural homotopy-level analogue of the set of isotropy groups which occur. This uses the geometric K-fixed point functor $X \longmapsto \Phi^K X$, which is the functor extending the K-fixed point functor on spaces, in the sense that $\Phi^K(\Sigma^\infty Y) = \Sigma^\infty(Y^K)$; it also enjoys similar formal properties to the space-level functor. We then define the set of isotropy groups of a spectrum X to be the set of subgroups K for which the geometric fixed point spectra $\Phi^K X$ are non-equivariantly essential. This gives the notion of a free $\mathbb{T}$-spectrum (alternatively characterized as a $\mathbb{T}$-spectrum X for which $E\mathbb{T}_+ \wedge X \longrightarrow X$ is an equivalence). We therefore suppose given a collection $\mathcal{H}$ of finite subgroups of $\mathbb{T}$, and we may consider the class of $\mathcal{H}$-free spectra (i.e. those with isotropy in $\mathcal{H}$), and the class of $\mathcal{H}$-semifree spectra (i.e. those with isotropy in $\mathcal{H} \cup \{\mathbb{T}\}$). The reader should concentrate on the case $\mathcal{H} = \{1\}$, which gives the usual classes of free and semifree spectra, and on the case $\mathcal{H} = \mathcal{F}$: the class of $\mathcal{F}$-semifree spectra is the class of all $\mathbb{T}$-spectra. However the additional generality makes the picture clearer, and the two special cases are representative of the two classes of examples: those with $\mathcal{H}$ finite, and those with $\mathcal{H}$ infinite. Analogous to the ring $H^*(B\mathbb{T}_+)$ we have the ring of operations

$$\mathcal{O}_{\mathcal{H}} = C(\mathcal{H}, \mathbb{Q})[c],$$

where $C(\mathcal{H}, \mathbb{Q})$ denotes the $\mathbb{Q}$-valued functions on the discrete set $\mathcal{H}$, and c is of degree -2. The notation is chosen to suggest that $\mathcal{O}_{\mathcal{H}}$ is a ring of functions on $\mathcal{H}$. This ring is Noetherian if $\mathcal{H}$ is finite and not otherwise. We let $e_H \in C(\mathcal{H}, \mathbb{Q}) = (\mathcal{O}_{\mathcal{H}})_0$ denote the idempotent with support $H \in \mathcal{H}$, and we let $c_H = e_H c$. Next we need the set $\mathcal{E} = \mathcal{E}_{\mathcal{H}}$ of Euler classes. If $\mathcal{H} = \{1\}$ this is simply the multiplicative subset $\{1, c_1, c_1^2, \dots\}$ of $\mathcal{O}_{\mathcal{H}}$ used for the localization theorem above, but in general it needs a little more explanation. For any finite subset $\phi \subseteq \mathcal{H}$ we have an associated idempotent $e_\phi \in \mathcal{O}_{\mathcal{H}}$, and we have an Euler class $c_\phi = e_\phi c + (1 - e_\phi)$, which is not a homogeneous element of $\mathcal{O}_{\mathcal{H}}$. The effect of c_ϕ on an $\mathcal{O}_\mathcal{H}$-module $M = e_\phi M \oplus (1-e_\phi)M$ is to multiply by c on the first factor and do nothing to the second: thus the result of inverting c_ϕ on M is again a graded module: $e_\phi M[c^{-1}] \oplus (1-e_\phi)M$. Thus our second ingredient is the multiplicative set

$$\mathcal{E}_\mathcal{H} = \langle c_\phi^k \mid \phi \subseteq \mathcal{H} \text{ finite, } k \geq 0\rangle$$

generated by these Euler classes. The category modelling semifree $\mathcal{H}$-spectra is then the category $\mathcal{A}_\mathcal{H}$ of $\mathcal{O}_\mathcal{H}$-modules N with a specified graded vector space V to act as a basis of $\mathcal{E}^{-1}N$: we call $\mathcal{A}_\mathcal{H}$ the *standard model.* It is convenient to package this as saying that we are given a *basing map*

$$\beta : N \longrightarrow (\mathcal{E}^{-1}\mathcal{O}_\mathcal{H}) \otimes V$$

which becomes an isomorphism when $\mathcal{E}$ is inverted. This makes clear that a morphism in $\mathcal{A}_\mathcal{H}$ is a diagram

$$\begin{array}{ccc} M & \xrightarrow{\theta} & N \\ \alpha \downarrow & & \downarrow \beta \\ (\mathcal{E}^{-1}\mathcal{O}_\mathcal{H}) \otimes U & \xrightarrow{1\otimes\phi} & (\mathcal{E}^{-1}\mathcal{O}_\mathcal{H}) \otimes V. \end{array}$$

We refer to N as the *nub* and V as the *vertex.* We also refer to an $\mathcal{O}_\mathcal{H}$-module with specified basing map as a *based* $\mathcal{O}_\mathcal{H}$-module, and to a morphism $\theta : M \longrightarrow N$ for which there is a compatible map ϕ as a *based map.* Note that if $\mathcal{H}$ is a singleton the existence of a basing isomorphism $\mathcal{E}^{-1}N \cong \mathcal{E}^{-1}\mathcal{O}_\mathcal{H} \otimes V$ for some V is automatic, but in general it puts a restriction on the modules N.

The connection with topology arises since $\mathcal{O}_\mathcal{F}$ is the ring of self maps $[E\mathcal{F}_+, E\mathcal{F}_+]_*^\mathbb{T}$. Thus $\mathcal{O}_\mathcal{F}$ acts on the homotopy groups of spectra of the form $E\mathcal{F}_+ \wedge X$ or $DE\mathcal{F}_+ \wedge X$ for any X, where $D(\cdot) = F(E\mathcal{F}_+, \cdot)$ denotes functional duality. From a $\mathbb{T}$-spectrum X we may therefore form an object of $\mathcal{A}$ as follows. Firstly we take $\pi_*^\mathbb{T}(DE\mathcal{F}_+ \wedge X)$ to be the nub, and $\pi_*(\Phi^\mathbb{T} X)$ as the vertex. Now

$$\pi_*^\mathbb{T}(DE\mathcal{F}_+ \wedge \tilde{E}\mathcal{F}) = \mathcal{E}^{-1}\pi_*^\mathbb{T}(DE\mathcal{F}_+) = \mathcal{E}^{-1}\mathcal{O}_\mathcal{F},$$

and so

$$\pi_*^\mathbb{T}(DE\mathcal{F}_+ \wedge \tilde{E}\mathcal{F} \wedge X) \cong \mathcal{E}^{-1}\mathcal{O}_\mathcal{F} \otimes \pi_*(\Phi^\mathbb{T} X).$$

The basing map is obtained by applying $\pi_*^\mathbb{T}$ to the map

$$DE\mathcal{F}_+ \wedge X \longrightarrow DE\mathcal{F}_+ \wedge \tilde{E}\mathcal{F} \wedge X \simeq DE\mathcal{F}_+ \wedge \tilde{E}\mathcal{F} \wedge \Phi^\mathbb{T} X$$

with cofibre $DE\mathcal{F}_+ \wedge \Sigma E\mathcal{F}_+ \wedge X \simeq \Sigma E\mathcal{F}_+ \wedge X$. Furthermore, since c is of negative degree and any element of $\pi_*^\mathbb{T}(E\mathcal{F}_+ \wedge X)$ is supported on a finite subspectrum, it follows that $\mathcal{E}^{-1}\pi_*^\mathbb{T}(E\mathcal{F}_+ \wedge X) = 0$, and so the resulting object

$$\pi_*^\mathcal{A}(X) := \left(\pi_*^\mathbb{T}(DE\mathcal{F}_+ \wedge X) \longrightarrow \pi_*^\mathbb{T}(DE\mathcal{F}_+ \wedge \tilde{E}\mathcal{F} \wedge \Phi^\mathbb{T} X) = \pi_*^\mathbb{T}(DE\mathcal{F}_+ \wedge \tilde{E}\mathcal{F}) \otimes \pi_*(\Phi^\mathbb{T} X)\right)$$

is therefore an object of $\mathcal{A}_\mathcal{H}$. If X is $\mathcal{H}$-semifree, the $\mathcal{O}_\mathcal{F}$-module structure factors through the projection $\mathcal{O}_\mathcal{F} \longrightarrow \mathcal{O}_\mathcal{H}$.

Now we may state our main classification theorem.

Classification Theorem: For any collection $\mathcal{H}$ of finite subgroups of the circle $\mathbb{T}$, the above invariant induces bijections
(i)

$$\{\mathcal{H}\text{-free rational spectra}\} / \simeq \;\longleftrightarrow\; \{\text{Euler-torsion } \mathcal{O}_{\mathcal{H}}\text{-modules}\} / \cong$$

where $\simeq$ denotes homotopy equivalence, and $\cong$ denotes isomorphism, and
(ii)

$$\{\mathcal{H}\text{-semifree rational spectra}\} / \simeq \;\longleftrightarrow\; \{\text{based } \mathcal{O}_{\mathcal{H}}\text{-modules}\} / \cong$$

where $\simeq$ denotes homotopy equivalence, and $\cong$ denotes isomorphism. In particular, *rational $\mathbb{T}$-equivariant cohomology theories are in bijective correspondence to isomorphism classes of based $\mathcal{O}_{\mathcal{F}}$-modules.*

In practice this is derived as a corollary of a theorem identifying the entire categories of spectra in algebraic terms. More precisely, recall that the derived category of a graded abelian category is the category of differential graded objects with homology isomorphisms inverted, although for practical purposes a more concrete construction is essential. The theorem identifies the categories of spectra as the derived category of the associated algebraic category:

$$\mathcal{H}\text{-free } \mathbb{T}\text{-spectra} \simeq D(\text{Euler torsion } \mathcal{O}_{\mathcal{H}}\text{-modules})$$

and

$$\mathcal{H}\text{-semifree } \mathbb{T}\text{-spectra} \simeq D(\text{based } \mathcal{O}_{\mathcal{H}}\text{-modules}).$$

Furthermore, cofibre sequences of spectra correspond to triangles under these equivalences. The point here is that both algebraic categories turn out to be abelian and one dimensional, so that morphisms in the derived category can be calculated from a short exact sequence involving Hom and Ext in the abelian category.

It is sometimes more practical to identify the place of a spectrum X in the classification by a different route. This amounts to identifying first the free part $E\mathcal{F}_+ \wedge X$ and its geometric fixed points $\Phi^{\mathbb{T}}X$, and then the map

$$q_X : \tilde{E}\mathcal{F} \wedge \Phi^{\mathbb{T}}X \simeq \tilde{E}\mathcal{F} \wedge X \longrightarrow \Sigma E\mathcal{F}_+ \wedge X$$

of which X is the fibre. It is not enough to identify the effect of q_X in homotopy: one must also take into account the twisting given by representations, and in general this requires both primary and secondary information. Nonetheless, there is a second model for semifree $\mathcal{H}$-spectra based on this approach, which we call the *torsion model.* We show it is equivalent to the standard model described above, and it is often the easiest route to placing a spectrum in the classification.

There are really three stages to the proof of these theorems. Firstly one shows, using idempotents in the Burnside rings of finite subgroups, that for $\mathcal{F}$-free spectra it is essentially enough to deal with the case of free spectra. Next, one constructs an Adams spectral sequence for free spectra, which collapses to a short exact sequence and gives a means of calculation. Because of the particularly simple algebraic behaviour of $\mathcal{O}_1 = \mathbb{Q}[c_1]$ this is enough to identify the entire triangulated category. The final stage is to take this work and process it: this stage is essentially formal.

Once we have algebraic models for various categories of spectra we naturally want to understand familiar topological constructions in algebraic terms. We begin in Part II with the change of groups functors. These we consider can all be modelled functorially, and we shall discuss only the full category of $\mathbb{T}$-spectra. The forgetful functor to non-equivariant spectra and its left and right adjoints, induction and coinduction, are straightforward. Similarly the geometric fixed point functor $X \longmapsto \Phi^{\mathbb{T}}X$ is the passage-to-vertex functor given as part of the structure. The first interesting functor is the geometric fixed point functor $\Phi^K : \mathbb{T} - \text{spectra} \longrightarrow \mathbb{T}/K - \text{spectra}$ for a finite subgroup K. This turns out to be easy to describe: we simply let $e_{[\supseteq K]} \in C(\mathcal{F}, \mathbb{Q})$ denote the idempotent supported on the set $[\supseteq K]$ of subgroups containing K. The algebraic model of Φ^K is multiplication by $e_{[\supseteq K]}$; this make sense since $e_{[\supseteq K]}\mathcal{O}_{\mathcal{F}}$ is the product $\prod_{H \supseteq K} \mathbb{Q}[c_H]$, which is naturally identified with the product indexed by the finite subgroups H/K of $\mathbb{T}/K$, ie the ring of operations $\overline{\mathcal{O}}_{\overline{\mathcal{F}}}$ for $\overline{\mathbb{T}} = \mathbb{T}/K$. Here we have used $\overline{\mathcal{F}}$ to denote the family of finite subgroups of $\overline{\mathbb{T}}$. As usual, the Lewis-May fixed point functor Ψ^K : $\mathbb{T}$-spectra $\longrightarrow$ $\mathbb{T}/K$-spectra (the spectrum $\Psi^K X$ is written X^K in [**19**]) is much harder to understand, and we only describe its behaviour here for $\mathcal{F}$-free and $\mathcal{F}$-contractible spectra, referring the reader to Chapters 9 and 10 for details of how these are spliced. Note that the $\mathbb{T}$-space $\tilde{E}\mathcal{F}$ may be viewed as the inflation of a $\mathbb{T}/K$-space, $\tilde{E}\overline{\mathcal{F}}$; we may then say that on $\mathcal{F}$-contractible spectra $X \simeq \tilde{E}\mathcal{F} \wedge \Phi^{\mathbb{T}}X$, we have $\Psi^K(X) = \tilde{E}\mathcal{F} \wedge \Phi^{\mathbb{T}}X$, so this is easy. We have seen that an $\mathcal{F}$-spectrum X is modelled by an Euler-torsion $\mathcal{O}_{\mathcal{F}}$-module N; from the form of Euler classes it follows that this is equivalent to specifying the function

$$\begin{array}{rcl} [N] : \mathcal{F} & \longrightarrow & \text{torsion } \mathbb{Q}[c]\text{-modules} \\ H & \longmapsto & e_H N. \end{array}$$

The Lewis-May fixed point functor groups these modules together according to the behaviour of the subgroup on passage to quotient. More precisely, we observe that passage to quotient $q : \mathbb{T} \longrightarrow \mathbb{T}/K = \overline{\mathbb{T}}$ defines a map $q_* : \mathcal{F} \longrightarrow \overline{\mathcal{F}}$ on finite subgroups. If the function $[N]$ models the $\mathcal{F}$-free spectrum X then the function $[\Psi^K N]$ modelling $\Psi^K X$ is the map

$$\begin{array}{rcl} \overline{\mathcal{F}} & \longrightarrow & \text{torsion } \mathbb{Q}[c]\text{-modules} \\ \overline{H} & \longmapsto & \bigoplus_{q_*(H)=\overline{H}} [N](H). \end{array}$$

A little thought shows that it is not a trivial matter to see how the $\mathcal{F}$-free and $\mathcal{F}$-contractible parts should be spliced together. Because the Lewis-May fixed point functor is so complicated, we actually approach it via its left adjoint, the inflation map $\inf_{\mathbb{T}/K}^{\mathbb{T}}$: $\mathbb{T}/K$-spectra $\longrightarrow$ $\mathbb{T}$-spectra. This is the functor given by regarding a $\mathbb{T}/K$ spectrum as a $\mathbb{T}$-spectrum by pullback along the quotient, and then building in representations (it is written $q^{\#}$ in [**19**], but more commonly i_* by abuse of notation; we shall stick to the more descriptive notation). From our description of Lewis-May fixed points it is easy to deduce inflation on $\overline{\mathcal{F}}$-contractible and $\overline{\mathcal{F}}$-free spectra. On $\overline{\mathcal{F}}$-contractible spectra $Y \simeq \tilde{E}\overline{\mathcal{F}} \wedge \Phi^{\overline{\mathbb{T}}}Y$ we have $\inf_{\mathbb{T}/K}^{\mathbb{T}} Y = \tilde{E}\mathcal{F} \wedge \Phi^{\overline{\mathbb{T}}}Y$. If $[P]$ is the model of the $\overline{\mathcal{F}}$-spectrum Y then the model $[\inf_{\mathbb{T}/K}^{\mathbb{T}} P]$ of $\inf_{\mathbb{T}/K}^{\mathbb{T}} Y$ is the composite

$$\mathcal{F} \xrightarrow{q_*} \overline{\mathcal{F}} \xrightarrow{[P]} \text{torsion } \mathbb{Q}[c]\text{-modules}.$$

In cases where N is Euler-torsion, the right adjoint of the inflation map is also its left adjoint; it therefore also gives a model for the topological quotient when X is K-free.

In Part III we apply the general theory of Parts I and II to several independent examples of particular interest. The first chapter of Part III is devoted to ordinary cohomology and its variants. After Eilenberg and Steenrod we define a cohomology theory to be ordinary if its coefficients are non-zero only in degree 0, and similarly in homology. For each integer q, an equivariant cohomology theory $F^*_G(\cdot)$ specifies a contravariant additive functor $G/H_+ \longmapsto F^q_G(G/H_+) = F^q_H$ on the stable category of orbits; such a functor is called a Mackey functor. As in the classical case, ordinary cohomology theories are classified by their non-zero Mackey functor M in degree 0, and we write $H^*_G(\cdot; M)$ for this theory and HM for its representing spectrum. Similarly, for each integer q a homology theory $F^G_*(\cdot)$ defines a covariant additive functor $G/H_+ \longmapsto F^G_q(G/H_+)$ on the stable category of orbits; such a functor is called a coMackey functor. Ordinary homology theories are classified by their associated coMackey functors N, and we write $H^G_*(\cdot; N)$ for this functor and JN for the representing spectrum. For finite groups G, the stable orbit category is self-dual, so that a coMackey functor can also be viewed as a Mackey functor; in this case the ordinary homology theory classified by a Mackey functor M is also represented by HM. However, for positive dimensional groups such as the circle, the functor given by a homology theory cannot usually be viewed as a Mackey functor.

Our first task is to identify objects of the form HM and JN in our model; we find that they are well behaved but by no means trivial. Finally, whenever one has an injective Mackey functor I one may consider the cohomology theory defined by Brown-Comenetz I-duality

$$hI^q_G(X) = \mathrm{Hom}(\underline{\pi}^G_q(X), I),$$

and its representing spectrum hI. Again, in the case of a finite group all rational Mackey functors are injective, and $HM = JM = hM$. Indeed, this is the basis of a simple proof that all rational cohomology theories are ordinary for finite groups. However, for the circle group the spectrum hI is rather complicated, and in particular it is unbounded; we identify it exactly in our model.

The next chapter is may perhaps have the widest appeal: in it we answer a number of obvious general questions. To begin with, we relate the model we have used to the use of Postnikov towers and the use of cells. In fact, we can understand the Atiyah-Hirzebruch spectral sequence $H^*_{\mathbb{T}}(X; \underline{K}^*_{\mathbb{T}}) \Longrightarrow K^*_{\mathbb{T}}(X)$ for $\mathcal{F}$-free spectra X completely, in terms of our model. It collapses at the E_2 page if and only if $K^{\mathbb{T}}_*(E\mathcal{F}_+)$ is injective over $\mathcal{O}_{\mathcal{F}}$. The latter condition holds for complex K-theory, so we recover McClure's theorem that the Atiyah-Hirzebruch spectral sequence for the rational K-theory of an $\mathcal{F}$-space collapses at E_2. However, in general there are arbitrarily long differentials. The contrast with the simplicity of the one dimensional nature of the category of Euler-torsion $\mathcal{O}_{\mathcal{F}}$-modules suggests that the Postnikov tower is a poor way to study $\mathbb{T}$-spectra. On the other hand, because of the simplicity of the graded maps between cells, we can contemplate homological algebra over the category of cells and graded maps. It is easy to construct a convergent spectral sequence based on cellular resolutions with a calculable E_2 term. Unfortunately the spectral sequence does not appear to be useful in general.

We do not have the means to detect purely unstable phenomena, but the splitting theorem of Segal and tom Dieck shows that suspension spectra of $\mathbb{T}$-spaces are very special, and we briefly comment on the implications of this for their algebraic models.

Finally we return to complex K-theory and identify its algebraic model. It is simple to describe in terms of representation theory, and is well behaved algebraically ('formal' in the torsion model). However there remain many interesting questions that we have not treated. Firstly, a qualitative comparison of the $\mathcal{F}$-spectrum Euler classes and the K-theory Euler classes is sufficient for our purpose, but an exact comparison using the Chern character, along the lines of Crabb's work [**5**], would be illuminating. Secondly, it would be interesting to compare our model with that of Brylinski [**3**]. Presumably these questions would be useful preparation for the more substantial project of modelling $\mathbb{T}$-equivariant elliptic cohomology as constructed by Grojnowski [**16**] and Ginzburg-Kapranov-Vaserrot [**6**].

The other motivating problem was that of understanding the $\mathbb{T}$-equivariant analogue of the Segal conjecture. Prior to the present work, there was the ironic situation that the harder profinite part was understood by virtue of work on the Segal conjecture for finite groups, whilst the easier rational part remained mysterious. Using the model described here, it is now an easy exercise to identify $DE\mathbb{T}_+$ in the torsion model as the composite

$$\mathcal{E}^{-1}\mathcal{O}_{\mathcal{F}} \otimes_{\mathbb{Q}} \mathcal{E}^{-1}\mathcal{O}_{\mathcal{F}} \longrightarrow \mathcal{E}^{-1}\mathcal{O}_{\mathcal{F}} \longrightarrow \mathcal{E}^{-1}\mathcal{O}_{\mathcal{F}}/\mathcal{O}_{\mathcal{F}} \longrightarrow \mathbb{Q}[c_1, c_1^{-1}]/\mathbb{Q}[c_1]$$

where the first map is the product. This object is formal, and the counterpart in the standard model is the inclusion of the kernel in the domain.

Turning to more specialized examples, we reach Tate cohomology theories in the sense of [**14**]. This construction on $\mathbb{T}$-spectra corresponds precisely to Tate cohomology in commutative algebra in the sense of [**9**]. Perhaps more interesting is our study of the *integral* Tate spectrum of *integral* complex equivariant K-theory: we write $K\mathbb{Z}$ for the representing spectrum by way of emphasis. We are able to identify the exact homotopy types of both $t(K\mathbb{Z}) \wedge E\mathcal{F}_+$ and $t(K\mathbb{Z}) \wedge \tilde{E}\mathcal{F}$ and the map q of which $t(K\mathbb{Z})$ is the fibre: the first is rational, and identified using our general theory, and the second is formed from K-theory with suitable coefficients by inflating and smashing with $\tilde{E}\mathcal{F}$.

Finally we turn to examples gaining their importance from algebraic K-theory. The motivation for the notion of a cyclotomic spectrum comes from the free loop space $\Lambda X = map(\mathbb{T}, X)$ on a $\mathbb{T}$-fixed space X. This has the property that if we take K-fixed points for a finite subgroup K we obtain the $\mathbb{T}/K$-space $map(\mathbb{T}/K, X)$, and if we identify the circle $\mathbb{T}$ with the circle $\mathbb{T}/K$ by the $|K|$th root isomorphism we recover ΛX. For spectra one also needs to worry about the indexing universe, but a cyclotomic spectrum is basically one whose geometric fixed point spectrum $\Phi^K X$, regarded as a $\mathbb{T}$-spectrum, is the original $\mathbb{T}$-spectrum X. After the suspension spectrum of a free loop space, the principal example comes from the topological Hochschild homology of $THH(F)$ of a functor F with smash products. Given such a cyclotomic spectrum X one may construct the topological cyclic spectrum $TC(X)$ of Bökstedt-Hsiang-Madsen [**2**], which is a non-equivariant spectrum. An intermediate construction of some interest is the $\mathbb{T}$-spectrum $TR(X)$. Although these constructions are principally of interest profinitely, it is instructive to identify the cyclotomic spectra in our model and follow the constructions through. In fact we show that cyclotomic spectra, are those spectra X such that the function $[N] : \mathcal{F} \longrightarrow$ torsion $\mathbb{Q}[c]$-modules modelling $E\mathcal{F}_+ \wedge X$ is constant, so that the structure map $\mathcal{E}^{-1}\mathcal{O}_\mathcal{F} \otimes V \longrightarrow \Sigma N$ commutes with any translation of the finite subgroups. Therefore the structure map factors through $\mathcal{E}^{-1}\mathcal{O}_\mathcal{F} \otimes V \longrightarrow (\mathcal{E}^{-1}\mathcal{O}_\mathcal{F})/\mathcal{O}_\mathcal{F} \otimes V$, and the map $(\mathcal{E}^{-1}\mathcal{O}_\mathcal{F})/\mathcal{O}_\mathcal{F} \otimes V \longrightarrow \Sigma N$ is a direct sum of copies of $\mathbb{Q}[c,c^{-1}]/\mathbb{Q}[c] \otimes V \longrightarrow \Sigma[N](1)$. Furthermore, we may recover Goodwillie's theorem that for any cyclotomic spectrum X we have $TC(X) = X^{h\mathbb{T}}$: topological cyclic cohomology coincides with cyclic cohomology in the rational setting.

Finally we come to Part IV in which we study the smash product and function spectrum constructions. These are by far the most complicated examples, and require more algebraic machinery than any of the other examples we consider. Furthermore, their complexity means that we are not able to show that our description is functorial, and the approach is necessarily indirect. This highlights a shortcoming of our method: the correct proof of our results would follow that used by Quillen in modelling rational homotopy of simply connected spaces. The functorial identification of smash products and function spectra would then be automatic. At present, such a proof is not accessible, but the present results strongly suggest that such a proof exists.

In any case, the model of the smash product is essentially the left derived tensor product, and the model of function spectra is its right adjoint. There are two warnings here: in the categories of $\mathcal{H}$-free spectra, there are not enough flat objects, so the left derived tensor product must be calculated in a larger category; it results in an Euler-torsion object since it coincides with the suspension of the right derived torsion product. With this caveat, if the spectra X and Y are modelled by M and N respectively then

$$X \wedge Y \text{ is modelled by } M \otimes^L N.$$

There is a corresponding caveat for function objects. It is convenient to consider the larger algebraic category in which no condition is placed on the behaviour of Euler classes. For $\mathcal{H}$-free spectra this is the category of *all* $\mathcal{O}_\mathcal{H}$-modules, and for $\mathcal{H}$-semifree spectra it is the category of *all* maps $N \longrightarrow \mathcal{E}^{-1}\mathcal{O}_\mathcal{H} \otimes V$. It turns out that the internal Hom functor in the abelian category is the composite functor $\Gamma\, \hat{\mathbf{Hom}}(M,N)$, where $\hat{\mathbf{Hom}}(M,N)$ is an object in the category with no condition on behaviour under inversion of Euler classes, and where Γ is the right adjoint to the inclusion of the smaller category. For example, in the case of $\mathcal{H}$-free spectra $\hat{\mathbf{Hom}}(M,N)$ is simply the $\mathcal{O}_\mathcal{H}$-module of $\mathcal{O}_\mathcal{H}$-morphisms, and for an arbitrary $\mathcal{O}_\mathcal{H}$-module M', the Euler-torsion module ΓM is defined to be the kernel of $M \longrightarrow \mathcal{E}^{-1}M$. In the semifree case, both functors are harder to describe (see Chapters 18 and 22), but formally much better behaved. The right adjoint of the functor of $M \longmapsto M \otimes^L N$ is $P \longmapsto R\Gamma\, R\hat{\mathbf{Hom}}(N,P)$. However, we warn that when there are not enough flat objects, this can be different from the right derived functor of $P \longmapsto \Gamma\, \hat{\mathbf{Hom}}(N,P)$.

Thus if the spectra Y and Z are modelled by N and P, then

$$\text{The internal function spectrum of maps from } Y \text{ to } Z \text{ is modelled by } R\Gamma\, R\hat{\mathbf{Hom}}(N,P).$$

An essential step in identifying the function spectrum on objects is to give a functorial identification of the product. In these terms we may say that if X_i is modelled by M_i then

$$\text{The internal product of the spectra } X_i \text{ is modelled by } R\Gamma \prod_i M_i,$$

and this model is functorial.

Finally, with the machinery of internal Hom, we can give give an alternative description of the inflation functor and the Lewis-May K-fixed point functor on the standard model. We again use bars to denote objects associated to the quotient $\overline{\mathbb{T}} = \mathbb{T}/K$:

$$\inf_{\overline{\mathcal{A}}}^{\mathcal{A}}(B) = S^0_{\mathcal{A}} \otimes_{\overline{\mathcal{O}}_{\overline{\mathcal{F}}}} B \text{ and } \Psi^K(C) = R\mathbf{IntHom}_{\overline{\mathcal{O}}_{\overline{\mathcal{F}}}}(S^0_{\mathcal{A}}, C).$$

This summarizes the contents of the body. There are also a number of appendices. Appendix A gives the structure of rational Mackey functors, and is of independent interest: in particular the category is of projective and injective dimension 1. Appendix B gives Quillen closed model category structures on the algebraic categories.

The author is grateful to the Nuffield Foundation for its support, and to the Universities of Chicago and of Georgia (Athens) for their hospitality. The author also thanks L.Hesselholt, J.P.May and N.P.Strickland for useful conversations, and J.P.May and L.Scull for many comments on an earlier draft.

Reading guide.

It is appropriate to comment briefly on reading this document, because the needs different readers will be different. Formally, Part I is the basis of all that follows, and is cumulative. *All remaining chapters in Parts II and III depend on Part I, but are otherwise independent* (with the two exceptions (i) Chapter 10 (which depends on the purely algebraic Chapter 9), and (ii) Chapter 15 (which depends on results about fixed point functors from Chapters 8 and 10)). Part IV depends only on Part I, but is again essentially cumulative. This is probably clearer in the table of chapter dependence opposite.

We expect that after reading Part I most readers will only want to read a selection from Part III, starting with Chapter 13. Indeed, the author recommends this.

Part II will be useful to those who (like the author) want to develop their intuition about change of groups functors. Readers interested in cyclotomic spectra and topological cyclic cohomology (Chapter 15) will also be forced to read much of Part II. Part IV will be of most interest to the minority of readers who want a more detailed understanding of the internal workings of the category of $\mathbb{T}$-spectra.

Finally we suggest all readers glance at Appendix C summarizing our conventions. There is an index of definitions, an index of notations, and a summary of the two main models: these should particularly assist selective readers.

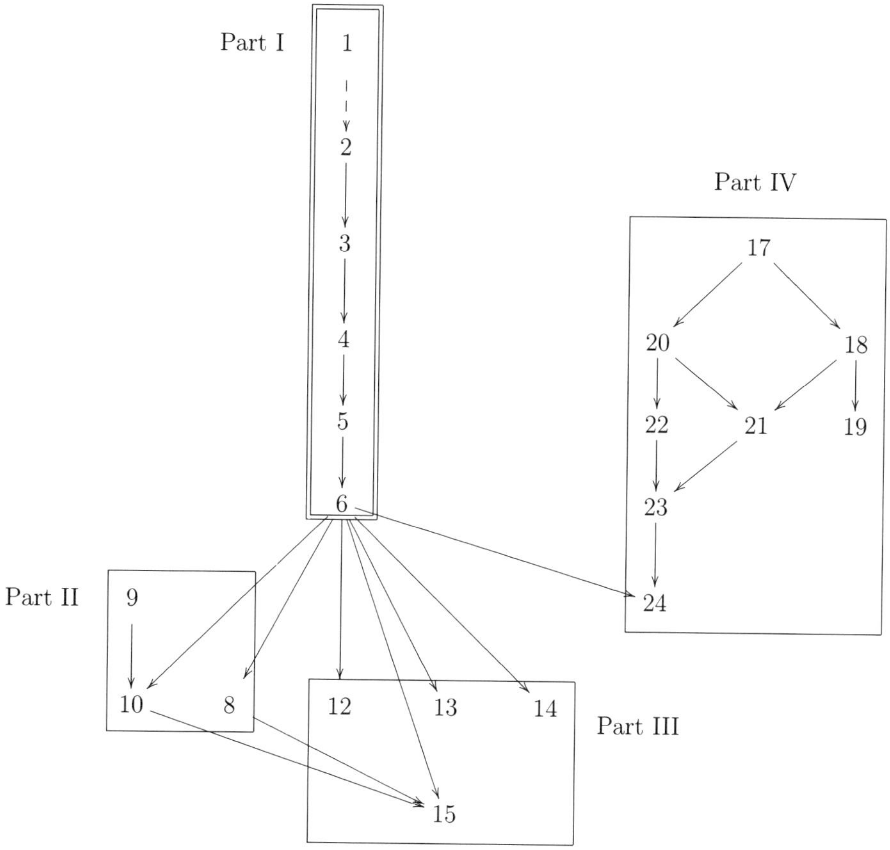

Chapter dependence

Part I

The algebraic model of rational $\mathbb{T}$-spectra.

CHAPTER 1

Introduction to Part I.

This chapter motivates Part I and provides a map for it. In Section 1.1 we explain the strategy used in Part I to analyse the category of rational $\mathbb{T}$-spectra, and in Section 1.2 give a brief guide to help readers with particular interests. This is followed in Sections 1.3 and 1.4 by accounts of Haeberly's example and a generalization of McClure's theorem: this is designed to show there is a need for analysis and some hope of achieving it.

1.1. Outline of the algebraic models.

The main business of Part I is to construct a complete algebraic model of the category of rational $\mathbb{T}$-spectra. Since spectra represent cohomology theories, this gives a complete algebraic classification of rational $\mathbb{T}$-equivariant cohomology theories. Having given the overview in the General Introduction, we concentrate here on the practical approach. In fact, we lead the reader through the investigative process to the algebraic model of $\mathbb{T}$-spectra. This should help explain how geometric information is packaged in the model, and how the algebraic model can be used.

The main problem in analysing $\mathbb{T}$-spectra is to choose basic objects which are easy to work with and which give theorems of practical use. We explained in the introduction that the building blocks familiar from finite groups of equivariance are not suitable: Eilenberg-MacLane spectra, Moore spectra and Brown-Comenetz spectra form distinct classes. This means that different methods must be used.

The redeeming feature of the circle group $\mathbb{T}$ is that there is no complication at all from representation theory since the Weyl groups are all connected. This means we can return to geometric intuition and concentrate on isotropy groups. It is appropriate for our present purpose to think of $\mathbb{T}$-spectra as generalized stable spaces. It is standard practice in transformation groups to consider various fixed point spaces X^H of a space X. In particular, spaces with a free action are especially approachable. One reason for this is that only one subgroup occurs as an isotropy group. In the rational case the behaviour at each finite subgroup is reasonably similar and reasonably simple. Therefore it is common to consider spaces X all of whose isotropy groups are finite. These are variously called $\mathcal{F}$-spaces, $\mathcal{F}$-free spaces, almost free spaces, or spaces without fixed points. We shall call them $\mathcal{F}$-spaces, and concentrate on the fact that they are equivalent to spaces constructed from cells $G/H \times E^n$ with H finite.

In any case, our analysis follows this time-honoured pattern, by breaking any object X into into $\mathcal{F}$-free and $\mathcal{F}$-contractible parts by the isotropy separation cofibering

$$X \longrightarrow X \wedge \tilde{E}\mathcal{F} \xrightarrow{q_X} X \wedge \Sigma E\mathcal{F}_+.$$

Here $E\mathcal{F}$ is the universal $\mathcal{F}$-free space, from which $E\mathcal{F}_+$ is formed by adding a disjoint basepoint, and $\tilde{E}\mathcal{F}$ is formed by taking the unreduced suspension. We thus consider X in two parts: the $\mathcal{F}$-contractible object $X(\mathbb{T}) = X \wedge \tilde{E}\mathcal{F}$ and the $\mathcal{F}$-free object $X(\mathcal{F}) = X \wedge E\mathcal{F}_+$. The object $X(\mathbb{T})$ is determined by its $\mathbb{T}$-homotopy groups as rational vector spaces. The main content of the analysis is therefore in understanding $\mathcal{F}$-objects such as $X(\mathcal{F})$, and how they may be stuck to $\mathcal{F}$-contractible objects $X(\mathbb{T})$. By use of idempotents in Burnside rings it is easy to see that $X(\mathcal{F})$ splits as a wedge of objects $X(H)$, one for each finite subgroup H, where only the isotropy group H is relevant to $X(H)$. The category of these will be called the category of $\mathbb{T}$-spectra over H and denoted $\mathbb{T}$-Spec$/H$; the mathematical core of the whole enterprise is the analysis of this category of objects $X(H)$. It turns out that $\pi^{\mathbb{T}}_*(X(H))$ is a torsion module over the ring $\mathcal{O}_H = \mathbb{Q}[c_H]$, in which c_H is an Euler class, and of degree -2, and that the category $\mathbb{T}$-Spec$/H$ of objects $X(H)$ is equivalent to the derived category of differential graded torsion $\mathbb{Q}[c_H]$-modules. The object $X(\mathcal{F})$ is thus determined by the torsion module $\pi^{\mathbb{T}}_*(X(\mathcal{F})) = \bigoplus_H \pi^{\mathbb{T}}_*(X(H))$ over $\mathcal{O}_{\mathcal{F}} = \prod_H \mathbb{Q}[c_H]$. Because we are working rationally it is not difficult to calculate homotopy groups of any precisely described spectrum, so this description is of practical use.

Finally we must determine the assembly map $q_X : X(\mathbb{T}) \longrightarrow \Sigma X(\mathcal{F})$. Note first that $\pi^{\mathbb{T}}_*(X(\mathbb{T}))$ is not naturally a module over $\mathcal{O}_{\mathcal{F}}$, and also that $\pi^{\mathbb{T}}_*(q_X)$ may be zero without q_X being zero. The answer is to take into account the twisting available from representations of $\mathbb{T}$. This twisting is measured by Euler classes, and since there are Thom isomorphisms for arbitrary $\mathcal{F}$-spectra we may consider the ring $\mathcal{E}^{-1}\mathcal{O}_{\mathcal{F}}$ formed from $\mathcal{O}_{\mathcal{F}}$ by inverting all Euler classes. We denote this ring $t^{\mathcal{F}}_*$, since it is in fact the $\mathcal{F}$-Tate cohomology of S^0 in the sense of [**14**]. It turns out that $t^{\mathcal{F}}_*$ is $\bigoplus_H \mathbb{Q}$ in positive even degrees and $\prod_H \mathbb{Q}$ in even degrees ≤ 0. By construction, $t^{\mathcal{F}}_*$ is a $\mathcal{O}_{\mathcal{F}}$-module, and q_X determines a map

$$\hat{q}_X : t^{\mathcal{F}}_* \otimes \pi^{\mathbb{T}}_*(X(\mathbb{T})) \longrightarrow \pi^{\mathbb{T}}_*(X(\mathcal{F})) = \bigoplus_H \pi^{\mathbb{T}}_*(X(H))$$

in the derived category of differential graded $\mathcal{O}_{\mathcal{F}}$-modules. It transpires that $\hat{q}_X$ is a complete invariant of q_X, so that X is determined by the rational vector space $\pi^{\mathbb{T}}_*(X(\mathbb{T}))$, the torsion $\mathbb{Q}[c_H]$-modules $\pi^{\mathbb{T}}_*(X(H))$, and the derived $\mathcal{O}_{\mathcal{F}}$-map $\hat{q}_X$. Continuing from this stage, it is not hard to identify which triples $(\pi^{\mathbb{T}}_*(X(\mathbb{T})), \pi^{\mathbb{T}}_*(X(\mathcal{F})), \hat{q}_X)$ occur, and to identify the relevant algebraic triangulated category.

In fact we may define an abelian category designed to include the maps $\hat{q}_X$ described in the previous paragraph. This is the *torsion model* category $\mathcal{A}_t$ whose objects are maps

$$q : t^{\mathcal{F}}_* \otimes V \longrightarrow T = \bigoplus_H T(H)$$

of $\mathcal{O}_{\mathcal{F}}$-modules, with $T(H)$ a torsion $\mathbb{Q}[c_H]$-module. It turns out that this category is abelian and of injective dimension 2. One may therefore consider differential graded objects in $\mathcal{A}_t$, and invert homology isomorphisms to form the derived category $D\mathcal{A}_t$. This category is equivalent to the category of rational $\mathbb{T}$-spectra, and provides the complete algebraic model we seek. However we prefer not to emphasize this model: the analysis is only possible by introducing a second model, which we call the *standard model.* This proves to be more convenient for most purposes. The real difficulty is that, since $\mathcal{A}_t$ is of dimension 2, it is rather hard to get a precise hold on morphisms in the derived category. On the other hand the standard model is of dimension 1. The identification of the standard model is the most important result of the analysis.

It will help to explain the construction of algebraic models for four triangulated categories of $\mathbb{T}$-spectra in increasing order of complexity. They are (i) the category of free $\mathbb{T}$-spectra, or more generally the category $\mathbb{T}$-Spec$/H$ of $\mathbb{T}$-spectra in which only the isotropy group H is important, (ii) the category of $\mathbb{T}$-Spec$/\mathcal{F}$ of $\mathcal{F}$-spectra, (iii) the category $\mathbb{T}$-Spec$_{sf}$ of semifree $\mathbb{T}$-spectra and (iv) the category of all rational $\mathbb{T}$-spectra. For each of these categories $\mathbb{C}$, we find an abelian category $\mathcal{A} = \mathcal{A}_{\mathbb{C}}$ of dimension 1, and a linearization functor $\pi_*^{\mathcal{A}} : \mathbb{C} \longrightarrow \mathcal{A}_{\mathbb{C}}$. Because the abelian category $\mathcal{A}_{\mathbb{C}}$ is so simple in each case, it is possible to reconstruct the original triangulated category $\mathbb{C}$ from it. Recall that the derived category of an abelian category $\mathcal{A}$ is the category formed from the category of differential graded objects by inverting homology isomorphisms; if $\mathcal{A}$ is of finite injective or projective dimension, the derived category may be constructed explicitly.

THEOREM 1.1.1. *If $\mathbb{C}$ is one of the above four categories of rational $\mathbb{T}$-spectra, there is a category $\mathcal{A} = \mathcal{A}_{\mathbb{C}}$ which is abelian and one dimensional so that there is an equivalence of triangulated categories*

$$\mathbb{C} \simeq D\mathcal{A},$$

where $D\mathcal{A}$ is the derived category of $\mathcal{A}$. Hence in particular, for any objects X and Y of $\mathbb{C}$, there is a natural short exact sequence

$$0 \longrightarrow \mathrm{Ext}_{\mathcal{A}}(\pi_*^{\mathcal{A}}(\Sigma X), \pi_*^{\mathcal{A}}(Y)) \longrightarrow [X,Y]_*^{\mathbb{T}} \longrightarrow \mathrm{Hom}_{\mathcal{A}}(\pi_*^{\mathcal{A}}(X), \pi_*^{\mathcal{A}}(Y)) \longrightarrow 0,$$

which splits unnaturally.

Before making the theorem explicit for the four categories we make some general remarks about the levels at which the theorem is useful. Firstly, every geometric object X of $\mathbb{C}$ has an algebraic model $\pi_*^{\mathcal{A}}(X)$ and there is a bijection between isomorphism classes in $\mathbb{C}$ and isomorphism classes in $\mathcal{A}$. Next, if we know the algebraic models of two objects X and Y, the short exact sequence allows us to use the algebra of the abelian category to calculate the group $[X,Y]_*^{\mathbb{T}}$ of maps between them. Finally, we may model all primary constructions (such as formation of cofibres, smash products, function spectra, composition of functions and calculation of Toda brackets) in the algebraic category. This much is internal to the category, but in addition, all homotopy functors of $\mathbb{T}$-spectra have their algebraic counterparts. It is very illuminating to identify the algebraic behaviour of various well known functors.

We now make Theorem 1.1.1 explicit in the four cases.

THEOREM 1.1.2. *If $\mathbb{C} = \mathbb{T}$-Spec$/H$ is the category of $\mathbb{T}$-spectra over H, then $\mathcal{A}$ is the category of torsion $\mathbb{Q}[c_H]$-modules. The functor $\pi_*^{\mathcal{A}}$ is simply $\mathbb{T}$-equivariant homotopy $\pi_*^{\mathbb{T}}$. This category $\mathcal{A}$ is abelian and one dimensional. Accordingly, for two $\mathbb{T}$-spectra X and Y over H there is a split short exact sequence*

$$0 \longrightarrow \mathrm{Ext}_{\mathbb{Q}[c_H]}(\pi_*^{\mathbb{T}}(\Sigma X), \pi_*^{\mathbb{T}}(Y)) \longrightarrow [X,Y]_*^{\mathbb{T}} \longrightarrow \mathrm{Hom}_{\mathbb{Q}[c_H]}(\pi_*^{\mathbb{T}}(X), \pi_*^{\mathbb{T}}(Y)) \longrightarrow 0. \quad \square$$

The proof of this will be completed in Section 4.4. The short exact sequence is Theorem 3.1.1, and it is the central result of the analysis of Part I.

THEOREM 1.1.3. *If* $\mathbb{C} = \mathbb{T}\text{-}Spec/\mathcal{F}$ *is the category of* $\mathcal{F}$*-spectra, then* $\mathcal{A}$ *is the full subcategory of* $\mathcal{O}_{\mathcal{F}}$*-modules* M *of the form* $M = \bigoplus_H M(H)$ *for torsion* $\mathbb{Q}[c_H]$*-modules* $M(H)$. *We refer to these as* $\mathcal{F}$*-finite torsion modules, and they may also be described as the* $\mathcal{O}_{\mathcal{F}}$*-modules annihilated by inverting all Euler classes. The functor* $\pi^{\mathcal{A}}_*$ *is simply* $\mathbb{T}$*-equivariant homotopy* $\pi^{\mathbb{T}}_*$. *The category of* $\mathcal{F}$*-finite torsion modules is abelian and one dimensional. Accordingly, for two* $\mathcal{F}$*-spectra* X *and* Y *there is a split short exact sequence*

$$0 \longrightarrow \mathrm{Ext}_{\mathcal{O}_{\mathcal{F}}}(\pi^{\mathbb{T}}_*(\Sigma X), \pi^{\mathbb{T}}_*(Y)) \longrightarrow [X,Y]^{\mathbb{T}}_* \longrightarrow \mathrm{Hom}_{\mathcal{O}_{\mathcal{F}}}(\pi^{\mathbb{T}}_*(X), \pi^{\mathbb{T}}_*(Y)) \longrightarrow 0. \quad \square$$

The proof of this will be completed in Section 4.5.

THEOREM 1.1.4. *If* $\mathbb{C} = \mathbb{T}\text{-}Spec_{sf}$ *is the category of semi-free spectra, then* $\mathcal{A}$ *is the category whose objects are morphisms* $M \longrightarrow \mathbb{Q}[c, c^{-1}] \otimes V$ *of* $\mathbb{Q}[c]$*-modules (for some graded vector space* V*) which become isomorphisms when* c *is inverted. This category* $\mathcal{A}$ *is abelian and one dimensional. The functor* $\pi^{\mathcal{A}}_*$ *is defined by*

$$\pi^{\mathcal{A}}_*(X) := \left(\pi^{\mathbb{T}}_*(X \wedge DE\mathbb{T}_+) \longrightarrow \pi^{\mathbb{T}}_*(X \wedge DE\mathbb{T}_+ \wedge \tilde{E}\mathcal{F})\right),$$

where $D(\cdot)$ *denotes functional duality. Accordingly, for two semifree* $\mathbb{T}$*-spectra there is a split short exact sequence*

$$0 \longrightarrow \mathrm{Ext}_{\mathcal{A}}(\pi^{\mathcal{A}}_*(\Sigma X), \pi^{\mathcal{A}}_*(Y)) \longrightarrow [X,Y]^{\mathbb{T}}_* \longrightarrow \mathrm{Hom}_{\mathcal{A}}(\pi^{\mathcal{A}}_*(X), \pi^{\mathcal{A}}_*(Y)) \longrightarrow 0. \quad \square$$

Finally the model of all rational $\mathbb{T}$-spectra is as follows.

THEOREM 1.1.5. *If* $\mathbb{C} = \mathbb{T}\text{-}Spec$ *then* $\mathcal{A}$ *is the category whose objects are morphisms* $M \longrightarrow t^{\mathcal{F}}_* \otimes V$ *of* $\mathcal{O}_{\mathcal{F}}$*-modules (for some graded vector space* V*) which become isomorphisms when all Euler classes are inverted (i.e. the kernel and cokernel are* $\mathcal{F}$*-finite torsion modules). This category* $\mathcal{A}$ *is abelian and one dimensional. The functor* $\pi^{\mathcal{A}}_*$ *is defined by*

$$\pi^{\mathcal{A}}_*(X) := \left(\pi^{\mathbb{T}}_*(X \wedge DE\mathcal{F}_+) \longrightarrow \pi^{\mathbb{T}}_*(X \wedge DE\mathcal{F}_+ \wedge \tilde{E}\mathcal{F})\right),$$

where $D(\cdot)$ *denotes functional duality. Accordingly, for two* $\mathbb{T}$*-spectra there is a split short exact sequence*

$$0 \longrightarrow \mathrm{Ext}_{\mathcal{A}}(\pi^{\mathcal{A}}_*(\Sigma X), \pi^{\mathcal{A}}_*(Y)) \longrightarrow [X,Y]^{\mathbb{T}}_* \longrightarrow \mathrm{Hom}_{\mathcal{A}}(\pi^{\mathcal{A}}_*(X), \pi^{\mathcal{A}}_*(Y)) \longrightarrow 0. \quad \square$$

The proof of this is given in Section 5.6. It should be emphasized that $\mathrm{Hom}_{\mathcal{A}}(M,N)$ and $\mathrm{Ext}_{\mathcal{A}}(M,N)$ are routinely computable, and that, because we are working rationally, there is usually no serious trouble in calculating $\pi^{\mathcal{A}}_*(X)$.

Part I begins with the concrete and moves towards the abstract in two steps. Thus we begin with the cohomology theories, move on to homotopy theory, pass to algebra by an Adams spectral sequence, and finally package this in categorical terms. Here is a more detailed outline of contents.

We begin with two sections which can be expressed in classical terms. These give evidence that there is some complexity in rational $\mathbb{T}$-equivariant cohomology theories, but not too much. In particular they give some evidence for the simplicity of $\mathcal{F}$-objects.

After this, the discussion is conducted in the Lewis-May **[19]** stable category of $\mathbb{T}$-spectra. The first step is to introduce the basic building blocks and the methods for breaking up general objects. This gives us the setting to construct an Adams spectral sequence, which provides the connection between topology and algebra. Once the Adams spectral sequence for $\mathbb{T}$-Spec$/H$ has been constructed, we need only do some algebra and certain formal manipulations to obtain and exploit all the algebraic models. We have taken the view that an abstract machine should only be introduced when there is a particular case on which its operation can be illustrated. Accordingly we have not described the transition from an Adams spectral sequence to an algebraic model (in Section 4.2) until we have constructed the simplest instance to which it applies. On the other hand Section 4.2 may be relevant in quite different settings, and it is written axiomatically so that it can be read and applied independently of the preceding sections.

1.2. Reading Guide for Part I.

Some readers may not wish to read all of the material in Part I, so we provide further guidance here.

Those only interested in the Atiyah-Hirzebruch spectral sequence for the K-theory of an $\mathcal{F}$-space will only need to read Sections 1.3, 1.4, 2.1, referring to Appendix A for the necessary facts about Mackey functors. Sections 1.3 and 1.4 are not used elsewhere in Part I. We shall return to the Atiyah-Hirzebruch spectral sequence in Section 13.1 of Part III, where we give more complete results.

Those interested in Mackey functors should read Section 2.1 and then refer to Appendix A. Mackey functors are not used until we consider ordinary cohomology theories in Chapter 12 from Part III.

The central material constructing the main Adams spectral sequence for the categories of $\mathcal{F}$-spectra and $\mathbb{T}$-spectra over H is to be found in Chapters 2 and 3. Maps from $\mathcal{F}$-contractible spectra to $\mathcal{F}$-free spectra are deduced in Sections 5.1 and 5.3. This is sufficient to answer most direct questions about particular $\mathbb{T}$-spectra, and may satisfy some readers. On the other hand some readers may wish to understand the shape of the algebraic models without reading these chapters: they should move directly to Chapters 5 and 6.

In Chapter 4, we explain the abstract process of reaching an algebraic model from an Adams spectral sequence and we illustrate it for $\mathbb{T}$-spectra over H. The goal of a full algebraic model is fulfilled in Chapter 5. We deduce the remaining topological input from the Adams spectral sequence in Sections 5.1 and 5.3, and construct the algebraic model in Section 5.4. It is then a simple matter to show in Section 5.6 that the algebra does indeed model the topology. Chapter 6 completes the circle by introducing the torsion model, closely following geometric intuition, and by showing that it gives a model equivalent to the standard model.

1.3. Haeberly's example.

We give two examples showing there is no Chern character isomorphism, for $\mathbb{T}$-equivariant K-theory: the first is a representation sphere, and the second is Haeberly's example [**17**]. Each example involves constructing a $\mathbb{T}$-space X whose equivariant K-theory is concentrated in even degrees, but whose ordinary cohomology with coefficients in the rationalized representation ring functor is nonzero in odd degrees. Since the homotopy functors of the K-theory spectrum are in even degrees, the K-theory cannot be a product of copies of ordinary cohomology. In the next section we give a proof of a generalization of McClure's result that there is a Chern isomorphism for $\mathbb{T}$-spaces X with $X^{\mathbb{T}}$ trivial.

First let c be the natural representation of $\mathbb{T}$ on the complex numbers, and let $X = S^{2c}$. By Bott periodicity $K^*_{\mathbb{T}}(S^{2c}) \cong K^*_{\mathbb{T}}(S^0)$, which is in even degrees. However, we have the cell structure

$$S^{2c} = S^0 \cup e^1 \wedge \mathbb{T}_+ \cup e^3 \wedge \mathbb{T}_+,$$

showing that the ordinary cohomology with coefficients in a Mackey functor M is the cohomology of the cochain complex

$$M(\mathbb{T}) \longrightarrow M(1) \longrightarrow 0 \longrightarrow M(1).$$

In particular the $H^3_{\mathbb{T}}(S^{2c}; M) = M(1)$. When M is the representation ring Mackey functor $M(1) = R(1) = \mathbb{Q}$.

Haeberly's example is a more complicated version of the same phenomenon. To explain, it is convenient to consider the group $\Gamma = \mathbb{T} \times \mathbb{T}'$ where both $\mathbb{T}$ and $\mathbb{T}'$ are copies of the circle group. The group Γ has a 3-dimensional complex representation $V = (1 \oplus t \oplus t^2) \otimes t'$, where t is the natural representation of $\mathbb{T}$ on $\mathbb{C}$, and similarly for $\mathbb{T}'$. We may consider the unit sphere $S(V)$ as a Γ-space, give it a disjoint basepoint and then form the $\mathbb{T}$-space $X = S(V)_+/\mathbb{T}'$. We could equally well describe X as a copy of $\mathbb{C}P^2_+$ on which $\mathbb{T}$ acts via $s(z_0 : z_1 : z_2) = (z_0 : sz_1 : s^2 z_2)$. From the first description it is easy to calculate the K-theory since we have $K^*_{\mathbb{T}}(X) = K^*_{\Gamma}(S(V)_+)$, because $S(V)$ is free as a $\mathbb{T}'$-space. Indeed, the cofibre sequence $S(V)_+ \longrightarrow S^0 \longrightarrow S^V$ of Γ-spaces gives an exact sequence

$$\cdots \longrightarrow K^i_{\Gamma}(S^V) \xrightarrow{\lambda(V)} K^i_{\Gamma}(S^0) \longrightarrow K^i_{\Gamma}(S(V)_+) \longrightarrow K^{i+1}_{\Gamma}(S^V) \longrightarrow \cdots .$$

Now by Bott periodicity $K^i_{\Gamma}(S^V)$ is $R(\Gamma)$ if i is even and 0 if i is odd, and because the degree 0 Euler class $\lambda(V) = (1-t')(1-tt')(1-t^2t')$ is not a zero divisor in $R(\Gamma) = \mathbb{Z}[t, t^{-1}, t', (t')^{-1}]$ we find

$$K^0_{\mathbb{T}}(X) = R(\Gamma)/\lambda(V) \text{ and } K^1_{\mathbb{T}}(X) = 0.$$

In particular the K-theory of X is entirely in even degrees.

On the other hand from the second description one may find a cell structure on X. Points $(z_0 : z_1 : z_2)$ with two coordinates zero are $\mathbb{T}$-fixed, points with $z_1 = 0$ have isotropy group C_2, and all other points have isotropy group 1. We may construct a cell structure with 1-skeleton $X^{(1)}$ the subspace of points with at least one coordinate zero. The three fixed points $(1 : 0 : 0)$, $(0 : 1 : 0)$, and $(0 : 0 : 1)$ are joined by three 1-cells. The tube of points $(z_0 : 0 : z_2)$ with $z_0 z_2 \neq 0$ forms a 1-cell with isotropy C_2; it is attached to $(1 : 0 : 0)$ at one end and $(0 : 0 : 1)$ at the other. Two similar tubes joining the other two pairs of fixed points form 1-cells with isotropy 1. The remaining points may be uniquely expressed in the form $(1 : z_1 : z_2)$ with $z_1 z_2 \neq 0$, and each orbit has a unique representative with z_1 a positive real number. Now decompose this space of representatives $\mathbb{R}_{>0} \times (\mathbb{C} \setminus \{0\})$ into the 2-cell of pairs of positive real numbers, and the 3-cell of other points; this completes the $\mathbb{T}$-CW decomposition of X. Note that the 2-cell $\mathbb{R}_{>0} \times \mathbb{R}_{>0}$ has its boundary naturally divided into three parts, each of which is attached to one of the 1-cells. Thus X may be given a $\mathbb{T}$-CW structure with two free 1-cells, one free 2-cell and one free 3-cell. Hence for any Mackey functor M we see that $H^*_{\mathbb{T}}(X; M)$ is the cohomology of a complex

$$3M(\mathbb{T}) \xrightarrow{d^0} M(C_2) \oplus 2M(1) \xrightarrow{d^1} M(1) \xrightarrow{d^2} M(1).$$

Furthermore, d^1 is surjective, since the attaching map of the 2-cell is of degree ± 1 to each of the free 1-cells. Thus $H^3(X; M) = M(1)$, and in particular if M is the rationalized representation ring Mackey functor this is the non-zero group $\mathbb{Q}$.

1.4. McClure's Chern character isomorphism for $\mathcal{F}$-spaces.

McClure has observed that the if X is an $\mathcal{F}$-space then the Atiyah-Hirzebruch spectral sequence for the K-cohomology of X does collapse at E_2. His proof involves appealing to unstable results and the work of Slominska. We shall give a proof of the corresponding statement for any cohomology theory whose homotopy functors are concentrated entirely in even degrees, and of the corresponding statement for homology theories. This applies in particular to K theory, by the Bott periodicity theorem. In Section 13.1 of Part III we shall give a necessary and sufficient condition for the collapse of the Atiyah-Hirzebruch spectral sequence for $\mathcal{F}$-spaces, which will give an alternative to the proof of this section.

Before stating the theorem, we recall that for each integer k it is appropriate to consider the entire system of homotopy groups $\pi^H_k(X) = [G/H_+ \wedge S^k, X]^{\mathbb{T}}$ as H runs through all subgroups of $\mathbb{T}$. It is appropriate to regard this as a functor $\underline{\pi}^{\mathbb{T}}_k(X) : G/H_+ \longmapsto [G/H_+ \wedge S^k, X]^{\mathbb{T}}$, on the category of stable orbits. An additive functor of this form is called a Mackey functor; we examine the algebraic structure of the category of rational Mackey functors in Appendix A, but for the present we only need the basic terminology. In line with the usual abbreviation, we write the coefficient functor $\underline{\pi}^{\mathbb{T}}_k(K)$ as $\underline{K}^{\mathbb{T}}_k$.

Since the orbits are the equivariant analogues of points, an ordinary cohomology theory is one for which the cohomology of each orbit is concentrated in degree zero. Thus ordinary cohomology theories correspond to Mackey functors M, and they are represented by Eilenberg-MacLane spectra HM.

THEOREM 1.4.1. *If K is any rational $\mathbb{T}$-spectrum with homotopy functors $\underline{K}^{\mathbb{T}}_m = 0$ for all odd integers m then for any $\mathcal{F}$-space X there are isomorphisms*
(a)

$$K^*_{\mathbb{T}}(X) \cong \prod_{n \in \mathbb{Z}} H^*_{\mathbb{T}}(\Sigma^{2n} X; \underline{K}^{\mathbb{T}}_{-2n})$$

and
(b)

$$K^{\mathbb{T}}_*(X) \cong \bigoplus_{n \in \mathbb{Z}} H^{\mathbb{T}}_*(\Sigma^{2n} X; \underline{K}^{\mathbb{T}}_{2n}).$$

This follows from a geometric statement.

THEOREM 1.4.2. *If K is any rational $\mathbb{T}$-spectrum with homotopy functors $\underline{K}^{\mathbb{T}}_m = 0$ for all odd integers m then*
(a)

$$F(E\mathcal{F}_+, K) \simeq F(E\mathcal{F}_+, \prod_{n\in\mathbb{Z}} \Sigma^{2n} H(\underline{K}^{\mathbb{T}}_{2n}))$$

and
(b)

$$K \wedge E\mathcal{F}_+ \simeq \bigvee_{n\in\mathbb{Z}} E\mathcal{F}_+ \wedge \Sigma^{2n} H(\underline{K}^{\mathbb{T}}_{2n}).$$

To see how Theorem 1.4.1 follows from 1.4.2 we use a lemma which is immediate from the definition of $E\mathcal{F}_+$ and its unreduced suspension $\tilde{E}\mathcal{F}$.

LEMMA 1.4.3. *For any $\mathcal{F}$-spectrum X,*
(a) $X \wedge \tilde{E}\mathcal{F} \simeq *$ and hence $X \simeq E\mathcal{F}_+ \wedge X$; also
(b) for any $\mathbb{T}$-spectrum Y we have $F(X, Y \wedge \tilde{E}\mathcal{F}) \simeq *$ and hence $F(X, Y \wedge E\mathcal{F}_+) \simeq F(X, Y)$. □

By 1.4.3 (a), Theorem 1.4.1 follows by applying $F(X, \ )$ to Part (a) of 1.4.2 and $X \wedge (\cdot)$ to Part (b) of 1.4.2 and taking homotopy groups.

Proof: We turn to the proof of 1.4.2. Note first that it is enough to prove Part (b); indeed, by 1.4.3 (b), Part (a) follows by applying $F(E\mathcal{F}_+, \cdot)$ to the equivalence of Part (b).

It is enough to construct a $\mathbb{T}$-map $\theta : K \wedge E\mathcal{F}_+ \longrightarrow E\mathcal{F}_+ \wedge \bigvee_{n\in\mathbb{Z}} \Sigma^{2n} H(\underline{K}_{2n})$ which is an H-equivalence for all finite subgroups H. By the Whitehead theorem it is sufficient that θ induces an isomorphism of π^H_* for all finite subgroups H. By 1.4.3 (b) again, it is equivalent to give the composite

$$\theta' : K \wedge E\mathcal{F}_+ \longrightarrow \bigvee_{n\in\mathbb{Z}} \Sigma^{2n} H(\underline{K}_{2n}),$$

and since this wedge is equivalent to the product we may specify θ' by giving its components. These are elements of the cohomology groups $[K \wedge E\mathcal{F}_+, HM]^*_{\mathbb{T}} = H^*_{\mathbb{T}}(K \wedge E\mathcal{F}_+; M)$ for various Mackey functors M. Accordingly we set about calculating the cohomology of $K \wedge E\mathcal{F}_+$.

The idea is to filter $E\mathcal{F}_+$ so that the subquotients are analogues of cells, but with all elements of finite order as isotropy groups. This extends the idea of [**8**]. Thus we note that if $H \subseteq L$ we have a projection $\mathbb{T}/H \longrightarrow \mathbb{T}/L$, and that the subgroups of finite order form a directed set. We may therefore let $\mathbb{T}/\!\!/\mathcal{F}_+ := \underset{\rightarrow H}{\text{holim}}\ \mathbb{T}/H_+$ where the limit is over all finite subgroups H (or over a cofinal sequence if that appears more comfortable). Analogously, if H is a finite subgroup of order n we may let $V(H)$ denote the representation t^n with kernel H, and there are maps $mV(H) \longrightarrow mV(L)$ (of degree $|L/H|^m$) for all m. We let $S(mV(\mathcal{F}))_+ := \underset{\rightarrow H}{\text{holim}}\ S(mV(H))_+$ for $0 \leq m \leq \infty$. The usefulness of these constructions is summarized in a lemma.

LEMMA 1.4.4. *The infinite sphere $S(\infty V(\mathcal{F}))_+$ is a model for $E\mathcal{F}_+$. We thus have a filtration*

$$* = S(0V(\mathcal{F}))_+ \subseteq S(1V(\mathcal{F}))_+ \subseteq S(2V(\mathcal{F}))_+ \subseteq \cdots \subseteq S(\infty V(\mathcal{F}))_+ = E\mathcal{F}_+$$

and the subquotients are generalized cells

$$S(mV(\mathcal{F}))_+/S((m-1)V(\mathcal{F}))_+ \simeq S^{2m-2} \wedge \mathbb{T}/\!\!/\mathcal{F}_+$$

for $1 \leq m < \infty$. Furthermore,

$$E\mathcal{F}_+ = \mathbb{T}/\!\!/\mathcal{F}_+ \cup \mathbb{T}/\!\!/\mathcal{F}_+ \wedge e^2 \cup \mathbb{T}/\!\!/\mathcal{F}_+ \wedge e^4 \cup \mathbb{T}/\!\!/\mathcal{F}_+ \wedge e^6 \cup \cdots .$$

Proof: Since $(S(mV(H)))^L = \emptyset$ if $L \not\subseteq H$ or $S(mV(H))$ if $L \subseteq H$ the fact that $S(\infty V(\mathcal{F}))_+$ is a universal space is clear. By definition $S(V(\mathcal{F}))_+ = \mathbb{T}/\!\!/\mathcal{F}_+$. To identify the higher quotients and give the 'cell' structure we use the fact that for $m \geq 2$ the cofibre sequences

$$S^{2m-3} \wedge \mathbb{T}/H_+ \longrightarrow S((m-1)V(H))_+ \longrightarrow S(mV(H))_+$$

fit into a direct system. □

Thus, for any spectrum K, we may form the spectral sequence of the filtered spectrum $K \wedge E\mathcal{F}_+$ which will have the form

$$E_1^{s,t} = H_{\mathbb{T}}^{s+t}(K \wedge (E\mathcal{F}_+^{(s)}/E\mathcal{F}_+^{(s-1)}); M) \Rightarrow H_{\mathbb{T}}^{s+t}(K \wedge E\mathcal{F}_+; M).$$

Indeed, from the form of the filtration, we find the spectral sequence is concentrated in the first quadrant in terms with even s where we have

$$E_1^{2m,t} = H_{\mathbb{T}}^t(K \wedge \mathbb{T}/\!\!/\mathcal{F}_+; M).$$

Using the change of groups isomorphism $H_{\mathbb{T}}^*(K \wedge \mathbb{T}/H_+; M) = H_H^*(K; M)$, we have a Milnor exact sequence

$$0 \longrightarrow \underset{\leftarrow H}{\lim}{}^1\, H_H^{t-1}(K; M) \longrightarrow H_{\mathbb{T}}^t(K \wedge \mathbb{T}/\!\!/\mathcal{F}_+; M) \longrightarrow \underset{\leftarrow H}{\lim}\, H_H^t(K; M) \longrightarrow 0.$$

It is in the analysis of this exact sequence that it is essential we are working rationally. Indeed, because H is finite, every rational H-spectrum is a product of Eilenberg-MacLane spectra and these are necessarily also Moore spectra. It now follows that, provided K has its homotopy functors in even degrees, the groups $H_H^t(K; M)$ are only nonzero for even t. The collapse of the spectral sequence is thus ensured once we show the $\underset{\leftarrow}{\lim}{}^1$ terms vanish. In fact the restriction maps

$$H_L^t(K; M) \longrightarrow H_H^t(K; M)$$

are surjective. Perhaps the quickest way to see this is to note that $H_H^*(HM'; M) = [HM', HM]_H^* = \mathrm{Hom}_H(M', M)$, for any Mackey functors M' and M. We may then use the corresponding fact for Mackey functors, that

$$\mathrm{Hom}_L(M', M) \longrightarrow \mathrm{Hom}_H(M', M)$$

is surjective. This surjectivity is due to the fact that all Weyl groups are connected, and it is easily deduced from Appendix A.

We conclude that if K has all its homotopy functors in even degrees then

$$H^*_{\mathbb{T}}(K \wedge E\mathcal{F}_+; M) = \prod_{m \in \mathbb{Z}} \lim_{\leftarrow H} H^*_H(\Sigma^{2m} K; M),$$

and in particular we can find a map

$$\theta'_{2m} : K \wedge E\mathcal{F}_+ \longrightarrow \Sigma^{2m} H(\underline{K}^{\mathbb{T}}_{2m})$$

inducing the identity in $\pi^H_{2m}(\bullet)$ for all finite subgroups H. The map

$$\theta' : K \wedge E\mathcal{F}_+ \longrightarrow \bigvee_{n \in \mathbb{Z}} \Sigma^{2n} H(\underline{K}^{\mathbb{T}}_{2n})$$

is thus an $\mathcal{F}$-equivalence and hence θ is a homotopy equivalence as required. □

In Section 13.1 of Part III we shall complete the picture of Atiyah-Hirzebruch spectral sequences for $\mathcal{F}$-spaces by giving an analysis without hypothesis on the rational cohomology theory. We characterize those theories $K^*_{\mathbb{T}}(\cdot)$ for which the spectral sequence always collapses at E_2, show that arbitrarily high differentials occur in general, and give a geometric explanation of them in terms of universal examples. The behaviour of the spectral sequence for arbitrary spaces X is much more complicated.

CHAPTER 2

Topological building blocks.

This chapter introduces all the basic ingredients for our analysis. In Section 2.1 we summarize what is known about the homotopy groups of the cells $\mathbb{T}/H_+$, and use idempotents of the Burnside rings to break them up; in Section 2.2 we show this extends to a complete splitting of the category of $\mathcal{F}$-spectra as a product of categories, one for each finite subgroup. In Section 2.3 we analyse the summands $E\langle H\rangle$ of $E\mathcal{F}_+$; their simplicity is fundamental to the success of our programme. The ring of self-maps of $E\langle H\rangle$ evidently acts as an algebra of natural operations on the homotopy groups of $\mathbb{T}$-spectra over H; in Section 2.4 we identify it as the principal ideal domain $\mathbb{Q}[c_H]$.

2.1. Natural cells and basic cells.

The idea is that in any context one should understand complicated objects by first understanding the building blocks and the way they can be stuck together. In practice this involves a complete description of the full subcategory on the basic building blocks. There are inevitably many different choices for the set of basic objects, and for practical purposes it is useful to choose those for which the basic subcategory is as simple as possible. The ideal is typified by representation theory over the complex numbers where where every morphism between simple modules is a scalar multiplication.

The natural building blocks for locally nice G-spaces are the G-cells $G/H_+ \wedge S^n$. The full category on these is complicated because of the group theory (which enters via the fixed point spaces $(G/H)^K$) and because of the topology (since we need to know the homotopy groups of spheres). Working over the rationals brings the topology under reasonable control, but the group theoretic complication is still considerable. However, if we work stably, we obtain a reasonable supply of idempotents which we use to chop up the cells into simpler pieces: the residual group theory and topology is much simpler for these pieces.

It is time to be quite definite, and we return to the case $G = \mathbb{T}$. This case is particularly favourable for several reasons. Firstly $\mathbb{T}$ is abelian, secondly there is only one infinite subgroup, and finally all Weyl groups are connected, so that there is hardly any group theoretic complication at all. There are two types of natural cells: the $\mathbb{T}$-fixed cells S^n and the cells $\mathbb{T}/H_+ \wedge S^n$ with finite isotropy group. The stable rational maps between them are not hard to calculate, using the standard change of groups isomorphisms and tom Dieck splitting. We shall have to refer to the rationalized Burnside ring $A(H)$ of finite H-sets; this is relevant because of Segal's result that $A(H) \cong [S^0, S^0]^H$ [**19**, V.2.11]. Here and elsewhere we shall always refer to homological grading unless otherwise specified, and we use the homology suspension $(\Sigma^t M)_n = M_{n-t}$.

LEMMA 2.1.1. *(a) The graded self maps of the fixed cell S^0 are*

$$[S^0, S^0]^{\mathbb{T}}_* = \mathbb{Q} \oplus \Sigma(\mathbb{Q}\mathcal{F}\,[d])$$

where $\mathbb{Q}\mathcal{F}$ denotes a rational vector space with basis $\mathcal{F}$, and d is an element of degree 2.
(b) The other maps involving the fixed cell are

$$[S^0, \mathbb{T}/H_+]^{\mathbb{T}}_* = \Sigma A(H)$$

and

$$[\mathbb{T}/H_+, S^0]^{\mathbb{T}}_* = A(H)$$

(c) Maps between two cells with finite isotropy groups are

$$[\mathbb{T}/H_+, \mathbb{T}/K_+]^{\mathbb{T}}_* = A(H \cap K) \oplus \Sigma A(H \cap K)$$

For the present the element d is simply a notational convenience, but we shall eventually introduce an operation c of degree -2 so that $cd^0 = 0$ and $cd^k = d^{k-1}$ for $k \geq 1$.

Proof: The tom Dieck splitting theorem [**19**, V.9.1] states that for any $\mathbb{T}$-space X

$$\pi^{\mathbb{T}}_*(X) = \pi_*(X) \oplus \bigoplus_H \pi_*(E(\mathbb{T}/H)_+ \wedge_{\mathbb{T}/H} \Sigma X^H).$$

Part (a) follows, as does Part (b)(i).

For Part (b)(ii) and Part (c) we use the change of groups isomorphism

$$[\mathbb{T} \wedge_H X, Y]^{\mathbb{T}}_* \cong [X, Y]^H_*$$

and the Wirthmüller isomorphism [**19**, II.6.5]

$$[Y, \mathbb{T} \wedge_H X]^{\mathbb{T}}_* \cong [Y, S^1 \wedge X]^H_*,$$

where X is an H-spectrum and Y is a $\mathbb{T}$-spectrum. To complete the proof of (c) we note that as an H-spectrum $\mathbb{T}/K_+$ is $S(U)_+$ where U is a representation with kernel $L = H \cap K$. Choosing an H-orbit in $S(U)$ for an equivariant 0-cell, we see there is a cofibre sequence of H-spaces

$$\Sigma H/L_+ \longrightarrow \Sigma H/L_+ \longrightarrow \Sigma S(U)_+$$

where the first map is $1 - g$ for some generator g of the cyclic group H/L. Since the group is abelian, g induces the identity in π^H_1, and Part (c) follows. □

The point to notice is that there is vertical complication (i.e. there are nontrivial maps of positive degree) and there is horizontal complication (i.e. the maps of degree zero do not form a simple ring). The vertical complexity is topological in origin, coming from the facts that $\mathbb{T}$ is of dimension 1 and that $B\mathbb{T}$ has positive dimensional rational cohomology. This is unavoidable, and indeed it is the source of the interesting mathematics. On the other hand we may eliminate the horizontal complexity simply by using simpler building blocks.

For this we need idempotents from the Burnside rings. Here we use the isomorphism $A(H) = [S^0, S^0]^H$ and the ring isomorphism $\phi : A(H) \xrightarrow{\cong} \prod_{K \subseteq H} \mathbb{Q}$ whose Kth coordinate takes $f : S^0 \longrightarrow S^0$ to the degree of the fixed point map $\Phi^H(f) : S^0 \longrightarrow S^0$, where Φ^H is the geometric H-fixed point functor [**19**, V.2]. We let e_K denote the idempotent which is 1 on the Kth factor and 0 on the others. This notation is consistent with restriction in the sense that if $K \subseteq H' \subseteq H$ then $res^H_{H'} e_K = e_K$. Notice that for any H-spectrum X we have $e_K X = (e_K S^0) \wedge X$ and hence we have non-equivariant equivalences $\Phi^L(e_K X) \simeq *$ if $L \neq K$, whilst $\Phi^K e_K X = \Phi^K X$.

DEFINITION 2.1.2. The *basic cells* are the $\mathbb{T}$-spectra

$$\sigma^n_{\mathbb{T}} = S^n$$

$$\sigma^n_H = \mathbb{T}_+ \wedge_H e_H S^n$$

for various integers n and finite subgroups H of $\mathbb{T}$.

We shall need a couple of elementary observations about basic cells.

LEMMA 2.1.3. *The geometric fixed points of a basic cell σ^n_H are non-equivariantly*

$$\Phi^K(\sigma^n_H) \simeq \begin{cases} S^1 \vee S^0 & \text{if } K = H \\ * & \text{if } K \neq H \end{cases}$$

Before giving a proof we record a useful lemma, which gives a substitute for the Mackey formula for the restriction of an induction.

LEMMA 2.1.4. *If $h \in H$ is a generator of H and Y is any H-spectrum, there is an H-equivariant cofibre sequence*

$$Y \xrightarrow{1-h} Y \longrightarrow \mathbb{T}_+ \wedge_H Y.$$

Proof: Apply $\wedge_H Y$ to the cofibre sequence $H_+ \xrightarrow{1-h} H_+ \longrightarrow \mathbb{T}_+$ of free left and right H-spaces. It is clear this makes sense and gives a cofibre sequence at the space level; it does not interact with the structure maps [**19**, II.4.1] and so applies also to spectra. □

We may now prove 2.1.3.

Proof: Since σ^n_H is a retract of $\mathbb{T}/H_+ \wedge S^n$ it can only have geometric fixed points for $K \subseteq H$. Applying the lemma we have a cofibre sequence

$$e_H S^n \xrightarrow{1-h} e_H S^n \longrightarrow \sigma^n_H.$$

If $K \neq H$ the result is clear since $\Phi^K e_H S^n \simeq *$, and for $K = H$ we use the fact that $h = 1$ on S^n. □

LEMMA 2.1.5. *(i) The duals of the basic cells are as follows: $D\sigma^0_{\mathbb{T}} = \sigma^0_{\mathbb{T}}$ and $D\sigma^0_H \simeq \Sigma^{-1}\sigma^0_H$.*
(ii) If $K \subseteq H$ there is an equivalence

$$\sigma^0_K \simeq \mathbb{T}_+ \wedge_H e_K S^0.$$

(iii) The natural cells can be obtained from the basic cells by the formula

$$\mathbb{T}/H_+ = \bigvee_{K \subseteq H} \sigma^0_K.$$

Note in particular that this means $\mathbb{T}/\!/\mathcal{F}_+ = \bigvee_H \sigma_H^0$, which completes the connection with Section 1.4 on McClure's theorem.

Proof: (i) The duality result follows from the fact that any map $f : S^0 \longrightarrow S^0$ is self-dual.
(ii) Consider the projection $H/K_+ \longrightarrow H/H_+ = S^0$; we construct a comparison map

$$\sigma_K^0 = \mathbb{T}_+ \wedge_K e_K S^0 \longrightarrow \mathbb{T}_+ \wedge_H e_K S^0$$

by applying $\mathbb{T} \wedge_H e_K(\cdot)$ and using compatibility of e_K under restriction. The cofibre is $\mathbb{T}_+ \wedge_H e_K \widetilde{H/K}$, where the tilde denotes unreduced suspension. Since this cofibre has no geometric L-fixed points unless $L \subseteq H$, it suffices to show it is H-contractible. By 2.1.4, $\mathbb{T}_+ \wedge_H e_K \widetilde{H/K}$ is the mapping cone of $1-h : e_K H/K \longrightarrow e_K H/K$, so it suffices to show $1-h$ is an equivalence in K-fixed points as a self-map of $\widetilde{H/K}$. Since $\widetilde{H/K}$ is K-equivariantly a wedge of circles, this follows from the Snake Lemma applied to the diagram

$$\begin{array}{ccccccc} 0 & \longrightarrow & J(H/K) & \longrightarrow & \mathbb{Z}[H/K] & \longrightarrow & \mathbb{Z} & \longrightarrow & 0 \\ & & 1-h\downarrow & & 1-h\downarrow & & 1-h\downarrow & & \\ 0 & \longrightarrow & J(H/K) & \longrightarrow & \mathbb{Z}[H/K] & \longrightarrow & \mathbb{Z} & \longrightarrow & 0 \end{array}$$

(iii) For the splitting we use the rational mark isomorphism $A(H) \xrightarrow{\cong} \prod_K \mathbb{Q}$ to see that $1 = \Sigma_K e_K$, and hence that the natural map $S^0 \xrightarrow{\simeq} \bigvee_K e_K S^0$ is an H-equivalence. Now use induction to extend it to a $\mathbb{T}$-equivalence $\mathbb{T}/H_+ = \mathbb{T}_+ \wedge_H S^0 \xrightarrow{\simeq} \bigvee_K \mathbb{T}_+ \wedge_H e_K S^0$. Now use Part (ii). □

Obviously the analogue of 2.1.1 (a) for basic cells is identical, but Parts (b) and (c) are much simpler.

LEMMA 2.1.6. *(a) The graded self maps of the fixed cell $\sigma_{\mathbb{T}}^0$ are*

$$[\sigma_{\mathbb{T}}^0, \sigma_{\mathbb{T}}^0]_*^{\mathbb{T}} = \mathbb{Q} \oplus \Sigma(\mathbb{Q}\mathcal{F}\,[d])$$

where $\mathbb{Q}\mathcal{F}$ denotes a rational vector space with basis $\mathcal{F}$, and d is an element of degree 2.
(b) The other maps involving the fixed cell are

$$[\sigma_{\mathbb{T}}^0, \sigma_H^0]_*^{\mathbb{T}} = \Sigma\mathbb{Q}$$

and

$$[\sigma_H^0, \sigma_{\mathbb{T}}^0]_*^{\mathbb{T}} = \mathbb{Q}$$

(c) Maps between two cells with finite isotropy groups are

$$[\sigma_H^0, \sigma_H^0]_*^{\mathbb{T}} = \mathbb{Q} \oplus \Sigma\mathbb{Q}$$

and

$$[\sigma_H^0, \sigma_K^0]_*^{\mathbb{T}} = 0$$

if $H \neq K$.

For the present the element d is simply a notational convenience, but we shall eventually introduce an operation c of degree -2 so that $cd^0 = 0$ and $cd^k = d^{k-1}$ for $k \geq 1$.

Proof: The result for $[\sigma_H^0, \sigma_{\mathbb{T}}^0]_*^{\mathbb{T}}$ is clear from the definitions, and the result for $[\sigma_{\mathbb{T}}^0, \sigma_H^0]_*^{\mathbb{T}}$ follows by duality.

The proof of Part (c) is essentially obtained by applying the idempotent e_H to the proof of 2.1.1. We change groups to see $[\sigma_H^0, \sigma_K^0]_*^{\mathbb{T}} = [e_H S^0, \sigma_K^0]_*^H$. If $K \neq H$ this is zero since σ_K^0 only has geometric L-fixed points if $L = K$, and so $e_H \sigma_K^0$ has no non-trivial geometric fixed points and is thus H-contractible. If $K = H$ we apply 2.1.4, together with the fact that $1 - h$ is the zero map of S^0. □

There are certain features worth special attention. The crudest is that there are no non-trivial maps $\sigma_H^0 \longrightarrow \sigma_K^0$ if $H \neq K$. Next, notice that except for degree zero the graded ring $[\sigma_{\mathbb{T}}^0, \sigma_{\mathbb{T}}^0]_*^{\mathbb{T}}$ is entirely in odd degrees, and hence its product structure is trivial. Finally, every degree zero map from $\sigma_{\mathbb{T}}^0$ to another basic cell is either zero or an isomorphism in homotopy: this is responsible for the finite dimensionality of the category of $\mathbb{T}$ -Mackey functors.

The curious reader may wish to read Appendix A; this will complete the justification of Section 1.4, but will otherwise not be needed until Part II. We also pursue the analysis of the graded category of stable maps between cells a little further in Section 13.2.

2.2. Separating isotropy types.

Building on the observations on cells in the Section 2.1, we pull the whole category of rational $\mathbb{T}$-spectra apart on similar lines. Obviously the fixed cell $\sigma_{\mathbb{T}}^0$ behaves quite differently from the others, so it is appropriate to separate $\mathbb{T}$ -fixed phenomena from $\mathcal{F}$-phenomena. The success of the present method depends on the fact that, in addition, the effects of different finite subgroups are completely decoupled. This idea is firmly based in geometric intuition since we are separating a spectrum X into parts $X(H)$ which only have geometric fixed points at H (i.e. $\Phi^K X(H)$ is non-equivariantly contractible if $K \neq H$).

Let $\mathbb{T}$-Spec be the stable homotopy category of rational $\mathbb{T}$-spectra, and consider the full subcategory $\mathbb{T}$-Spec$/\mathbb{T}$ of $\mathcal{F}$-contractible $\mathbb{T}$-spectra X (i.e. with $\Phi^H X$ non-equivariantly contractible for all finite subgroups H), and the full subcategory $\mathbb{T}$-Spec$/\mathcal{F}$ of $\mathcal{F}$-spectra. We view the category $\mathbb{T}$-Spec as being generated by these two subcategories. More formally we consider the isotropy separation cofibering

$$E\mathcal{F}_+ \wedge X \longrightarrow X \longrightarrow \tilde{E}\mathcal{F} \wedge X :$$

the last term $\tilde{E}\mathcal{F} \wedge X$ lies in $\mathbb{T}$-Spec$/\mathbb{T}$ since $\Phi^H(\tilde{E}\mathcal{F} \wedge X) \simeq (\Phi^H \tilde{E}\mathcal{F}) \wedge (\Phi^H X)$, which is non-equivariantly contractible for all finite subgroups H, and the first term $E\mathcal{F}_+ \wedge X$ is visibly an $\mathcal{F}$-spectrum. We thus let $X(\mathbb{T}) = \tilde{E}\mathcal{F} \wedge X$ and $X(\mathcal{F}) = E\mathcal{F}_+ \wedge X$. Our approach is to analyse the categories $\mathbb{T}$-Spec$/\mathbb{T}$ and $\mathbb{T}$-Spec$/\mathcal{F}$ rather completely and then study how to assemble them to make $\mathbb{T}$-Spec. The analysis of $\mathbb{T}$-Spec$/\mathcal{F}$ forms the main content of the work, and it begins by observing that the decoupling implies that $\mathbb{T}$-Spec$/\mathcal{F} \simeq \prod_H \mathbb{T}$-Spec$/H$ where $\mathbb{T}$-Spec$/H$ is the full subcategory on $\mathbb{T}$ -spectra X with $\Phi^K X$ non-equivariantly contractible if $K \neq H$. In particular $X(\mathcal{F}) \simeq \bigvee_H X(H)$ in $\mathbb{T}$-Spec$/\mathcal{F}$, for suitable spectra $X(H)$ in $\mathbb{T}$-Spec$/H$. In a sense to be made precise below $\mathbb{T}$-Spec$/H$ has global dimension 1 and is thus easily controlled.

The analysis of $\mathbb{T}$-Spec/$\mathbb{T}$ is straightforward obstruction theory, together with the classical analysis of the nonequivariant category of rational spectra. The obstruction theory is easily done integrally. This is standard at the space level and may be deduced for spectra by passage to limits; alternatively it may be deduced from [**19**, II.9].

PROPOSITION 2.2.1. *(i) $\Phi^H X$ is non-equivariantly contractible for all finite subgroups H if and only if the inclusion $X = X \wedge S^0 \longrightarrow X \wedge \tilde{E}\mathcal{F}$ is a $\mathbb{T}$-homotopy equivalence.*
(ii) For any $\mathbb{T}$-spectrum X there is a natural equivalence

$$\tilde{E}\mathcal{F} \wedge X \simeq \tilde{E}\mathcal{F} \wedge \Phi^{\mathbb{T}} X$$

(see Appendix C.1 to interpret the right hand side).
(iii) For any $\mathbb{T}$-spectrum X we have the natural isomorphism

$$[X, \tilde{E}\mathcal{F} \wedge Y]^{\mathbb{T}}_* = [\Phi^{\mathbb{T}} X, \Phi^{\mathbb{T}} Y]_*. \quad \square$$

For a vector space W of dimension d we let $X[W]$ be a d-fold wedge of copies on X, and if W is a graded vector space we let $X[W] = \bigvee_n \Sigma^n X[W_n]$. Combining the parts of 2.2.1 we see

$$X(\mathbb{T}) = \tilde{E}\mathcal{F} \wedge X \simeq \tilde{E}\mathcal{F} \wedge \Phi^{\mathbb{T}} X \simeq \tilde{E}\mathcal{F} \wedge S^0[\pi_*(\Phi^{\mathbb{T}} X)],$$

since the nonequivariant rational spectrum $\Phi^{\mathbb{T}} X$ is equivalent to a wedge $S^0[\pi_*(\Phi^{\mathbb{T}} X)]$ of spheres. For morphisms, we see that if X and Y lie over $\mathbb{T}$, we have $[X, Y]^{\mathbb{T}}_* = [\Phi^{\mathbb{T}} X, \Phi^{\mathbb{T}} Y]_*$, and the required reduction to algebra follows.

COROLLARY 2.2.2. *The functor $\Phi^{\mathbb{T}}$ induces an equivalence of categories*

$$\mathbb{T}\text{-}Spec/\mathbb{T} \simeq \text{Graded vector spaces.}$$

Proof:
If the equivalence is also to preserve cofibre sequences, one must interpret the right hand side as the derived category of differential graded vector spaces.

Now return to the analysis of $\mathcal{F}$-spectra. The idea is that any $\mathcal{F}$-spectrum may be formed from natural cells with finite isotropy groups and hence from basic cells with finite isotropy groups. Since basic cells with one isotropy type cannot be non-trivially attached to those of another, the splitting follows.

THEOREM 2.2.3. (Decoupling of finite isotropy groups.)
(i) For any $\mathcal{F}$-spectrum X and each finite subgroup H of $\mathbb{T}$ there is a spectrum $X(H)$ in $\mathbb{T}$-Spec/H formed from basic cells with isotropy H so that there is an equivalence

$$X \xrightarrow{\simeq} \bigvee_H X(H).$$

(ii) Given two $\mathcal{F}$-spectra X and Y the natural map

$$\prod_H [X(H), Y(H)]^{\mathbb{T}} \xrightarrow{\cong} [X, Y]^{\mathbb{T}}$$

wedging together components is an isomorphism.

Before giving the proof, let us make explicit the universal example $E\mathcal{F}_+$: because of its importance we give this in some detail.

LEMMA 2.2.4. *Assuming there is a splitting of $X = E\mathcal{F}_+$ as in the theorem, the summand $E\mathcal{F}_+(H)$ is the suspension spectrum of a certain space $E\langle H\rangle$. This space has the property that there are non-equivariant equivalences*

$$E\langle H\rangle^K \simeq \begin{cases} * & \text{if } K \neq H \\ S^0 & \text{if } K = H. \end{cases}$$

In fact $E\langle 1\rangle = E\mathbb{T}_+$, but $E\langle H\rangle$ is not usually of the form Y_+ for any space Y. Furthermore, $E\langle H\rangle$ is equipped with a stable map $E\langle H\rangle \longrightarrow S^0$, which is a non-equivariant equivalence in geometric H-fixed points.

Proof: To construct $E\langle H\rangle$, we let $\mathcal{F}^k$ denote the family of finite subgroups of order $\leq k$, and hence obtain the filtration

$$E\mathbb{T}_+ = E\mathcal{F}^1_+ \subseteq E\mathcal{F}^2_+ \subseteq E\mathcal{F}^3_+ \subseteq \cdots \subseteq \bigcup_k E\mathcal{F}^k_+ = E\mathcal{F}_+.$$

When H is of order k we define $E\langle H\rangle$ to be the cofibre of $E\mathcal{F}^{k-1}_+ \longrightarrow E\mathcal{F}^k_+$. The content of 2.2.3 is that the filtration splits rationally (see [**11**] for the generalization to arbitrary compact Lie groups). By construction, the map $E\mathcal{F}^{k-1}_+ \longrightarrow E\mathcal{F}^k_+$ is an equivalence in K-fixed points whenever $K \neq H$. Since $E\mathcal{F}^{k-1}_+$ has no H-fixed points, we see $E\langle H\rangle^H \simeq (E\mathcal{F}^k_+)^H \simeq S^0$.

We now have the string of equivalences

$$E\langle H\rangle \overset{1}{\simeq} E\langle H\rangle \wedge E\mathcal{F}_+ \overset{2}{\simeq} E\langle H\rangle \wedge E\mathcal{F}_+(H) \overset{3}{\simeq} E\mathcal{F}^k_+ \wedge E\mathcal{F}_+(H) \overset{4}{\simeq} E\mathcal{F}_+ \wedge E\mathcal{F}_+(H) \overset{5}{\simeq} E\mathcal{F}_+(H).$$

These hold for the following reasons: (1) since $E\langle H\rangle$ is an $\mathcal{F}$-space, (2) since by considering geometric fixed points, $E\langle H\rangle \wedge E\mathcal{F}_+(K) \simeq *$ if $K \neq H$, (3) since $E\mathcal{F}^{k-1}_+ \wedge E\mathcal{F}_+(H) \simeq *$, (4) since $E\mathcal{F}^k_+ \longrightarrow E\mathcal{F}_+$ is an equivalence in H-fixed points, and (5) since $E\mathcal{F}_+(H)$ is an $\mathcal{F}$-spectrum.

The stable map to S^0 is the composite $E\langle H\rangle \longrightarrow E\mathcal{F}_+ \longrightarrow S^0$, where the first map comes from the splitting. The equivalence in geometric H-fixed points follows since $E\langle H\rangle$ is the only contributor to the H-fixed points of $E\mathcal{F}_+$. □

REMARK 2.2.5. We shall see in 2.4.1 that the map $E\langle H\rangle \longrightarrow S^0$ is unique up to multiplication by a non-zero rational. We normalize it so that the canonical generators of the zeroth homology of H-fixed points correspond.

Assuming there is a splitting as in Part (i) of the theorem we use the universal example $E\mathcal{F}_+$ to identify $X(H)$ exactly and deduce the association $X \longmapsto X(H)$ is functorial. Indeed, since any $\mathcal{F}$-spectrum X is equivalent to $E\mathcal{F}_+ \wedge X$ and $E\mathcal{F}_+ \simeq \bigvee_H E\langle H\rangle$, one then finds $X \simeq \bigvee_H E\langle H\rangle \wedge X$. It follows that

$$X(H) \simeq E\langle H\rangle \wedge X;$$

the equivalences $X(H) = S^0 \wedge X(H) \simeq E\langle H\rangle \wedge X(H) \simeq E\langle H\rangle \wedge X$ follow by the geometric fixed point Whitehead theorem.

In any case we obtain the categorical consequence we need.

COROLLARY 2.2.6. *There is an equivalence of triangulated categories*

$$\mathbb{T}\text{-}Spec/\mathcal{F} \simeq \prod_H \mathbb{T}\text{-}Spec/H$$

induced by the functors $X \longmapsto (X(H))_H$ *and* $(X(H))_H \longmapsto \bigvee_H X(H)$. □

We turn to the proof of 2.2.3. The basic tool is the following.

LEMMA 2.2.7. *If* H *is a finite subgroup of* $\mathbb{T}$ *and if* X *is formed from basic cells* σ_H^n *with isotropy* H *and if* Y *is a* $\mathcal{F}$*-spectrum formed using basic cells* σ_K^n *with isotropy* $K \neq H$ *then*

$$[X, Y]^{\mathbb{T}} = 0 \text{ and } X \wedge Y \simeq *.$$

Proof: The fact that $[\sigma_H^0, Y]^{\mathbb{T}} = 0$ follows from 2.1.6 by induction on the number of cells of Y and passage to direct limits. The fact that $[X, Y]^{\mathbb{T}} = 0$ now follows by passage to wedges, cofibres and direct limits in the X variable.

To see that $X \wedge Y \simeq *$ note that $D(\sigma_H^0) \simeq \Sigma^{-1}\sigma_H^0$ so that $\sigma_H^0 \wedge \sigma_K^0 \simeq *$ and argue by induction and limits. □

Proof of 2.2.3: If 2.2.3 (i) holds for X, we shall say X splits, and we note first that Part (ii) of 2.2.3 holds for any pair of $\mathcal{F}$-spectra X and Y which both split. Indeed we have

$$[\bigvee_H X(H), \bigvee_K Y(K)]^{\mathbb{T}} = \coprod_H [X(H), \bigvee_K Y(K)]^{\mathbb{T}}.$$

Since $[X(H), \bigvee_{K \neq H} Y(K)]^{\mathbb{T}} = 0$ by 2.2.7 we conclude

$$[\bigvee_H X(H), \bigvee_K Y(K)]^{\mathbb{T}} = \prod_H [X(H), Y(H)]^{\mathbb{T}}$$

as required.

We now turn to the proof of Part (i), in which the basic ingredients are the splitting of cells as in 2.1.5, and an inductive step as follows.

LEMMA 2.2.8. *If* $X \longrightarrow Y \longrightarrow Z$ *is a cofibre sequence of* $\mathcal{F}$*-spectra and both* X *and* Y *split, so too does* Z.

Before proving 2.2.8, we note how the theorem follows. Indeed any $\mathcal{F}$-spectrum has an inductive filtration so that the subquotients are natural $\mathcal{F}$-cells. In other words $X \simeq \underset{\rightarrow\, n}{\text{holim}}\ X_n$ where $X_0 = *$ and where there are cofibre sequences $C_n \longrightarrow X_n \longrightarrow X_{n+1}$ where C_n is a wedge of natural cells with finite isotropy. The result therefore follows for X_n by induction on n, and since there is a cofibre sequence

$$\bigvee_n X_n \longrightarrow \bigvee_n X_n \longrightarrow \underset{\rightarrow\, n}{\text{holim}}\ X_n$$

it also follows for X. It thus remains only to prove 2.2.8.

Proof of 2.2.8: By hypothesis X and Y both split, so we may form the diagram

$$\begin{array}{ccccc} X & \longrightarrow & Y & \longrightarrow & Z \\ \simeq\downarrow & & \simeq\downarrow & & \\ \bigvee_H X(H) & \xrightarrow{f} & \bigvee_K Y(K) & & . \end{array}$$

By Part (ii) of the theorem $f \simeq \bigvee_H f(H)$ for suitable maps $f(H) : X(H) \longrightarrow Y(H)$. It follows that $Z \simeq \bigvee_H Z(H)$ where $Z(H)$ is the cofibre of $f(H)$. To see the cofibre may be formed from basic cells we may use the usual cellular approximation argument since any map from a basic cell to a higher dimensional basic cell is null-homotopic. □

This completes the proof of 2.2.3. □

2.3. The single strand spectra $E\langle H\rangle$.

It turns out that $E\langle H\rangle$ and its skeleta are very basic objects of $\mathbb{T}$-Spec$/H$. Furthermore, since $Y \simeq Y \wedge E\langle H\rangle$ for any object Y of $\mathbb{T}$-Spec$/H$, it follows that $[E\langle H\rangle, E\langle H\rangle]^{\mathbb{T}}_*$ acts as an algebra of operations on objects of $\mathbb{T}$-Spec$/H$. It is therefore fitting to spend a little time considering maps between the skeleta of $E\langle H\rangle$. Topologists may like to think of $E\langle H\rangle$ as a version of $\mathbb{C}P^\infty_+$ massively simplified by the facts that $\eta^2 = 0$, and James periodicity is trivial.

From the construction of $E\langle H\rangle$ as $E\mathcal{F}_+(H)$ we find its basic structure.

LEMMA 2.3.1. *The spectrum $E\langle H\rangle$ may be constructed stably with a single basic cell σ_H^{2m} in each even dimension:*

$$E\langle H\rangle = \sigma^0_H \cup \sigma^0_H \wedge e^2 \cup \sigma^0_H \wedge e^4 \cup \ldots .$$

Proof: Consider the construction of $E\mathcal{F}_+$ by attaching cells $\mathbb{T}/\!\!/\mathcal{F}_+ \wedge e^n$ given in 1.4.4, and apply the cellular approximation argument to the splitting. □

REMARK 2.3.2. This is an intrinsically stable result since the basic cells do not exist unstably.

Using this cell structure we may consider the skeleta $E\langle H\rangle^{(2m)}$. These are the fundamental objects of $\mathbb{T}$-Spec$/H$: in Chapter 3 we shall see that they are the only locally finite indecomposables. Since we only have contributions from $\mathbb{T}$-isotropy and H-isotropy we should identify the H-homotopy types.

PROPOSITION 2.3.3. *(a) $E\langle H\rangle$ is H-equivariantly equivalent to $e_H S^0$.*
(b) $E\langle H\rangle^{(2m)}$ is H-equivariantly equivalent to $e_H(S^0 \vee S^{2m+1})$

Proof: For Part (a) we apply e_H to the H-equivalences $S^0 \simeq E\mathcal{F}_+|_H \simeq \bigvee_K E\langle K\rangle|_H$; the result follows since $e_H E\langle K\rangle \simeq *$ if $H \neq K$.

Now for Part (b) we argue by induction on m. Indeed the case $m = 0$ states that σ^0_H is H-equivariantly $e_H(S^0 \vee S^1)$, which follows by applying e_H to the H-equivalence

$\mathbb{T}/H_+|_H \simeq S^0 \vee S^1$ (which is true at the unstable level). Suppose the result is proved for $E\langle H\rangle^{(2m)}$, and consider the attaching map of the $(2m+2)$-cell. It must be an H-equivariant equivalence on the common copy of $e_H S^{2m+1}$ by (a), since no higher dimensional cell can cancel it. □

For objects X of $\mathbb{T}$-Spec$/H$ the non-trivial part of the Mackey functor $\underline{\pi}^{\mathbb{T}}_n(X)$ comes from $\pi^{\mathbb{T}}_n(X) = [S^0, X]^{\mathbb{T}}_n$ and $\pi^H_n(X)$. Indeed by 2.1.5

$$\pi^K_n(X) = [\mathbb{T}/K_+, X]^{\mathbb{T}}_n = [\bigvee_{L\subseteq K} \sigma^0_K, X]^{\mathbb{T}}_n$$

and this is $[\sigma^0_H, X]^{\mathbb{T}}_n$ if $H \subseteq K$ and 0 otherwise. A form of the Whitehead theorem follows.

LEMMA 2.3.4. (Naive Whitehead theorem in $\mathbb{T}$-Spec$/H$.) *If X and Y are $\mathcal{F}$-spectra over H, a map $f : X \longrightarrow Y$ is an equivalence provided it is a π^H_* isomorphism.*

Proof: Since X and Y are $\mathcal{F}$-spectra it is enough to verify that f induces an isomorphism in π^K_* for all finite subgroups K; by the above discussion this follows from the particular case $H = K$. □

We record the homotopy groups of the skeleta of $E\langle H\rangle$.

LEMMA 2.3.5. *(a) For any natural number k*

$$\pi^{\mathbb{T}}_*(E\langle H\rangle^{(2k)}) = \bigoplus_{0\leq i\leq k} \Sigma^{2i+1}\mathbb{Q} \text{ and } \pi^H_*(E\langle H\rangle^{(2k)}) \cong \mathbb{Q}\oplus\Sigma^{2k+1}\mathbb{Q}.$$

(b)

$$\pi^{\mathbb{T}}_*(E\langle H\rangle) = \bigoplus_{i\geq 0} \Sigma^{2i+1}\mathbb{Q} \text{ and } \pi^H_*(E\langle H\rangle) \cong \mathbb{Q}.$$

Proof: The values of π^H_* are immediate from 2.3.3. The $\mathbb{T}$-homotopy groups are easily calculated from the cell structure by induction, using the fact that $\pi^{\mathbb{T}}_*(\sigma^0_H) = \Sigma\mathbb{Q}$. □

It follows that each of the attaching maps

$$\sigma^{2m+1}_H \longrightarrow E\langle H\rangle^{(2m)} \longrightarrow E\langle H\rangle^{(2m)}/E\langle H\rangle^{(2m-2)} = \sigma^{2m}_H$$

is nontrivial in $[\sigma^{2m+1}_H, \sigma^{2m}_H]^{\mathbb{T}} = \mathbb{Q}$.

The following uniqueness result will have fundamental implications.

PROPOSITION 2.3.6. *Let k be a natural number or ∞, and let X be a $\mathbb{T}$-spectrum of the form $X = \sigma^0_H \cup \sigma^0_H \wedge e^2 \cup \cdots \cup \sigma^0_H \wedge e^{2k}$. If the attaching maps $\sigma^{2n+1}_H \longrightarrow X^{2n} \longrightarrow X^{2n}/X^{2n-2} = \sigma^{2n}_H$ are nontrivial for all n, then X is equivalent to $E\langle H\rangle^{(2k)}$.*

Three particular cases are of special interest.

COROLLARY 2.3.7. *(a)* (Thom): *If $V(L)$ denotes a one dimensional representation with kernel L containing H, there is an equivalence*

$$E\langle H\rangle \wedge S^{V(L)} \simeq E\langle H\rangle \wedge S^2.$$

(b) (James): *For any natural number* m*, and for* $0 \leq k \leq \infty$ *there is an equivalence*

$$E\langle H\rangle^{(2m+2k)}/E\langle H\rangle^{(2m-2)} \simeq \Sigma^{2m} E\langle H\rangle^{(2k)}.$$

(c) (Atiyah): *For any natural number* m *there is an equivalence*

$$D(E\langle H\rangle^{(2m)}) \simeq \Sigma^{-2m-1} E\langle H\rangle^{(2m)}.$$

Proof: For Part (a) note that applying $S^{V(L)}\wedge$ preserves cofibre sequences. Thus we need only remark that $S^{V(L)} \wedge \sigma^0_H \simeq S^2 \wedge \sigma^0_H$. This follows from the calculation

$$S^{V(L)} \wedge \mathbb{T}_+ \wedge_H e_H S^0 \simeq \mathbb{T}_+ \wedge_H (S^{V(L)} \wedge e_H S^0) \simeq \mathbb{T}_+ \wedge_H S^2 \wedge e_H S^0 \simeq S^2 \wedge \mathbb{T}_+ \wedge_H e_H S^0,$$

where the middle equivalence follows since $S^{V(L)}$ is H-fixed. Since $S^{V(L)}$ is invertible it is clear that $S^{V(L)}\wedge$ preserves non-trivial maps.

Parts (b) follows from 2.1.5(i) and the fact that duality is idempotent.

Part (c) is clear. □

Part (a) is the statement that there are universal Thom isomorphisms for $\mathbb{T}$-spectra over H, Part (b) is the strong analogue of James periodicity, and Part (c) is the straightforward analogue of Atiyah duality. We warn that $DE\langle H\rangle$ is not an object of $\mathbb{T}$-Spec$/H$, and in particular its form is quite different from the pattern suggested in Part (c): see Section 13.5

Proof of 2.3.6 By the naive Whitehead theorem 2.3.4, it suffices to construct a map $f : X \longrightarrow E\langle H\rangle^{(2k)}$ which is an isomorphism in $[\sigma^0_H, \cdot]^{\mathbb{T}}_*$. We do this by obstruction theory, defining f on successive skeleta. In fact define f on $X^{(0)} = \sigma^0_H$ to be the inclusion of the 0-cell. By 2.3.3 $[\sigma^r_H, E\langle H\rangle^{(2k)}]^{\mathbb{T}} = 0$ for $0 < r < 2k+1$ and so this extends uniquely to a map f.

Now calculate $\pi^H_*(X^{2n})$ by induction on n, and note that $\pi^H_*(X) \cong \pi^H_*(E\langle H\rangle^{(2k)})$. The map f is certainly an isomorphism in π^H_0, and to see it is an isomorphism in π^H_{2k+1} we follow through the construction, supposing by induction on n that f_n gives an equivalence $X^{(2n)} \simeq E\langle H\rangle^{(2n)}$. One extension to f_{n+1} may be obtained by completing the map of cofibre sequences

$$\begin{array}{ccccc}
\sigma^{2n+1}_H & \longrightarrow & X^{(2n)} & \longrightarrow & X^{(2n+2)} \\
\downarrow & & \simeq\downarrow & & \\
\sigma^{2n+1}_H & \longrightarrow & E\langle H\rangle^{(2n)} & \longrightarrow & E\langle H\rangle^{(2n+2)},
\end{array}$$

in which the left hand vertical is an equivalence by hypothesis on X; by uniqueness this gives the only possibility. □

The main use of the Thom isomorphism is to give Euler classes. In fact we shall refer to the composite $E\langle H\rangle = E\langle H\rangle \wedge S^0 \stackrel{1\wedge e(V)}{\longrightarrow} E\langle H\rangle \wedge S^{V(H)} \simeq E\langle H\rangle \wedge S^2$ as the Euler class c_H. It induces a map $c_H : X \longrightarrow X \wedge S^2$ for any object X of $\mathbb{T}$-Spec$/H$. More exactly the operation c_H is the part of the Euler class appearing in $\mathbb{T}$-Spec$/H$; the Euler class $\chi(V(H))$ for $V(H)$ makes contributions to $\mathbb{T}$-Spec$/K$ for all subgroups $K \subseteq H$. If K is of order d it is perhaps best to view c_K as analogous to the dth cyclotomic polynomial $\Phi_d(z)$. In fact by analogy with $1 - z^n = \prod_{d|n} \Phi_d(z)$, if H is of order n we have $\chi(V(H))$ acting on the category of spectra over K as c_K if $K \subseteq H$, and as 1 otherwise. We shall discuss this global behaviour of Euler classes further in Section 4.6, but for the present we need only worry about the behaviour of Euler classes over one subgroup, and we now may make this more precise. Indeed, if we define $\tilde{E}\langle H\rangle$ by the cofibre sequence $E\langle H\rangle \longrightarrow S^0 \longrightarrow \tilde{E}\langle H\rangle$ it behaves quite analogously to $\tilde{E}\mathcal{F}$.

LEMMA 2.3.8. *There are equivalences*
(i) $S^{\infty V(H)} \wedge E\langle H\rangle \simeq *$
(ii) $S(\infty V(H))_+ \wedge E\langle H\rangle \simeq E\langle H\rangle$
(iii) $\tilde{E}\langle H\rangle \wedge S^{\infty V(H)} \simeq S^{\infty V(H)}$.

Proof: Part (i) is clear by considering the various Φ-fixed points, and the other parts follow from the cofibre sequences, $S(\infty V(H))_+ \longrightarrow S^0 \longrightarrow S^{\infty V(H)}$ and $E\langle H\rangle \longrightarrow S^0 \longrightarrow \tilde{E}\langle H\rangle$. □

The other useful fact is that the skeleton $\tilde{E}\langle H\rangle^{(2k-1)}$ behaves on $\mathbb{T}$-Spec$/H$ rather like the spheres $S^{kV(H)}$; we therefore use the notation

$$\sigma^{kV(H)} = \tilde{E}\langle H\rangle^{(2k-1)}.$$

COROLLARY 2.3.9. *(i) For any object Y of $\mathbb{T}$-Spec$/H$ and any $k \geq 0$, there is an equivalence $Y \wedge \sigma^{kV(H)} \simeq Y \wedge S^{2k}$.*
(ii) For each $k \geq 0$, the inclusion $\sigma^{kV(H)} \longrightarrow \sigma^{(k+1)V(H)}$ induces multiplication by c_H in the homotopy of any object of $\mathbb{T}$-Spec$/H$.

Proof: Both parts follow from the corresponding facts for $S^{kV(H)}$.
First note that we have a cofibre sequence $S(kV(H))_+ \longrightarrow S^0 \longrightarrow S^{kV(H)}$. The first term splits as $\bigvee_{K\subseteq H} E\langle K\rangle^{(2k-2)}$: indeed it is clear that $S(kV(H))_+(K) \simeq *$ unless $K \subseteq H$, and we claim $S(kV(H))_+(K) \simeq E\langle K\rangle^{(2k-2)}$. From the cell structure of $S(kV(H))_+$ it follows that $S(kV(H))_+(K) = \sigma^0_K \cup \sigma^0_K \wedge e^2 \cup \cdots \cup \sigma^0_K \wedge e^{2k-2}$, and the attaching maps must be non-trivial because of the known K-equivariant structure of $S(kV(H))_+$; the assertion follows from the uniqueness theorem, 2.3.6. Only the term $E\langle H\rangle^{(2k-2)}$ survives smashing with an object of $\mathbb{T}$-Spec$/H$. □

2.4. Operations: self-maps of $E\langle H\rangle$.

As remarked above, for any object X of $\mathbb{T}$-Spec$/H$, we have an equivalence $X \simeq X \wedge E\langle H\rangle$ and hence $[E\langle H\rangle, E\langle H\rangle]^{\mathbb{T}}_*$ acts on X, and on all homology and cohomology theories on $\mathbb{T}$-Spec$/H$. The best thing of all is that this ring of operations is the principal ideal domain $\mathbb{Q}[c_H]$, and is thus of global dimension 1. The aim of the next chapter is to construct an Adams spectral sequence based on these operations, which has the uncharacteristically simple form of a short exact sequence.

We begin by calculating the ring of operations.

THEOREM 2.4.1. (Homotopy operations in $\mathbb{T}$-Spec$/H$.) *The ring of self-maps of* $E\langle H\rangle$ *is*

$$[E\langle H\rangle, E\langle H\rangle]^{\mathbb{T}}_* = \mathbb{Q}[c_H].$$

where c_H *(of degree* -2*) is the Euler class of* $V(H)$ *in the sense that it is the composite* $E\langle H\rangle = E\langle H\rangle \wedge S^0 \stackrel{1\wedge e(V)}{\longrightarrow} E\langle H\rangle \wedge S^{V(H)} \simeq E\langle H\rangle \wedge S^2$.

Proof: We calculate the additive structure of $[E\langle H\rangle, E\langle H\rangle]^{\mathbb{T}}_*$ by passage to limits from the case of finite skeleta. Of course we know that $[\sigma^0_H, E\langle H\rangle]^{\mathbb{T}}_* = \mathbb{Q}$ so that by induction on k we see that $[E\langle H\rangle^{(2k)}, E\langle H\rangle]^{\mathbb{T}}_*$ is nonzero only in degrees $0, -2, -4, \ldots, -2k$, in each of which it is $\mathbb{Q}$. The additive result follows using the Milnor exact sequence.

Choose a generator $c_H : E\langle H\rangle \longrightarrow \Sigma^2 E\langle H\rangle$ of $[E\langle H\rangle, E\langle H\rangle]^{\mathbb{T}}_{-2}$; for the multiplicative statement it is sufficient to show that all composites c_H^n for $n \geq 0$ are essential. For this it is enough to show that c_H induces an isomorphism in $\pi^{\mathbb{T}}_{2k}$ for $k \geq 1$. This follows by the argument used in the proof of 2.3.6, since c_H is the unique extension of the composite $E\langle H\rangle^{(2)} \longrightarrow E\langle H\rangle^{(2)}/E\langle H\rangle^{(0)} \simeq \sigma^2_H \longrightarrow \Sigma^2 E\langle H\rangle$, in which the first map is the collapse map and the last is the inclusion of the bottom cell.

To see that c_H may be chosen to be the Euler class it is enough to observe that the Euler class is nontrivial in $\pi^{\mathbb{T}}_2$. But the cofibre of the Euler class map is $\Sigma E\langle H\rangle \wedge \mathbb{T}/H_+ \simeq \sigma^1_H$, which has $\pi^{\mathbb{T}}_*$ concentrated in degree 2. □

For us the fundamental object will be the $\mathbb{Q}[c_H]$-module $\pi^{\mathbb{T}}_*(X)$; notice that c_H decreases the degree of a homotopy element by 2.

THEOREM 2.4.2. *For any object* X *of* $\mathbb{T}$*-Spec$/H$, the* $\mathbb{Q}[c_H]$*-module* $\pi^{\mathbb{T}}_*(X)$ *is torsion in the sense that for any element* $x \in \pi^{\mathbb{T}}_*(X)$ *there is a natural number* n *so that* $(c_H)^n x = 0$.

Proof: In fact x is supported on a finite $\mathbb{T}$-Spec$/H$-subcomplex $X' \subseteq X$ in the sense that $x = i_* x'$ for some $x' \in \pi^{\mathbb{T}}_*(X')$ where i denotes the inclusion. Since $\pi^{\mathbb{T}}_*(X')$ is bounded below, it is immediate that $c^n_H x' = 0$ for sufficiently large n. □

Note in particular that this implies that $\mathbb{Q}[c_H]$ itself does not occur as $\pi^{\mathbb{T}}_*(X)$ for any object X of $\mathbb{T}$-Spec$/H$. Accordingly, the function spectrum $F(E\langle H\rangle, E\langle H\rangle)$ is not an object of $\mathbb{T}$-Spec$/H$, so that $\mathbb{T}$-Spec$/H$ does not have arbitrary products.

We should record the $\mathbb{Q}[c_H]$-modules for the spectra we have already studied.

LEMMA 2.4.3. *There are* $\mathbb{Q}[c_H]$*-module isomorphisms*

$$\pi^{\mathbb{T}}_*(E\langle H\rangle) = \Sigma\mathbb{I}(H),$$

where $\mathbb{I}(H) = \Sigma^{-2}\mathbb{Q}[c_H, c_H^{-1}]/\mathbb{Q}[c_H]$ *and*

$$\pi^{\mathbb{T}}_*(E\langle H\rangle^{(2k)}) = \Sigma(\mathbb{Q}[c_H]/c_H^{k+1}).$$

Proof: The first calculation was part of the proof of 2.4.1, and the second follows since the inclusion $E\langle H\rangle^{(2k)} \longrightarrow E\langle H\rangle$ induces an isomorphism of $\pi^{\mathbb{T}}_i$ for $i \leq 2k+1$. □

The fundamental fact here is that $\pi^{\mathbb{T}}_*(E\langle H\rangle)$ is $\mathbb{Q}[c_H]$-injective, and this explains the notation. Indeed the $\mathbb{Q}[c_H]$-module $\Sigma^2\mathbb{I}(H) = \mathbb{Q}[c_H, c_H^{-1}]/\mathbb{Q}[c_H]$ is c_H-divisible and hence injective. We have chosen to name $\mathbb{I}(H)$ rather than its double suspension, because $\mathbb{I}(H)$ is concentrated in degrees $0, 2, 4, 6, \ldots$.

CHAPTER 3

Maps between $\mathcal{F}$-free $\mathbb{T}$-spectra.

This chapter forms the core of Part I. In it, we construct an Adams spectral sequence for the category of $\mathbb{T}$-spectra over H, and observe that it collapses to a short exact sequence: this gives us control over morphisms between $\mathcal{F}$-spectra. In Section 3.1 we state the theorem and outline the strategy. In Section 3.2 we prove the Hurewicz and Whitehead theorems, which are basic ingredients. In Section 3.3 we calculate maps into wedges of suspensions of $E\langle H\rangle$, and then in Section 3.4 we show the algebra of $\mathbb{Q}[c_H]$-modules is so simple that this deals with all injectives. Finally, in Section 3.5 we are equipped to deduce the general case.

3.1. The Adams short exact sequence.

The purpose of this chapter is to prove the following central theorem.

THEOREM 3.1.1. (The Adams short exact sequence.) *For any spectra X and Y in $\mathbb{T}$-Spec/H there is a short exact sequence*

$$0 \longrightarrow \mathrm{Ext}_{\mathbb{Q}[c_H]}(\pi_*^{\mathbb{T}}(\Sigma X), \pi_*^{\mathbb{T}}(Y)) \longrightarrow [X, Y]_*^{\mathbb{T}} \longrightarrow \mathrm{Hom}_{\mathbb{Q}[c_H]}(\pi_*^{\mathbb{T}}(X), \pi_*^{\mathbb{T}}(Y)) \longrightarrow 0.$$

We note that one immediate consequence is a Whitehead theorem based on $\pi_*^{\mathbb{T}}$ for $\mathbb{T}$-Spec/H: any map inducing an isomorphism in $\pi_*^{\mathbb{T}}$ induces an isomorphism of $[X, \cdot]_*^{\mathbb{T}}$, and $[\cdot, Y]_*^{\mathbb{T}}$, and is thus an equivalence. Unlike the naive Whitehead theorem 2.3.4 based on π_*^H, this has real content. In fact, it will be one of the main step in the proof of 3.1.1. On the other hand, 3.1.1 itself allows us to be more quantitative and recover π_*^H from $\pi_*^{\mathbb{T}}$. Indeed we know $\pi_*^{\mathbb{T}}(\sigma_H^0) = \Sigma\mathbb{Q}$ so that a little homological algebra gives the following result.

COROLLARY 3.1.2. *For any spectrum Y in $\mathbb{T}$-Spec/H there is a short exact sequence*

$$0 \longrightarrow \pi_*^{\mathbb{T}}(Y)/c_H \longrightarrow \pi_*^H(Y) \longrightarrow \Sigma^{-1} ann(c_H, \pi_*^{\mathbb{T}}(Y)) \longrightarrow 0. \quad \square$$

3.2. The Whitehead and Hurewicz theorems for $\mathbb{T}$-spectra over H.

We remarked that the Whitehead theorem was a consequence of Theorem 3.1.1. In practice the Whitehead theorem is the basis for the proof, and this in turn depends on the Hurewicz theorem.

Before giving the statement it is worth preparing the reader for a potential source of confusion. A geometric object is said to be c-connected if it may be constructed with cells of dimension greater than c, and a graded abelian group is said to be c-connected if it is only non-zero in degrees greater than c. Accordingly the basic $(c+1)$-cell σ_H^{c+1} is c-connected. However, there is a dimension shift in the Adams isomorphism, so that

$\pi^{\mathbb{T}}_*(\sigma^{c+1}_H)$ is $(c+1)$-connected: the expected algebraic connectivity is one more than the topological connectivity.

THEOREM 3.2.1. (Hurewicz theorem for $\mathbb{T}$-Spec$/H$.) *If Z is an object of $\mathbb{T}$-Spec$/H$ and $\pi^{\mathbb{T}}_*(Z)$ is $(c+1)$-connected then Z is c-connected. If in addition $\pi^{\mathbb{T}}_{c+2}(Z) \neq 0$ then $\pi^H_{c+1}(Z) \neq 0$, so the result is sharp.*

Proof: By duality and the universal Thom isomorphism 2.3.7(a) we conclude $[S^{V(H)}, Z]^{\mathbb{T}}_* \cong [S^2, Z]^{\mathbb{T}}_*$. The cofibre sequence $\mathbb{T}/H_+ \longrightarrow S^0 \longrightarrow S^{V(H)}$ gives the Gysin sequence

3.2.2.
$$\cdots \longleftarrow \pi^{\mathbb{T}}_{n-1}(Z) \xleftarrow{c_H} \pi^{\mathbb{T}}_{n+1}(Z) \longleftarrow \pi^H_n(Z) \longleftarrow \pi^{\mathbb{T}}_n(Z) \xleftarrow{c_H} \pi^{\mathbb{T}}_{n+2}(Z) \longleftarrow \cdots .$$

It follows that $\pi^H_*(Z)$ is c-connected as required. □

THEOREM 3.2.3. (Whitehead theorem for $\mathbb{T}$-Spec$/H$.) *If X and Y are spectra in $\mathbb{T}$-Spec$/H$ and $f : X \longrightarrow Y$ induces an isomorphism in $\pi^{\mathbb{T}}_*$ then f is an equivalence.*

Proof: The hypothesis is equivalent to saying the cofibre Z of f is $\pi^{\mathbb{T}}_*$-acyclic, and it is enough to show Z is contractible. By the naive Whitehead theorem 2.3.4, it is enough to show that $\pi^H_*(Z) = 0$. This follows from the Hurewicz theorem 3.2.1. □

3.3. The injective case.

In this section we prove the result for a convenient collection of spectra Y with $\pi^{\mathbb{T}}_*(Y)$ injective. We then spend an algebraic interlude showing that there are enough injectives of a very simple sort.

LEMMA 3.3.1. *If the $\mathbb{Q}[c_H]$-module $\pi^{\mathbb{T}}_*(Y)$ is injective and the functor*
$$\pi^{\mathbb{T}}_* : [X, Y]^{\mathbb{T}}_* \longrightarrow \mathrm{Hom}_{\mathbb{Q}[c_H]}(\pi^{\mathbb{T}}_*(X), \pi^{\mathbb{T}}_*(Y))$$
is an isomorphism for $X = \sigma^0_H$ then it is an isomorphism for all objects X of $\mathbb{T}$-Spec$/H$.

Proof: Since $\pi^{\mathbb{T}}_*(Y)$ is $\mathbb{Q}[c_H]$-injective, the codomain is a cohomology theory of X, and $\pi^{\mathbb{T}}_*$ is a natural transformation of cohomology theories on $\mathbb{T}$-Spec$/H$. Since any object of $\mathbb{T}$-Spec$/H$ can be constructed from cells σ^n_H, the result follows. □

We show the hypothesis of the lemma holds in an important case.

LEMMA 3.3.2. *Passage to homotopy induces an isomorphism*
$$[\sigma^0_H, E\langle H\rangle]^{\mathbb{T}}_* \xrightarrow{\cong} \mathrm{Hom}_{\mathbb{Q}[c_H]}(\pi^{\mathbb{T}}_*(\sigma^0_H), \pi^{\mathbb{T}}_*(E\langle H\rangle)).$$

Proof: Both sides are simply $\mathbb{Q}$ in degree 0, and the calculation of $\pi^{\mathbb{T}}_*(E\langle H\rangle)$ in 2.4.3 showed that the inclusion $\sigma^0_H \longrightarrow E\langle H\rangle$ induces an isomorphism in $\pi^{\mathbb{T}}_0$. □

COROLLARY 3.3.3. *If Y is an arbitrary wedge of suspensions of $E\langle H\rangle$ then*

$$\pi_*^{\mathbb{T}} : [X,Y]_*^{\mathbb{T}} \xrightarrow{\cong} \mathrm{Hom}_{\mathbb{Q}[c_H]}(\pi_*^{\mathbb{T}}(X), \pi_*^{\mathbb{T}}(Y))$$

is an isomorphism for all spectra X in $\mathbb{T}$-Spec$/H$.

Proof: Recall that an object T in a category with sums is said to be *small* if the natural map

$$\bigoplus_i \mathrm{Hom}(T, Y_i) \longrightarrow \mathrm{Hom}(T, \bigoplus_i Y_i)$$

is always an isomorphism. The case when $X = \sigma_H^0$ follows from 3.3.2, since both σ_H^0 and $\pi_*^{\mathbb{T}}(\sigma_H^0) = \Sigma\mathbb{Q}$ are small.

Next observe that since $\mathbb{Q}[c_H]$ is Noetherian, an arbitrary sum of injective modules is injective. The case of arbitrary X follows by 3.3.1. □

3.4. Injectives in the category of torsion $\mathbb{Q}[c_H]$-modules.

In this section we classify the injectives in the category of torsion $\mathbb{Q}[c_H]$-modules: they are all formed by taking sums of suspensions of the single module $\mathbb{I}(H)$, which we already recognize as the homotopy of $\Sigma^{-1}E\langle H\rangle$ by 2.4.3.

PROPOSITION 3.4.1. *If M is a torsion $\mathbb{Q}[c_H]$-module then there is an embedding $M \longrightarrow \bigoplus_i \Sigma^{n_i}\mathbb{I}(H)$ for suitable indexing set and integers n_i. Furthermore, if M is injective the embedding e may be taken to be an isomorphism.*

Proof: We consider the filtration

$$0 = F_0M \subseteq F_1M \subseteq F_2M \subseteq \cdots \subseteq \bigcup_j F_jM = M$$

of M, where $F_jM = \{x \in M \mid c_H^j x = 0\}$; this exhausts M because M is a torsion module. Now the associated graded module GrM has trivial c_H-action, so it is natural to view it as a graded vector space. We shall construct an embedding $e : M \longrightarrow GrM \otimes \mathbb{I}(H)$. Since $\mathrm{Hom}_{\mathbb{Q}[c_H]}(M,N) = \lim_{\leftarrow j} \mathrm{Hom}_{\mathbb{Q}[c_H]}(F_jM, N)$ it is sufficient to give e one filtration at a time, using the notation

$$e_j : F_jM \longrightarrow \left(\bigoplus_{0\le i\le j} F_iM/F_{i-1}M \right) \otimes \mathbb{I}(H)$$

and starting with the unique homomorphism when $j = 0$. Now if we suppose that e_j has been defined, we form e_{j+1} as follows. The codomain is injective so e_j extends to a map $\hat{e}_j$ on $F_{j+1}M$ with the same codomain, and we use the canonical embedding $F_{j+1}M/F_jM \longrightarrow F_{j+1}M/F_jM \otimes \mathbb{I}(H)$ as the final coordinate. This final coordinate ensures that no element of $F_{j+1}M \setminus F_jM$ lies in the kernel of e.

Now, if M is injective, we show that e is also an epimorphism. Indeed, if e is not surjective we may choose an element y not in the image, with torsion order as small as possible. More precisely we suppose $c_H^k y = 0$ and that the image of e includes all elements annihilated by c_H^{k-1}; note that $k \ge 2$ by construction of e. By hypothesis, we may choose x

so that $e(x) = c_H y$, and since M is injective there is an element $\tilde{x}$ so that $c_H\tilde{x} = x$. Now, let $\tilde{y} = e(\tilde{x})$, so that $c_H(y - \tilde{y}) = 0$. By construction of e there is an x' with $e(x') = y - \tilde{y}$; hence $e(x' + \tilde{y}) = y$, contradicting the choice of k. □

3.5. Proof of Theorem 3.1.1.

We now have all the ingredients for the proof of Theorem 3.1.1.

PROPOSITION 3.5.1. *Any injective torsion $\mathbb{Q}[c_H]$-module I is realizable in the sense that there is an object $Y(I)$ of $\mathbb{T}$-Spec/H so that $\pi_*^{\mathbb{T}}(Y(I)) \cong I$. Furthermore, this determines the spectrum $Y(I)$ up to unique homotopy equivalence, and Theorem 3.1.1 holds with $Y = Y(I)$.*

Proof: We saw in Proposition 3.4.1 that

$$I \cong \bigoplus_i \Sigma^{n_i}\mathbb{I}(H)$$

for a suitable indexing set and collection of integers n_i. We may take

$$Y(I) = \bigvee_i \Sigma^{n_i-1}E\langle H\rangle,$$

and obtain $\pi_*^{\mathbb{T}}(Y(I)) \cong I$. Theorem 3.1.1 holds with $Y = Y(I)$ by 3.3.3.

Now if $Y'(I)$ is some other spectrum in $\mathbb{T}$-Spec/H with $\pi_*^{\mathbb{T}}(Y'(I)) \cong I$ there is a unique map $Y'(I) \longrightarrow Y(I)$ inducing the identity in $\pi_*^{\mathbb{T}}$ by 3.3.3. By the Whitehead Theorem 3.2.3 this map is an equivalence. □

Now suppose Y is an arbitrary $\mathbb{T}$-spectrum over H; by 3.4.1 we may form a short exact sequence

$$0 \longrightarrow \pi_*^{\mathbb{T}}(Y) \xrightarrow{i} I \xrightarrow{\delta} J \longrightarrow 0$$

of torsion $\mathbb{Q}[c_H]$-modules with I injective. Since $\mathbb{Q}[c_H]$ is of injective dimension 1 it follows that J is also injective.

Now by 3.5.1, there is a unique map $Y \longrightarrow Y(I)$ inducing i in homotopy. Since i is injective, the cofibre has homotopy J, and hence by 3.5.1 is equivalent to $Y(J)$. We have thus realized the short exact sequence by a cofibre sequence

$$Y \longrightarrow Y(I) \longrightarrow Y(J).$$

Applying $[X, \cdot]_*^{\mathbb{T}}$ we obtain a long exact sequence

$$\begin{array}{ccccccc} \cdots \longrightarrow & [X,Y]_*^{\mathbb{T}} & \longrightarrow & [X,Y(I)]_*^{\mathbb{T}} & \longrightarrow & [X,Y(J)]_*^{\mathbb{T}} & \longrightarrow \cdots \\ & & & \| & & \| & \\ & & & \mathrm{Hom}_{\mathbb{Q}[c_H]}(\pi_*^{\mathbb{T}}(X), I) & \xrightarrow{\delta_*} & \mathrm{Hom}_{\mathbb{Q}[c_H]}(\pi_*^{\mathbb{T}}(X), J) & \end{array}$$

Theorem 3.1.1 follows by definition of Hom and Ext. □

The Adams short exact sequence 3.1.1 and the realization of injectives 3.5.1 are enough to allow us to make algebraic models, both of $\mathbb{T}$-Spec/H and of $\mathbb{T}$-Spec/$\mathcal{F}$. In the next chapter, we establish a suitable abstract framework for this categorical reprocessing, using it to identify the models in Sections 4.4 and 4.5.

CHAPTER 4

Categorical reprocessing.

An Adams spectral sequence arises from a homology functor on a triangulated category with values in an abelian category. The algebraic archetype of a triangulated category is the derived category of an abelian category. In this chapter we show that the algebraic simplicity of certain categories allows us to deduce the structure of the triangulated categories of spectra from their abelianizations. An example of the type of model we obtain is Theorem 4.4.1: the category $\mathbb{T}$-Spec$/H$ is equivalent to the derived category of torsion $\mathbb{Q}[c_H]$-modules.

We begin in Section 4.1 by recalling the construction of the derived category in an appropriate form. In Section 4.2 we give abstract conditions (one dimensionality being the most important) which we show in Section 4.3 are sufficient to ensure an abelian category determines a triangulated category of which it is the abelianization. In Sections 4.4 and 4.5 we show this applies to the two cases we have encountered so far: the category $\mathbb{T}$-Spec$/H$, and the category of $\mathcal{F}$-spectra. The chapter ends with Section 4.6, which restores Euler classes to their rightful prominence in the model of the category of $\mathcal{F}$-spectra.

In the next chapter we show that these methods also applies to a suitable abelianization of the category of all $\mathbb{T}$-spectra.

4.1. Recollections about derived categories.

We motivate our approach by considering the homotopy category of differential graded R-modules for a differential graded algebra R, such as $\mathbb{Q}[c_H]$. Henceforth we often abbreviate 'differential graded' to 'dg'. The associated derived category $D(R)$ is obtained from the category of dg R-modules by forming the category of fractions in which homology isomorphisms are inverted. For a category of fractions to be useful, one must give a fairly concrete construction: this has the additional merit of proving its existence. The usual method is to take the objects to be dg R-modules, and maps in the derived category from M to N to be $[\Gamma M, N] = [M, \Delta N]$ where ΓM is projective as a R-module and equipped with a homology isomorphism $\Gamma M \longrightarrow M$, and ΔN is injective as a R-module and equipped with a homology isomorphism $N \longrightarrow \Delta N$. In the example $R = \mathbb{Q}[c_H]$ we want to restrict attention to torsion modules, and form the derived category of dg torsion $\mathbb{Q}[c_H]$-modules. Since there are no projective torsion $\mathbb{Q}[c_H]$-modules, we are forced to use the construction of the derived category in terms of injectives.

Since most references restrict themselves to derived categories of ungraded algebras, and since we need to discuss derived categories of other abelian categories, we shall spend some time outlining well-known constructions. In this section we concentrate on giving a very concrete approach, but the construction is a special case of Quillen's construction of the homotopy category of a closed model category [**22**]. We therefore use terminology consistent with Quillen's, and give details in Appendix B of how to regard our algebraic model categories as closed model categories.

We shall repeatedly use the following formality in constructing relevant derived categories. This is more familiar to topologists in its dual counterpart, where it is used to construct a category in which weakly equivalent spaces are isomorphic. The dual properties are then the cellular approximation theorem, and two theorems of J.H.C.Whitehead.

Suppose given a category $\mathbb{C}$, a collection S of morphisms which we intend to invert, and a collection $\mathcal{I}$ of fibrant objects (which have the character of injectives) such that the following conditions hold.

CONDITIONS 4.1.1. (i) For every object X of $\mathbb{C}$ there is an element $s : X \longrightarrow \Delta X$ of S with $\Delta X \in \mathcal{I}$
(ii) If $s : I \longrightarrow I'$ is an element of S with $I, I' \in \mathcal{I}$ then s is an isomorphism, and
(iii) For every object I of $\mathcal{I}$ and morphism $s : X \longrightarrow Y$ in S, the induced map $s^* : \mathbb{C}(Y, I) \longrightarrow \mathbb{C}(X, I)$ is a bijection.

We may then extend Δ to a functor $\Delta : \mathbb{C} \longrightarrow \mathbb{C}$ by filling in the diagram

$$\begin{array}{ccc} X & \longrightarrow & \Delta X \\ {\scriptstyle f}\downarrow & & \downarrow{\scriptstyle \Delta f} \\ Y & \longrightarrow & \Delta Y \end{array}$$

using (iii). Next define a category $S^{-1}\mathbb{C}$ by giving it the same objects as $\mathbb{C}$ and morphisms $S^{-1}\mathbb{C}(X, Y) = \mathbb{C}(\Delta X, \Delta Y)$, and a functor $l : \mathbb{C} \longrightarrow S^{-1}\mathbb{C}$ by using Δ.

LEMMA 4.1.2. *The morphism $l : \mathbb{C} \longrightarrow S^{-1}\mathbb{C}$ constructed above is the localization of the category $\mathbb{C}$ away from S.* □

Now suppose given a graded algebra R with zero differential, and let $\mathbb{C}$ be the homotopy category of differential graded R-modules. More generally, we may suppose given a graded abelian category $\mathbb{A}$, and let $\mathbb{C}$ be the category of differential graded objects of $\mathbb{A}$ (ie objects X of $\mathbb{A}$ equipped with a map $d : X \longrightarrow \Sigma X$ with $d^2 = 0$). For dg $\mathbb{A}$-objects X and Y, we shall denote the dg abelian group of $\mathbb{A}$-morphisms by $\mathrm{Hom}(X, Y)$, and the group of homotopy classes by $[X, Y]$; it is useful to note that $[X, Y] = H_0(\mathrm{Hom}(X, Y))$.

Of course, a short exact sequence of dg $\mathbb{A}$-objects induces a long exact sequence in homology. It follows that a cofibre sequence of dg $\mathbb{A}$-objects also induces a long exact sequence in homology, but cofibre sequences have the advantage that they are preserved by application of $\mathrm{Hom}(\cdot, Y)$ and $\mathrm{Hom}(X, \cdot)$, and hence they also induce long exact sequences on applying $[\cdot, Y]$ and $[X, \cdot]$. We can only expect short exact sequences of dg $\mathbb{A}$-objects to induce long exact sequences of $[\cdot, Y]$ if Y consists of injective $\mathbb{A}$-objects and of $[X, \cdot]$ if X consists of projective $\mathbb{A}$-objects.

Now let S be the class of homology isomorphisms and $\mathcal{I}$ be the collection of differential graded $\mathbb{A}$-objects which may be formed from injective $\mathbb{A}$-objects with zero differential by a finite number of cofibre sequences.

PROPOSITION 4.1.3. *If $\mathbb{A}$ is of finite injective dimension, then with these choices of S and $\mathcal{I}$, Conditions 4.1.1 hold and hence the derived category $D(\mathbb{A})$ may be formed from the homotopy category of dg $\mathbb{A}$-objects by inverting homology isomorphisms.*

Proof: The finiteness of injective dimension is only used to verify Condition (i), so we begin without assumption on $\mathbb{A}$.

LEMMA 4.1.4. *If J is an injective $\mathbb{A}$-object with zero differential and E is exact then $[E, J] = 0$.*

Proof: Suppose $f : E \longrightarrow J$ is a map of dg $\mathbb{A}$-objects. Since $d = 0$ on J it follows that $f = 0$ on $im(d) = ker(d)$. Hence f factors through $E/ker(d)$; by injectivity of J this map may be extended along the inclusion $E/ker(d) \subseteq E$, providing a map $h : E \longrightarrow J$ of $\mathbb{A}$-objects. Since $d = 0$ on J we find $dh + hd = f$, so that f is null-homotopic. □

COROLLARY 4.1.5. *Conditions 4.1.1 (ii) and (iii) hold for these choices of $\mathbb{C}$, $\mathcal{I}$ and S.*

Proof: We first show Condition (iii) holds. If $f : X \longrightarrow Y$ is a homology isomorphism, its cofibre E is exact. Since a cofibre sequence induces a long exact sequence in the first variable it is enough to show $[E, I] = 0$ for all objects I in $\mathcal{I}$. Since I may be formed from injective $\mathbb{A}$-objects using finitely many cofibre sequences, this follows from Lemma 4.1.4 using exact sequences in the second variable.

Now for Condition (ii) we must find an inverse t to $s : I \longrightarrow I'$. By Condition (iii) we have a bijection $s^* : [I', I] \xrightarrow{\cong} [I, I]$, and similarly with I' in the second variable. Now choose t so that $s^*t = id$: this ensures $ts = id$, and since $s^*(st) = s^*(id)$ we conclude that $st = id$ as well. □

Let us now turn to Condition (i) and begin with a useful observation.

LEMMA 4.1.6. *If $X \longrightarrow Y \longrightarrow Z$ is a cofibre sequence and two of the three objects admit homology isomorphisms to an object of $\mathcal{I}$, then so does the third.*

Proof: Since the property is invariant under isomorphism and suspension we may as well suppose that there are homology isomorphisms $e : X \longrightarrow I$ and $f : Y \longrightarrow J$; it is enough by Verdier's octahedral axiom and the 5-lemma to show that we may complete the square up to homotopy

$$\begin{array}{ccc} X & \longrightarrow & Y \\ {\scriptstyle e}\downarrow & & \downarrow{\scriptstyle f} \\ I & \dashrightarrow & J. \end{array}$$

But this is immediate from Condition (iii). □

The analogue for short exact sequences now follows.

COROLLARY 4.1.7. *If* $0 \longrightarrow X \xrightarrow{i} Y \xrightarrow{q} Z \longrightarrow 0$ *is a short exact sequence and two of the three objects admit homology isomorphisms to an object of* $\mathcal{I}$*, then so does the third.*

Proof: The main point is that by Condition (iii), if we have a homology isomorphism $A \longrightarrow B$ and one of the two objects admits a homology isomorphism to an object of $\mathcal{I}$, so too does the other.

If X is the term not known to admit a homology isomorphism to an object of $\mathcal{I}$, we note that there is a homology isomorphism $X \longrightarrow F(q)$, where $F(q)$ denotes the mapping fibre of q, and apply 4.1.6. Similarly for Z we use the homology isomorphism $C(i) \longrightarrow Z$ where $C(i)$ denotes the mapping cone of i.

Finally for Y we use the short exact sequence $0 \longrightarrow Y \longrightarrow C(i) \longrightarrow \Sigma X \longrightarrow 0$, the homology isomorphism $C(i) \longrightarrow Z$ and the first case. □

LEMMA 4.1.8. *If* $\mathbb{A}$ *is of finite homological dimension then Condition 4.1.1 (i) also holds.*

Proof: An arbitrary dg $\mathbb{A}$-object X has dg sub-object dX of boundaries, and both dX and X/dX have trivial differential. Accordingly, by 4.1.7 and the above remarks, it is sufficient to deal with the case that X has zero differential.

For this we argue by induction on the length of injective resolution. Indeed for an injective object we may use the identity, and any $\mathbb{A}$-object X of positive injective dimension lies in an exact sequence $0 \longrightarrow X \longrightarrow I \longrightarrow Q \longrightarrow 0$, with I injective and Q of lower injective dimension. □

This completes the proof of Proposition 4.1.3 □

Finally it will be convenient to begin the construction of an Adams spectral sequence for objects of $D(\mathbb{A})$. Consider the natural map

$$\theta : [X, Y]_{derived} \longrightarrow \mathrm{Hom}_{\mathbb{A}}(H_*(X), H_*(Y))$$

given by passage to homology.

PROPOSITION 4.1.9. *If* $H_*(Y)$ *is injective then* θ *is an isomorphism.*

Proof: First note that if I is an injective object of $\mathbb{A}$, regarded as a dg $\mathbb{A}$-object with zero differential, then

$$[X, I] = H_0(\mathrm{Hom}(X, I)) = \mathrm{Hom}(H_*(X), I).$$

It will be convenient later to prove a more precise statement in some detail.

LEMMA 4.1.10. *Suppose that* I *is a dg object with zero differential, and* L *is a dg* $\mathbb{A}$*-object with cycles* $zL = ker(d)$*. Consider the map*

$$H_* : [L, I] \longrightarrow \mathrm{Hom}(H_*(L), I).$$

(i) If $\mathrm{Ext}^1(L/zL, I) = 0$ *then* H_* *is surjective.*
(ii) If $\mathrm{Ext}^1(L/dL, I) = 0$ *then* H_* *is injective.*

Proof: (i) Given $\phi : zL/dL = H_*(L) \longrightarrow I$ we may extend it to a map $\hat{\phi} : L/dL \longrightarrow I$ by the given condition. The resulting composite $L \longrightarrow L/dL \longrightarrow I$ is a dg map since d is zero on I, and it visibly induces ϕ on homology.
(ii) Given dg maps $\tilde{\phi}, \tilde{\psi} : L \longrightarrow I$, both are zero on dL and hence give maps $\hat{\phi}, \hat{\psi} : L/dL \longrightarrow I$. If they agree in homology, the map $\delta := \hat{\phi} - \hat{\psi}$ is zero on zL/dL, and hence gives a map $\overline{\delta} : L/zL \longrightarrow I$. Now d gives an embedding $L/zL \longrightarrow L$, and by hypothesis, $\overline{\delta}$ extends along this to give a map $h : L \longrightarrow I$. Now $hd + dh = hd = \tilde{\phi} - \tilde{\psi}$, so that the given maps are homotopic as required. □

Continuing with the proof of the proposition, note that by construction, $[X, I] = [X, I]_{derived}$ when I is injective. This proves the result if Y has zero differential. It also shows that if $H_*(Y)$ is injective, then there is a map $Y \longrightarrow H_*(Y)$ inducing the identity in homology. Thus $[X, Y]_{derived} = [X, H_*(Y)]_{derived}$, and the general case follows. □

Following the abuse of notation standard in topology we henceforth omit the subscript *derived*. Thus $[X, Y]$ now denotes maps in the derived category.

4.2. Split linear triangulated categories.

The aim of this section and the next is to show that a sufficiently simple triangulated category is determined by its abelianization. We treat this abstractly for the usual reasons: firstly it becomes apparent exactly what properties are being used, and secondly there are several instances of the phenomenon that we want to treat at once.

Thus we suppose given a triangulated category $\mathbb{D}$, which might be a derived category of some abelian category or a category of spectra; we shall use $X, Y, \ldots$ to denote objects of $\mathbb{D}$, and $[X, Y]$ to denote morphisms. We also suppose given a graded abelian category $\mathbb{A}$, and a functor $H_* : \mathbb{D} \longrightarrow \mathbb{A}$, exact in the sense that triangles are taken to long exact sequences; in the applications H_* is some form of homology. The idea is that H_* is a linearization, so that we require in particular the Inverse Function Theorem holds in the sense that homology isomorphisms are invertible (i.e. that H_* creates isomorphisms).

The example to bear in mind at present is the abelian category $\mathbb{A} = \text{tors-}\mathbb{Q}[c_H]\text{-mod}$, and the triangulated category $\mathbb{D} = \mathbb{T}\text{-Spec}/H$, with $\pi_*^{\mathbb{T}}$ providing the linearization.

One can hope to incorporate non-linear phenomena by using derived functors in $\mathbb{A}$; this is the situation when there is an Adams spectral sequence based on H_*. One could think of this as a formal power series description of $\mathbb{D}$ with coefficients in $\mathbb{A}$. However, if two or more derived functors are involved, there will usually be differentials in the Adams spectral sequence, and one cannot hope to recover $\mathbb{D}$ from $\mathbb{A}$ alone. Suppose then that $\mathbb{A}$ has injective dimension 1, so that the Adams spectral sequence takes the form of a short exact sequence

4.2.1.
$$0 \longrightarrow \mathrm{Ext}_{\mathbb{A}}(H_*(\Sigma X), H_*(Y)) \longrightarrow [X, Y] \longrightarrow \mathrm{Hom}_{\mathbb{A}}(H_*(X), H_*(Y)) \longrightarrow 0.$$

This happens in case $\mathbb{D} = \mathbb{T}\text{-Spec}/H$ by Theorem 3.1.1.

The Adams short exact sequence shows that objects X are classified by their images $H_*(X)$ in $\mathbb{A}$. Indeed, if $H_*(X) \cong H_*(Y)$ we may lift an algebraic isomorphism to a map $f : X \longrightarrow Y$ in $\mathbb{D}$; since it is a homology isomorphism by construction, it is an equivalence in $\mathbb{D}$. If we are to recover $\mathbb{D}$ from $\mathbb{A}$ we should therefore require that H_* is surjective on isomorphism classes of objects. It is sufficient to show enough injectives are in the image.

LEMMA 4.2.2. *If enough injectives are in the image of H_* then all objects of $\mathbb{A}$ are in the image of H_*.*

By Corollary 3.5.1 this applies when $\mathbb{D} = \mathbb{T}\text{-Spec}/H$.

Proof: First note that if I is a realizable injective then any retract of I is also realizable, so that all injectives are realizable by hypothesis. Indeed a retract of I is of the form eI for some idempotent $e : I \longrightarrow I$. Now if $H_*(X(I)) = I$ then we may lift e uniquely to a map $X(e) : X(I) \longrightarrow X(I)$ by the Adams short exact sequence, and $X(e)$ is also idempotent. Thus $H_*(X(e)X(I)) = eI$ as required, since I is injective.

Now an arbitrary object M of $\mathbb{A}$ has an injective resolution $0 \longrightarrow M \longrightarrow I \stackrel{f}{\longrightarrow} J \longrightarrow 0$, and we may realize I by $X(I)$, J by $X(J)$ and the map f by a unique map $X(f) : X(I) \longrightarrow X(J)$ by the Adams short exact sequence. Thus the fibre of $X(f)$ realizes M. □

However we still cannot expect to recover $\mathbb{D}$ from $\mathbb{A}$ without further hypothesis, because the extension 4.2.1 will not generally be split. For $\mathbb{A} = \text{tors-}\mathbb{Q}[c_H]\text{-mod}$, splitting of objects into even and odd graded parts will do what is required.

DEFINITION 4.2.3. (i) A one dimension graded abelian category $\mathbb{A}$ is *split* if every object M of $\mathbb{A}$ has a splitting $M = M_+ \oplus M_-$ so that $\text{Hom}(M_\delta, N_\epsilon) = 0$ and $\text{Ext}(M_\delta, N_\epsilon) = 0$ if $\delta \neq \epsilon$, and so that $(\Sigma M)_\epsilon = \Sigma(M_{\epsilon+1})$.
(ii) We say that a triangulated category $\mathbb{D}$ is *split linear* if there is a functor $H_* : \mathbb{D} \longrightarrow \mathbb{A}$, to a split one dimensional category $\mathbb{A}$, which (a) is surjective on isomorphism classes of objects, (b) creates isomorphisms, and (c) has the property that there are natural Adams short exact sequences 4.2.1.

We shall see in 4.3.2 that a split one dimensional abelian category $\mathbb{A}$ is the linearization of a split linear triangulated category $\mathbb{D}$ unique up to equivalence of categories. Furthermore, we show in 4.3.1, that if $\mathbb{D}$ arises as a homotopy category this equivalence preserves triangles.

We have motivated the above definitions with the example $\mathbb{D} = \mathbb{T}\text{-Spec}/H$, $\mathbb{A} = \text{tors-}\mathbb{Q}[c_H]\text{-mod}$ and $H_* = \pi^{\mathbb{T}}_*$, but there is a general example.

LEMMA 4.2.4. *If $\mathbb{A}$ is a split one dimensional abelian category then the derived category $D(\mathbb{A})$ is a split linear triangulated category with linearization $\mathbb{A}$.*

Proof: Homology gives a functor $H_* : D(\mathbb{A}) \longrightarrow \mathbb{A}$. This is surjective on objects since any object of $\mathbb{A}$ can be viewed as a dg object with zero differential. The Adams short exact sequence is established in the usual way. First, homology gives a natural map

$$[X, Y] \longrightarrow \text{Hom}_{\mathbb{A}}(H_*(X), H_*(Y)),$$

where $[X, Y]$ denotes maps in the derived category, which is an isomorphism when $H_*(Y)$ is injective, by 4.1.9. This gives the case of the Adams spectral sequence when $H_*(Y)$ is injective, and it only remains to prove that all objects have Adams resolutions by objects of this type.

For an arbitrary object Y in the derived category there is a triangle

$$Y \longrightarrow I \longrightarrow J$$

so that

$$0 \longrightarrow H_*(Y) \longrightarrow H_*(I) \longrightarrow H_*(J) \longrightarrow 0$$

is an injective resolution of $H_*(Y)$ in $\mathbb{A}$. To see this, find an embedding $e : H_*(Y) \longrightarrow I$ with I injective. Regard I as a dg object with zero differential and use the injective case to lift e to a map $Y \longrightarrow I$ in the derived category. The mapping cone of this map necessarily has injective homology, and hence qualifies as J. This completes the proof that $D(\mathbb{A})$ is a split linear triangulated category with linearization $\mathbb{A}$. □

4.3. The uniqueness theorem.

The main theorem of the section states that when $\mathbb{A}$ is split one dimensional, $D(\mathbb{A})$ is essentially the only category with abelianization $\mathbb{A}$.

THEOREM 4.3.1. *If $\mathbb{A}$ is a split one-dimensional abelian category, and $\mathbb{D}$ is (i) the homotopy category of a model category and (ii) a split linear triangulated category with linearization $\mathbb{A}$, then there is an equivalence*

$$\mathbb{D} \simeq D(\mathbb{A})$$

of triangulated categories.

We begin by establishing an equivalence of categories, and later show it preserves triangles.

PROPOSITION 4.3.2. *If $\mathbb{A}$ is a split one dimensional abelian category and $\mathbb{D}$ is a split linear triangulated category with linearization $\mathbb{A}$, then $\mathbb{D}$ is equivalent to $D(\mathbb{A})$. Furthermore, the Adams short exact sequence is split.*

Proof: Consider the two split linear triangulated categories, $\mathbb{D}$ and $\mathbb{E} = D(\mathbb{A})$. Each is equipped with an exact functor H_* to $\mathbb{A}$, which is essentially surjective on objects, and which creates isomorphisms. Furthermore we have an exact sequence 4.2.1 for objects of $\mathbb{D}$, and similarly for $\mathbb{E}$.

We shall construct a functor $p : \mathbb{D} \longrightarrow \mathbb{E}$ which is an equivalence of categories over $\mathbb{A}$. First note that the splitting of objects of $\mathbb{A}$ lifts to splittings of objects of $\mathbb{D}$ and $\mathbb{E}$. Indeed if $H_*(X) = H_+ \oplus H_-$ then, since H_* is essentially surjective, there is an object X_+ with $H_*(X_+) \cong H_+$. Furthermore, it follows from the Adams short exact sequence 4.2.1 that we may lift the projection $H_*(X) \longrightarrow H_+$ to a map $X \longrightarrow X_+$; arguing similarly with H_- we obtain a map $X \longrightarrow X_+ \vee X_-$ which is a homology isomorphism and hence an equivalence. We refer to objects X with $X = X_+$ or $X = X_-$ as *pure parity* objects; note that if X and Y are both of pure parity then the Adams short exact sequence either says $[X, Y] =$

$\mathrm{Hom}_{\mathbb{A}}(H_*(X), H_*(Y))$ if they are of the same parity, or $[X, Y] = \mathrm{Ext}_{\mathbb{A}}(H_*(\Sigma X), H_*(Y))$ if they are of opposite parity.

Following Adams [**1**], we may consider d and e invariants of a map $f : X \longrightarrow Y$ in $\mathbb{D}$. Any map f has a d invariant $d(f) := H_*(f) \in \mathrm{Hom}_{\mathbb{A}}(H_*(X), H_*(Y))$, and if $d(f) = 0$ then the class of the extension

$$0 \longrightarrow H_*(Y) \longrightarrow H_*(C(f)) \longrightarrow H_*(\Sigma X) \longrightarrow 0$$

gives the e invariant, $e(f) \in \mathrm{Ext}_{\mathbb{A}}(H_*(\Sigma X), H_*(Y))$. This much applies to any triangulated category, and in our case the e invariant identifies the position of a morphism in the Adams short exact sequence 4.2.1.

Now choose such a splitting for each object of $\mathbb{D}$ and $\mathbb{E}$, and a preferred representative from each isomorphism class of pure parity objects of $\mathbb{E}$. Define p on objects by $p(X) = p(X_+) \vee p(X_-)$ where $p(X_\delta)$ is the preferred representative of objects with homology $H_*(X_\delta)$. On morphisms $f : X \longrightarrow Y$ between pure parity objects we take $p(f)$ to be the map with the same d or e invariant as f. For arbitrary X and Y we view f as a 2×2 matrix of maps between the pure parity summands of X and Y; this gives a splitting of the Adams short exact sequence 4.2.1, so that any map $f : X \longrightarrow Y$ has a well defined e invariant. We define $p(f)$ by letting p act on each of the four matrix entries. By matrix multiplication, the functoriality of p on all morphisms will follow from that on morphisms between pure parity objects. There are four cases to consider.

Consider maps $X \xrightarrow{f} Y \xrightarrow{g} Z$ between pure parity objects. If X, Y and Z have the same parity $d(gf) = d(g)d(f)$ since H_* is a functor, and hence $p(gf) = p(g)p(f)$. Similarly if either f or g is a map between objects of the same parity, then functoriality comes from the naturality of the e invariant in the sense that $e(gf) = e(g)d(f)$ or $e(gf) = d(g)e(f)$ as the case may be. Finally if both f and g are maps between objects with opposite parities then f and g both have Adams filtration 1, so gf has Adams filtration 2 and $gf = 0$ because $\mathbb{A}$ is one dimensional; $p(g)p(f) = 0$ for analogous reasons. This completes the construction of a functor $p : \mathbb{D} \longrightarrow \mathbb{E}$; a functor in the reverse direction can be constructed in the same way, and their composites are isomorphic to the relevant identity functors. Thus p is an equivalence as required. □

It remains to consider the behaviour of p on triangles, and we begin the discussion without further assumptions on the categories. Suppose given a triangle $X \xrightarrow{f} Y \xrightarrow{g} Z \xrightarrow{h} \Sigma X$ in $\mathbb{D}$. First note that, because H_* is exact, p commutes with suspensions. The ideal outcome would be to show that $p(X) \xrightarrow{p(f)} p(Y) \xrightarrow{p(g)} p(Z) \xrightarrow{p(h)} \Sigma p(X)$ is a triangle in $\mathbb{E}$. To do this, we would have to show that a map determines the triangle it lies in, up to automorphism, using only categorical constructions on its d and e invariants.

LEMMA 4.3.3. *Let $f : X \longrightarrow Y$ be a morphism in $\mathbb{D}$.*
The d and e invariants of f determine (a) its mapping cone Z up to isomorphism and (b) the d invariants of the maps $g : Y \longrightarrow Z$ and $h : Z \longrightarrow \Sigma X$.

Proof: For (a) we note that it is only necessary to determine the isomorphism class of $H_*(Z)$. Indeed it lies in an extension

$$0 \longrightarrow C \longrightarrow H_*(Z) \longrightarrow \Sigma K \longrightarrow 0,$$

where C is the cokernel and K the kernel of $H_*(f)$. In fact the extension class is the image of $e(f)$ under the composite

$$\mathrm{Ext}_{\mathbb{A}}(H_*(\Sigma X), H_*(Y)) \longrightarrow \mathrm{Ext}_{\mathbb{A}}(H_*(\Sigma X), C) \longrightarrow \mathrm{Ext}_{\mathbb{A}}(\Sigma K, C).$$

The construction clearly identifies $d(g)$ and $d(h)$. □

To proceed further we wish to complete the diagram

$$\begin{array}{ccccccccc} p(X) & \xrightarrow{p(f)} & p(Y) & \xrightarrow{g'} & Z' & \xrightarrow{h'} & \Sigma p(X) & \xrightarrow{-\Sigma p(f)} & \Sigma p(Y) \\ =\downarrow & & =\downarrow & L & \alpha\vdots & R & =\downarrow & & =\downarrow \\ p(X) & \xrightarrow{p(f)} & p(Y) & \xrightarrow{p(g)} & p(Z) & \xrightarrow{p(h)} & \Sigma p(X) & \xrightarrow{-\Sigma p(f)} & \Sigma p(Y) \end{array}$$

in which the first row is obtained by embedding $p(f)$ into a triangle in $\mathbb{E}$. Since p is functorial, $p(g)p(f) = p(gf) = 0$ and so a map α exists such that the square L commutes; the problem is to choose the map to make the square R commute. If this is possible, then a 5-lemma argument in homology shows that α is an equivalence in $\mathbb{E}$, and so the second row is isomorphic to the first, and hence is a triangle as required.

The standard method of discussing such problems is that of Toda brackets, but these are not available in an arbitrary triangulated category. Since they are available in the cases of interest to us we shall impose further conditions. Note in particular that $D(\mathbb{A})$ is the homotopy category of the category of dg $\mathbb{A}$-objects so that it does have Toda brackets.

Proof of 4.3.1: We showed in 4.2.4 that the derived category $\mathbb{E} = D(\mathbb{A})$ of differential graded objects of $\mathbb{A}$ is a split linear triangulated category with linearization $\mathbb{A}$, and it was constructed as a homotopy category. We have also shown in 4.3.2 that there is an equivalence $\mathbb{D} \simeq D(\mathbb{A})$ of categories. It remains to show that if $\mathbb{D}$ is a homotopy category, the equivalence preserves triangles.

Consider the diagram of differential graded objects of $\mathbb{A}$ and homomorphisms (i.e. not up to homotopy)

$$\begin{array}{ccccccccc} A & \xrightarrow{\theta} & B & \xrightarrow{i} & C & \xrightarrow{\pi} & \Sigma A & \xrightarrow{-\Sigma\theta} & \Sigma B \\ =\downarrow & & =\downarrow & L & \alpha\vdots & R & =\downarrow & & =\downarrow \\ A & \xrightarrow{\theta} & B & \xrightarrow{\phi} & C' & \xrightarrow{\psi} & \Sigma A & \xrightarrow{-\Sigma\theta} & \Sigma B \end{array}$$

in which C is the mapping cone of θ, and i and π are the canonical maps.

A map α exists so that L strictly commutes if and only if $\phi\theta \simeq 0$, and choices of α correspond exactly to null-homotopies H of $\phi\theta$. The condition that R commutes up to homotopy is then that $\pi \simeq \psi\alpha$, which is equivalent to the existence of a null-homotopy K

of $\psi\phi$ so that $1 \simeq \psi H - K\theta$. In other words, there is a solution α so that both L and R commute if and only if the Toda bracket

$$\langle \psi, \phi, \theta \rangle \subseteq [\Sigma A, \Sigma A]$$

exists and contains the identity.

Exactly similar arguments apply to the category $\mathbb{D}$ (see [**22**, I.3]), so that the result will follow once we show that the map p preserves Toda brackets. This follows from the discussion in [**1**, Section 5], and specifically from diagram (5.1) there. In fact we form the analogue of Adams' diagram (5.1) for calculating the Toda bracket $\langle h, g, f \rangle$ in the model category whose homotopy category is $\mathbb{D}$. Since f, g, h are successive maps in a triangle, 1 is an element of the Toda bracket and there are choices of null-homotopies so that the composite $HF \simeq 1$ in Adams' notation. Now apply p to the resulting diagram in $\mathbb{D}$, and conclude that $1 = p(1) \in \langle p(h), p(g), p(f) \rangle$. □

4.4. The algebraicization of the category of $\mathbb{T}$-spectra over H.

In this section we quickly complete the proof of the following theorem.

THEOREM 4.4.1. *There is an equivalence of triangulated categories between $\mathbb{T}$-Spec/H and the derived category of dg torsion $\mathbb{Q}[c_H]$-modules:*

$$\mathbb{T}\text{-}Spec/H \simeq D(tors\text{-}\mathbb{Q}[c_H]).$$

Proof: By the uniqueness theorem, 4.3.1, it is only necessary to observe that $\mathbb{T}$-Spec/H is a split linear triangulated category arising as the homotopy category of a model category.

We take $\mathbb{D} = \mathbb{T}$-Spec/H, $\mathbb{A} =$ tors-$\mathbb{Q}[c_H]$-mod, and $H_*(X) = \pi_*^{\mathbb{T}}(X)$. It is easily verified that tors-$\mathbb{Q}[c_H]$-mod is abelian and has injective dimension 1. The split condition arises since $\mathbb{Q}[c_H]$ is concentrated in even degrees, so that for any $\mathbb{Q}[c_H]$-module M we may take M_+ to be the even-graded part of M and M_- to be the odd graded part. The Adams short exact sequence 3.1.1 shows that $\mathbb{A}$ is the linearization of $\mathbb{D}$.

Finally, it follows from the construction of Lewis-May that $\mathbb{T}$-Spec/H is the homotopy category of a model category. The model category we have in mind is the category of $\mathcal{F}$-spectra, with weak equivalences created by the functor $X \longmapsto e_H\pi_*^{\mathbb{T}}(X)$. Since $\pi_*^{\mathbb{T}}(X) = e_H\pi_*^{\mathbb{T}}(X)$ for objects of $\mathbb{T}$-Spec/H,we see that homology is well defined. We conclude that Theorem 4.3.1 applies to show there is an equivalence

$$p : \mathbb{T}\text{-Spec}/H \xrightarrow{\simeq} D(\text{tors-}\mathbb{Q}[c_H])$$

of triangulated categories; this completes the proof of Theorem 4.4.1. □

It may be instructive to give an example illustrating how maps are represented in $D(\text{tors-}\mathbb{Q}[c_H])$.

EXAMPLE 4.4.2. Consider a non-trivial map $f : \sigma_H^1 \longrightarrow \sigma_H^0$. First we note that $\pi_*^{\mathbb{T}}(\sigma_H^0) = \Sigma\mathbb{Q}$, and there is an exact sequence

$$0 \longrightarrow \Sigma\mathbb{Q} \longrightarrow \Sigma\mathbb{I}(H) \xrightarrow{c_H} \Sigma^3\mathbb{I}(H) \longrightarrow 0.$$

Thus $p(\sigma_H^0) = F(c_H)$, which we may illustrate:

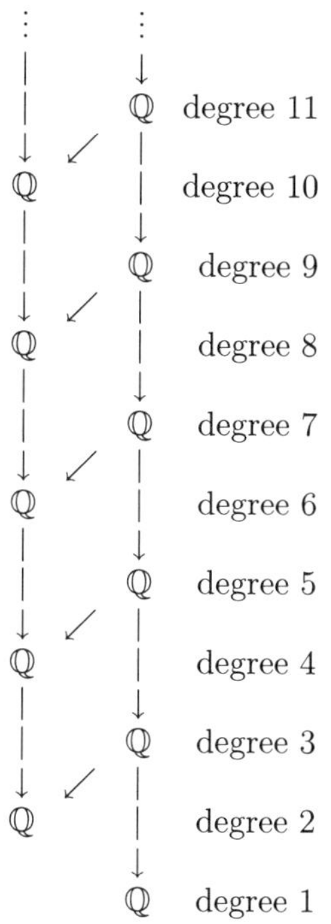

Here degree is displayed vertically, the vertical arrows are multiplication by c_H, and the differential is displayed diagonally. Now the map $f : \sigma_H^1 \longrightarrow \sigma_H^0$ has zero d-invariant, and its e-invariant is a non-zero element of $\mathbb{Q} \cong \mathrm{Ext}(\Sigma^3\mathbb{Q}, \Sigma\mathbb{Q})$. Thus $p(f)$ is non-trivial in all even degrees and zero in all odd degrees. The cofibre of f is equivalent to $E\langle H\rangle^{(2)}$, and it is natural to choose an Adams resolution corresponding to the minimal injective resolution $0 \longrightarrow \pi_*^{\mathbb{T}}(E\langle H\rangle^{(2)}) \longrightarrow \Sigma\mathbb{I}(H) \longrightarrow \Sigma^5\mathbb{I}(H) \longrightarrow 0$. The reader may find it instructive to show that $C(p(f))$ is quasi-isomorphic to $p(E\langle H\rangle^{(2)})$. □

4.5. The algebraicization of the category of $\mathcal{F}$-spectra.

By the Decoupling Theorem 2.2.3 the step from $\mathbb{T}$-spectra over H to $\mathcal{F}$-spectra is relatively small. However, before we state the theorem, we must introduce the relevant algebra.

If Y is any $\mathcal{F}$-spectrum then $Y \simeq E\mathcal{F}_+ \wedge Y$ so $\pi^{\mathbb{T}}_*(Y)$ is a module over the ring of operations $\mathcal{O}_{\mathcal{F}} := [E\mathcal{F}_+, E\mathcal{F}_+]^{\mathbb{T}}_*$. Using 2.2.3, it follows from 2.4.1 that $\mathcal{O}_{\mathcal{F}}$ is the product $\prod_H \mathbb{Q}[c_H]$. Let $e_H \in (\mathcal{O}_{\mathcal{F}})_0$ denote the idempotent which is projection onto the Hth factor, and let $c \in (\mathcal{O}_{\mathcal{F}})_{-2}$ denote the total Chern class c with Hth coordinate c_H (i.e. $c_H = e_H c$ in $e_H\mathcal{O}_{\mathcal{F}} = \mathbb{Q}[c_H]$).

Arbitrary modules over this ring may be quite unpleasant, but by the Decoupling Theorem 2.2.3, the modules $\pi^{\mathbb{T}}_*(Y)$ for an $\mathcal{F}$-spectrum Y which occur geometrically are all rather well behaved. Indeed they all decompose as the sum

$$\pi^{\mathbb{T}}_*(Y) = \bigoplus_H \pi^{\mathbb{T}}_*(Y(H))$$

where $\pi^{\mathbb{T}}_*(Y(H)) = e_H\pi^{\mathbb{T}}_*(Y)$ is actually a torsion module over $\mathbb{Q}[c_H] = e_H\mathcal{O}_{\mathcal{F}}$.

We therefore consider the following two conditions on an $\mathcal{O}_{\mathcal{F}}$-module M.

CONDITION 4.5.1. M is *$\mathcal{F}$-finite* if it is the direct sum of its submodules $M(H) := e_H M$:

$$M = \bigoplus_H M(H).$$

and

CONDITION 4.5.2. M is a *torsion* $\mathcal{O}_{\mathcal{F}}$-module if for each $x \in M$ there is a number N so that $c^N x = 0$.

We refer to the category of full subcategory of $\mathcal{O}_{\mathcal{F}}$-modules satisfying Conditions 4.5.1 and 4.5.2 as *$\mathcal{F}$-finite torsion modules*, and denote it tors-$\mathcal{O}^f_{\mathcal{F}}$-mod.

THEOREM 4.5.3. *There is an equivalence*

$$p : \mathbb{T}\text{-}Spec/\mathcal{F} \xrightarrow{\simeq} D(tors\text{-}\mathcal{O}^f_{\mathcal{F}})$$

of triangulated categories.

Proof: We take $\mathbb{D} = \mathbb{T}$-Spec$/\mathcal{F}$ and $\mathbb{A} =$ tors-$\mathcal{O}^f_{\mathcal{F}}$-mod and apply the uniqueness theorem 4.3.1. We have motivated our definitions by the fact that homotopy gives a functor

$$\pi^{\mathbb{T}}_* : \mathbb{T}\text{-Spec}/\mathcal{F} \longrightarrow \text{tors-}\mathcal{O}^f_{\mathcal{F}}\text{-mod}.$$

Furthermore it is clear from the Lewis-May construction that $\mathbb{T}$-Spec$/\mathcal{F}$ is the homotopy category of the model category of $\mathcal{F}$-spectra with weak equivalences defined as usual. By the Whitehead Theorem 3.2.3 for $\mathbb{T}$-Spec$/H$, weak equivalences of $\mathcal{F}$-spectra may equally well be characterized as $\pi^{\mathbb{T}}_*$-equivalences.

Since the idempotents e_H show that the category tors-$\mathcal{O}^f_{\mathcal{F}}$-mod splits to give an equivalence

$$\text{tors-}\mathcal{O}^f_{\mathcal{F}}\text{-mod} \simeq \prod_H \text{tors-}\mathbb{Q}[c_H]\text{-mod},$$

it follows that tors-$\mathcal{O}^f_{\mathcal{F}}$-mod is a split one dimensional graded abelian category. The Decoupling Theorem 2.2.3 showed that there was a similar splitting of $\mathbb{T}$-Spec$/\mathcal{F}$, and the map $\pi^{\mathbb{T}}_*$ is compatible with these splittings. Hence from 3.1.1 we obtain Adams short exact sequences of the appropriate form to show that $\mathbb{A}$ is the linearization of $\mathbb{D}$. □

4.6. Euler classes revisited.

It is instructive to have an alternative description of tors-$\mathcal{O}^f_{\mathcal{F}}$-mod in terms of Euler classes. First we need some notation.

There is an Euler class associated to any representation V of $\mathbb{T}$ with $V^{\mathbb{T}} = 0$. This is based on the map $e(V) : S^0 \longrightarrow S^V$, and exists when we have Thom isomorphisms. We let $\chi(V)$ denote the associated Euler class. Provided the requisite Thom isomorphisms exist, the fact that $e(V \oplus W) = (1 \wedge e(V)) \circ e(W)$ shows they are multiplicative in the sense that $\chi(V \oplus W) = \chi(V)\chi(W)$. Evidently, if $V^{\mathbb{T}} \neq 0$ then $\chi(V) = 0$, since $e(V)$ is null.

Now consider the case of $\mathcal{F}$-spectra. So far we have restricted attention to the effect on $\mathbb{T}$-spectra over H when $V = V(H)$, but now we need to take all isotropy groups into account. Thus we consider the map $e(V) \wedge 1 : S^0 \wedge E\mathcal{F}_+ \longrightarrow S^V \wedge E\mathcal{F}_+$, and note that, if we split $E\mathcal{F}_+$ into its summands $E\langle H\rangle$, we obtain $c_H^{v(H)} : E\langle H\rangle \longrightarrow \Sigma^{2v(H)}E\langle H\rangle$ where $v(H) = dim_{\mathbb{C}}(V^H)$. For any complex representation V with $V^{\mathbb{T}} = 0$ the Euler class $\chi(V)$ is multiplication by $c_H^{v(H)}$ on the H'th summand; note that $V^H = 0$ for almost all H. If $V^{\mathbb{T}} \neq 0$ then $\chi(V) = 0$. To avoid undue dependence on representations we will use an alternative notation.

DEFINITION 4.6.1. If $\phi \subseteq \mathcal{F}$ we let e_ϕ be the idempotent with support ϕ, and we let

$$c_\phi = e_\phi c + (1 - e_\phi).$$

Thus, if V is one dimensional with kernel H, we have $\chi(V) = c_{[\subseteq H]}$ where $[\subseteq H]$ is the set of subgroups of H.

More generally, if $v : \mathcal{F} \longrightarrow \mathbb{Z}_{\geq 0}$ is any function, we may let $c^v \in \mathcal{O}_{\mathcal{F}}$ be the non-homogeneous element which is $c_H^{v(H)}$ over H. For any representation V we may associate the function $v(H) = dim_{\mathbb{C}}(V^H)$, with finite support: we then have $\chi(V) = c^v$.

Finally, we let $\mathcal{E}$ denote the multiplicative set of all Euler classes:

$$\mathcal{E} = \{c^v \mid v : \mathcal{F} \longrightarrow \mathbb{Z}_{\geq 0} \text{ with finite support}\}.$$

REMARK 4.6.2. The case in which ϕ is the singleton $\{H\}$ gives $e_H c + (1 - e_H)$. It is natural to identify this with the element c_H regarded as an element of $\mathcal{O}_{\mathcal{F}}$, although we have so far paid little attention to c_H outside $\mathbb{Q}[c_H]$.

It can be convenient to deal with a smaller collection of Euler classes, especially if we are inverting them.

LEMMA 4.6.3. *The multiplicative set $\mathcal{E}$ is generated by either*
(i) the elements c_H with H finite or
(ii) the elements c_ϕ with $\phi \subset \mathcal{F}$ finite.

Proof: For Part (i) we just note that $c^v = \prod_H c_H^{v(H)}$, where the product is finite since v has finite support. Part (ii) follows. □

REMARK 4.6.4. Although the Euler classes are not homogeneous, if M is a graded $\mathcal{O}_{\mathcal{F}}$-module, $\mathcal{E}^{-1}M$ is still naturally graded. Indeed, an element c_ϕ acts on $M = e_\phi M \oplus (1-e_\phi)M$ by c on the first factor, and by 1 on the second.

LEMMA 4.6.5. *Inverting the set $\mathcal{E}$ is equivalent to inverting the set of Euler classes of representations.*

Proof: We need only remark that if v is any function with finite support, we can find a subgroup H containing all supporting subgroups. Then if k is the maximum value of v, the Euler class c^v divides $c^k_{[\subseteq H]} = \chi(kV(H))$. □

We may now characterize $\mathcal{F}$-finite torsion modules in terms of Euler classes.

PROPOSITION 4.6.6. *The following conditions on a $\mathcal{O}_{\mathcal{F}}$-module M are equivalent:*
(i) M is a $\mathcal{F}$-finite torsion module.
(ii) M is Euler torsion in the sense that $\mathcal{E}^{-1}M = 0$.

Proof: It is obvious that any $\mathcal{F}$-finite torsion module is Euler torsion.

For any module M, the map $\bigoplus_H M(H) \longrightarrow M$ is injective. If $x \in M$ has $c^k_\phi x = 0$ then $x \in e_\phi M = \bigoplus_{H \in \phi} M(H)$, so that if M is Euler torsion it is $\mathcal{F}$-finite. It is then clearly also c-torsion. □

To complete the analogy with the topological case we introduce algebraic suspension functors analogous to topological suspensions by a representation. The idea is that by analogy with the Thom isomorphism, these suspend the summands of a module for different subgroups by different amounts.

DEFINITION 4.6.7. If $v : \mathcal{F} \longrightarrow \mathbb{Z}$ is a function with finite support, and if ϕ_i is the set of subgroups on which v takes the value i, we define the suspension functor

$$\Sigma^v : \mathcal{O}_{\mathcal{F}}\text{-mod} \longrightarrow \mathcal{O}_{\mathcal{F}}\text{-mod}.$$

by $\Sigma^v M = \bigoplus_i \Sigma^i e_{\phi_i} M$, noting that the sum has only finitely many terms.

This has the familiar properties of suspension.

LEMMA 4.6.8. *The generalized suspension on $\mathcal{O}_{\mathcal{F}}$-modules has the properties*
(i) Σ^0 is the identity functor.
(ii) $\Sigma^v \Sigma^w = \Sigma^{v+w}$.
(iii) Σ^v is an invertible functor. □

Note that the Euler class c^v can be regarded as a degree zero map

$$c^v : M \longrightarrow \Sigma^v M$$

for any module M. Hence in particular

$$\mathcal{E}^{-1}M = \lim_{\rightarrow v} (\Sigma^v M, c^*).$$

CHAPTER 5

Assembly and the standard model.

So far, we have worked entirely locally: one finite subgroup at a time. By the Decoupling Theorem 2.2.3, that is sufficient to deal with arbitrary $\mathcal{F}$-spectra. However, finite isotropy is related to $\mathbb{T}$-isotropy, so, if we want to take the $\mathbb{T}$-fixed part into account, we must relate the parts lying over various finite subgroups. In this chapter, we show how to assemble the finite and infinite isotropy into a single algebraic object giving a complete invariant.

We begin in Sections 5.1 and 5.3 by showing how to deduce Adams spectral sequences for the semifree and general case from the free and $\mathcal{F}$-free cases. Section 5.2 introduces the ring $t^{\mathcal{F}}_*$, which encodes the twisting by representations in relating the $\mathcal{F}$-free and $\mathcal{F}$-contractible parts. The Adams spectral sequence then suggests an obvious algebraic candidate for the abelianization, which we call the *standard model*: this is introduced formally in Section 5.4, and its homological algebra is investigated in Section 5.5. This equips us to apply the criteria of the uniqueness theorem 4.3.1, and we show in Section 5.6 that the triangulated category of $\mathbb{T}$-spectra is equivalent to the derived category of the standard model category. We conclude with four sections for later use: Section 5.7 describing all maps between injective spectra, Section 5.8 discussing cells, spheres and Euler classes, Section 5.9 introducing some explicit notation and giving some important examples, and Section 5.10 formalizing a useful trick for dealing with $\mathcal{O}_{\mathcal{F}}$-modules.

5.1. Assembly.

Now that we understand the categories $\mathbb{T}$-Spec/$\mathbb{T}$ and $\mathbb{T}$-Spec/$\mathcal{F}$, it is time to return to the question of how to fit them together. If X is an object of $\mathbb{T}$-Spec/$\mathbb{T}$, and Y is an object of $\mathbb{T}$-Spec/$\mathcal{F}$ then $[Y,X]^{\mathbb{T}}_*=0$. This follows from 2.2.1 (iii) since $X\simeq \tilde{E}\mathcal{F}\wedge X$ by 2.2.1 (i), and $\Phi^{\mathbb{T}}Y\simeq *$. It is therefore enough to understand $[X,Y]^{\mathbb{T}}_*$. Now, as remarked above $X\simeq \tilde{E}\mathcal{F}\wedge X\simeq \tilde{E}\mathcal{F}\wedge \Phi^{\mathbb{T}}X$ (where we have suppressed notation for inflation on the right). Since the nonequivariant spectrum $\Phi^{\mathbb{T}}X$ splits as a wedge of spheres, the essential case is $[\tilde{E}\mathcal{F},Y]^{\mathbb{T}}_*$. The obstacle to using our available machinery is that $\pi^{\mathbb{T}}_*(\tilde{E}\mathcal{F})$ is not a $\mathbb{Q}[c_H]$-module.

We begin with the simpler case when Y has only one relevant isotropy group i.e. Y is an object of $\mathbb{T}$-Spec/H.

THEOREM 5.1.1. *If $X\simeq X\wedge\tilde{E}\mathcal{F}$ and Y is in $\mathbb{T}$-Spec/H there is a short exact sequence*

$$0\longrightarrow \mathrm{Ext}_{\mathbb{Q}[c_H]}(\mathbb{Q}[c_H,c_H^{-1}]\otimes\pi^{\mathbb{T}}_*(\Sigma X),\pi^{\mathbb{T}}_*(Y))\longrightarrow$$
$$[X,Y]^{\mathbb{T}}_*\longrightarrow \mathrm{Hom}_{\mathbb{Q}[c_H]}(\mathbb{Q}[c_H,c_H^{-1}]\otimes\pi^{\mathbb{T}}_*(X),\pi^{\mathbb{T}}_*(Y))\longrightarrow 0.$$

Proof: As in the construction of the Adams short exact sequence, it is enough to construct a natural transformation

$$\phi : [X,Y]^{\mathbb{T}}_* \longrightarrow \mathrm{Hom}_{\mathbb{Q}[c_H]}(\mathbb{Q}[c_H, c_H^{-1}] \otimes \pi^{\mathbb{T}}_*(X), \pi^{\mathbb{T}}_*(Y))$$

and show it is an isomorphism when $\pi^{\mathbb{T}}_*(Y)$ is an injective $\mathbb{Q}[c_H]$-module. Furthermore, by the splitting in 4.3.1 we may also suppose $\pi^{\mathbb{T}}_*(Y)$ is concentrated in even degrees.

We first explain how to define ϕ. The problem is to get $\mathbb{Q}[c_H]$ acting naturally on some functor of $\pi^{\mathbb{T}}_*(X)$. Observe that for any object Y of $\mathbb{T}$-Spec$/H$ and any other spectrum Z, $\mathbb{Q}[c_H]$ does act on $F(Z,Y)$, even though it is not usually an $\mathcal{F}$-spectrum. We thus define ϕ on a morphism $f : X \longrightarrow Y$, by forming $f \wedge 1 : X \wedge DE\langle H\rangle \longrightarrow Y \wedge DE\langle H\rangle$ and then applying homotopy:

$$\phi(f) = (f_* : \pi^{\mathbb{T}}_*(X \wedge DE\langle H\rangle) \longrightarrow \pi^{\mathbb{T}}_*(Y \wedge DE\langle H\rangle)).$$

The next two lemmas show that the domain and codomain have the correct form.

For the codomain we need a variant on a well known observation.

LEMMA 5.1.2. *There is an equivalence*

$$E\langle H\rangle \wedge F(E\langle H\rangle, X) \simeq E\langle H\rangle \wedge X.$$

Proof: The cofibre of the natural map from right to left is $E\langle H\rangle \wedge F(\tilde{E}\langle H\rangle, X)$, which is contractible. This follows from the fact that $E\langle H\rangle$ may be formed from cells σ^0_H, whilst $\sigma^0_H \wedge F(\tilde{E}\langle H\rangle, X) \simeq F(D\sigma^0_H \wedge \tilde{E}\langle H\rangle, X)$, and $D\sigma^0_H \wedge \tilde{E}\langle H\rangle$ is contractible. □

Since $Y \simeq E\langle H\rangle \wedge Y$, we have an equivalence $Y \wedge DE\langle H\rangle \simeq Y$, and the codomain of $\phi(f)$ is $\pi^{\mathbb{T}}_*(Y)$ as required.

LEMMA 5.1.3. *There is a natural isomorphism of $\mathbb{Q}[c_H]$-modules*

$$\pi^{\mathbb{T}}_*(\tilde{E}\mathcal{F} \wedge X \wedge DE\langle H\rangle) \cong \mathbb{Q}[c_H, c_H^{-1}] \otimes \pi_*(\Phi^{\mathbb{T}} X).$$

Proof: First note that $\tilde{E}\mathcal{F}$ may be replaced by $\tilde{E}\langle H\rangle$, since the cofibre of $\tilde{E}\langle H\rangle \longrightarrow \tilde{E}\mathcal{F}$ is the cofibre of $E\langle H\rangle \longrightarrow E\mathcal{F}_+$, namely the wedge of spectra $E\langle K\rangle$ with $K \neq H$. When $K \neq H$ we have $E\langle K\rangle \wedge DE\langle H\rangle \simeq *$ by 2.2.7.

Next, we have $\pi^{\mathbb{T}}_*(\tilde{E}\mathcal{F} \wedge X \wedge DE\langle H\rangle) \cong \pi_*(\Phi^{\mathbb{T}}(X \wedge DE\langle H\rangle))$. Since $\Phi^{\mathbb{T}}$ commutes with smash products it is enough to deal with the case $\Phi^{\mathbb{T}} X = S^0$. But then $\pi^{\mathbb{T}}_*(\tilde{E}\mathcal{F} \wedge DE\langle H\rangle) = \pi^{\mathbb{T}}_*(\tilde{E}\langle H\rangle \wedge DE\langle H\rangle) = \pi^{\mathbb{T}}_*(DE\langle H\rangle)[c_H^{-1}]$, by 2.3.9, and we have seen in 2.4.1 that $\pi^{\mathbb{T}}_*(DE\langle H\rangle) \cong \mathbb{Q}[c_H]$. □

This shows the domain of $\phi(f)$ has the required form, so ϕ is now defined and we may turn to the proof of the theorem.

We already have the machinery to calculate $[X,Y]^{\mathbb{T}}_*$ as a rational vector space when $\pi^{\mathbb{T}}_*(Y)$ is injective. Indeed, since $\Phi^{\mathbb{T}} X$ splits as a wedge of spheres, $X \simeq \bigvee_i \Sigma^{n_i} \tilde{E}\mathcal{F}$, and so $[X,Y]^{\mathbb{T}} \cong \prod_i [\Sigma^{n_i} \tilde{E}\mathcal{F}, Y]^{\mathbb{T}}_*$. It is thus enough to deal with the special case $X = \tilde{E}\mathcal{F}$. The cofibre sequence $E\mathcal{F}_+ \longrightarrow S^0 \longrightarrow \tilde{E}\mathcal{F}$ thus gives an exact sequence

$$\cdots \longleftarrow [S^0, Y]^{\mathbb{T}}_* \longleftarrow [\tilde{E}\mathcal{F}, Y]^{\mathbb{T}}_* \longleftarrow [\Sigma E\mathcal{F}_+, Y]^{\mathbb{T}}_* \longleftarrow \cdots .$$

Since $\pi_*^{\mathbb{T}}(S^0 \wedge DE\langle H\rangle) \cong \mathbb{Q}[c_H]$ and $\pi_*^{\mathbb{T}}(\Sigma E\mathcal{F}_+ \wedge DE\langle H\rangle) \cong \pi_*^{\mathbb{T}}(\Sigma E\langle H\rangle) \cong \mathbb{Q}[c_H, c_H^{-1}]/\mathbb{Q}[c_H]$, the transformation ϕ compares this with the sequence

$$\mathrm{Hom}_{\mathbb{Q}[c_H]}(\mathbb{Q}[c_H], \pi_*^{\mathbb{T}}(Y)) \longleftarrow \mathrm{Hom}_{\mathbb{Q}[c_H]}(\mathbb{Q}[c_H, c_H^{-1}], \pi_*^{\mathbb{T}}(Y)) \longleftarrow \mathrm{Hom}_{\mathbb{Q}[c_H]}(\mathbb{Q}[c_H, c_H^{-1}]/\mathbb{Q}[c_H], \pi_*^{\mathbb{T}}(Y)),$$

which is in fact short exact. It is tautologous that ϕ identifies the first terms

$$[S^0, Y]_*^{\mathbb{T}} \cong \mathrm{Hom}_{\mathbb{Q}[c_H]}(\mathbb{Q}[c_H], \pi_*^{\mathbb{T}}(Y))$$

For the third, ϕ is simply passage to homotopy, so, since $\pi_*^{\mathbb{T}}(Y)$ is injective, 3.1.1 shows that ϕ gives an isomorphism

$$[\Sigma E\mathcal{F}_+, Y]_*^{\mathbb{T}} \cong \mathrm{Hom}_{\mathbb{Q}[c_H]}(\mathbb{Q}[c_H, c_H^{-1}]/\mathbb{Q}[c_H], \pi_*^{\mathbb{T}}(Y)).$$

Since $\pi_*^{\mathbb{T}}(Y)$ is concentrated in even degrees, both terms are concentrated in even degrees, and the geometric sequence is also short exact. By the Five Lemma, ϕ gives an isomorphism

$$[\tilde{E}\mathcal{F}, Y]_*^{\mathbb{T}} \cong \mathrm{Hom}_{\mathbb{Q}[c_H]}(\mathbb{Q}[c_H, c_H^{-1}], \pi_*^{\mathbb{T}}(Y)),$$

at least as rational vector spaces. However if $Z = Z' \wedge S^{\infty V(H)}$ then c_H acts invertibly on $F(Z, Y)$ since c_H is the Euler class of $V(H)$ on spectra over H. Since $\mathbb{Q}[c_H, c_H^{-1}]$ is a graded field, the $\mathbb{Q}[c_H]$-module structure on $\pi_*^{\mathbb{T}}(F(Z, Y))$ follows from its structure as a vector space. □

5.2. The ring $t_*^{\mathcal{F}}$.

The proof in the previous section generalizes readily to $\mathcal{F}$-spectra Y which are nontrivial at only a finite number of subgroups. We now need to discuss the case where Y has nontrivial isotropy at an infinite number of subgroups.

As discussed in Section 4.4 above, for an $\mathcal{F}$-spectrum Y we view $\pi_*^{\mathbb{T}}(Y)$ as a module over the ring $\mathcal{O}_{\mathcal{F}} = \prod_H \mathbb{Q}[c_H]$. and $c \in (\mathcal{O}_{\mathcal{F}})_{-2}$ denotes the total Chern class c with Hth coordinate c_H. This section introduces a ring which will play a pivotal role.

We continue by analogy with the previous section. The complication is that, since $\mathcal{F}$ is infinite, sums and products do not coincide. Once again a central role is played by $\pi_*^{\mathbb{T}}(\tilde{E}\mathcal{F} \wedge DE\mathcal{F}_+ \wedge X)$; since $\mathcal{O}_{\mathcal{F}}$ acts on $DE\mathcal{F}_+$, this is a $\mathcal{O}_{\mathcal{F}}$-module. Particularly important is the case $X = S^0$:

$$t_*^{\mathcal{F}} := \pi_*^{\mathbb{T}}(\tilde{E}\mathcal{F} \wedge DE\mathcal{F}_+).$$

The notation $t_*^{\mathcal{F}}$ is chosen since it is the coefficient ring of $\mathcal{F}$-Tate homology for S^0 in the sense of [**14**]; this will play no part in what follows, but suggests a number of parallels worth further investigation.

Since $\tilde{E}\mathcal{F} = \varinjlim_{V^{\mathbb{T}}=0} S^V$ we see that $t_*^{\mathcal{F}}$ is obtained by inverting all Euler classes in $\mathcal{O}_{\mathcal{F}}$:

$$t_*^{\mathcal{F}} = \mathcal{E}^{-1}\mathcal{O}_{\mathcal{F}}$$

(see Section 4.6 for information about Euler classes). The canonical warning is that since Euler classes have finite support, $t_*^{\mathcal{F}}$ is not simply $\mathcal{O}_{\mathcal{F}}[c^{-1}]$. Although it is easy to deduce the additive structure of $t_*^{\mathcal{F}}$ directly, we give an alternative approach.

LEMMA 5.2.1. *As a graded vector space* $t_*^{\mathcal{F}} = \pi_*^{\mathbb{T}}(\tilde{E}\mathcal{F} \wedge DE\mathcal{F}_+)$ *is* $\prod_H \mathbb{Q}$ *in degrees* $0, -2, -4, -6, \ldots$ *and* $\bigoplus_H \mathbb{Q}$ *in degrees* $2, 4, 6, \ldots$. *Furthermore there is a short exact sequence*

$$0 \longrightarrow \mathcal{O}_{\mathcal{F}} \longrightarrow t_*^{\mathcal{F}} \xrightarrow{\epsilon} \Sigma^2 \mathbb{I} \longrightarrow 0$$

of $\mathcal{O}_{\mathcal{F}}$*-modules, and the map*

$$c : t_2^{\mathcal{F}} = \bigoplus_H \mathbb{Q} \longrightarrow \prod_H \mathbb{Q} = t_0^{\mathcal{F}}$$

describing the extension is the natural inclusion of the sum in the product.

Proof: The additive structure of $t_*^{\mathcal{F}}$ follows by applying $\pi_*^{\mathbb{T}}(\bullet \wedge DE\mathcal{F}_+)$ to the cofibre sequence $E\mathcal{F}_+ \longrightarrow S^0 \longrightarrow \tilde{E}\mathcal{F}$; indeed the homotopy groups of $E\mathcal{F}_+ \wedge DE\mathcal{F}_+ \simeq E\mathcal{F}_+$ follow from 2.3.5 and are in positive degrees, whilst those of $DE\mathcal{F}_+ \simeq \prod_H DE\langle H \rangle$ follow from 2.4.1, and are in non-positive degrees. This also gives the short exact sequence.

To determine the effect of $c : t_2^{\mathcal{F}} \longrightarrow t_0^{\mathcal{F}}$ it is enough to consider the square

$$\begin{array}{ccc} t_2^{\mathcal{F}} & \xrightarrow{c} & t_0^{\mathcal{F}} \\ \uparrow & & \uparrow \\ \pi_2^{\mathbb{T}}(F(E\langle H \rangle, S^0) \wedge \tilde{E}\mathcal{F}) & \xrightarrow{c_H} & \pi_0^{\mathbb{T}}(F(E\langle H \rangle, S^0) \wedge \tilde{E}\mathcal{F}), \end{array}$$

where the verticals are induced by the projection $E\mathcal{F}_+ \longrightarrow E\langle H \rangle$ onto a direct factor. We saw in the previous section that the bottom horizontal was c_H. □

Although the occurrence of both sums and products might appear unattractive, both seem to play an essential role. However, they do not prevent a rather surprising example of Tate duality.

LEMMA 5.2.2. (Tate duality.) *There is an isomorphism*

$$\mathrm{Hom}_{\mathcal{O}_{\mathcal{F}}}(t_*^{\mathcal{F}}, \Sigma^2 \mathbb{I}) \cong t_*^{\mathcal{F}}$$

of $\mathcal{O}_{\mathcal{F}}$*-modules.*

Proof: We apply $\mathrm{Hom}_{\mathcal{O}_{\mathcal{F}}}(\cdot, \Sigma^2 \mathbb{I})$ to the short exact sequence of 5.2.1; the result is again short exact since $\Sigma^2 \mathbb{I}$ is injective amongst $\mathcal{F}$-finite torsion modules, so that $\mathrm{Ext}_{\mathcal{O}_{\mathcal{F}}}(\mathbb{I}, \mathbb{I}) = 0$:

$$0 \longrightarrow \mathcal{O}_{\mathcal{F}} \longrightarrow \mathrm{Hom}_{\mathcal{O}_{\mathcal{F}}}(t_*^{\mathcal{F}}, \Sigma^2 \mathbb{I}) \longrightarrow \Sigma^2 \mathbb{I} \longrightarrow 0.$$

To identify the extension, apply the idempotent e_H, and note that c_H acts invertibly. □

We may also record the broader relevance of $t_*^{\mathcal{F}}$ to topology.

LEMMA 5.2.3. *There is a natural isomorphism of* $\mathcal{O}_{\mathcal{F}}$*-modules*

$$\pi_*^{\mathbb{T}}(\tilde{E}\mathcal{F} \wedge X \wedge DE\mathcal{F}_+) \cong t_*^{\mathcal{F}} \otimes \pi_*(\Phi^{\mathbb{T}} X).$$

Proof: Note that $\pi_*^{\mathbb{T}}(\tilde{E}\mathcal{F} \wedge X \wedge DE\mathcal{F}_+) \cong \pi_*(\Phi^{\mathbb{T}}(X \wedge DE\mathcal{F}_+))$. Since $\Phi^{\mathbb{T}} X \simeq \bigvee_i S^{n_i}$, and $\Phi^{\mathbb{T}}$ commutes with smash products, it is enough to deal with the case $\Phi^{\mathbb{T}} X = S^0$, which is true by definition. □

5.3. Global assembly.

In this section we calculate maps from an $\mathcal{F}$-contractible spectrum to an $\mathcal{F}$-spectrum. First we need to begin by introducing the relevant algebra. The ring $\mathcal{O}_{\mathcal{F}}$ is fairly complicated, but so far it has sufficed to restrict attention to the $\mathcal{F}$-finite torsion modules in the sense of 4.5.2 and 4.5.1 (namely those occurring as $\pi^{\mathbb{T}}_*(Y)$ for an $\mathcal{F}$-spectrum Y). We must now include modules like the twisting module $t^{\mathcal{F}}_*$ introduced in the previous section.

In particular we need to get the homological algebra under control. First, note that for an arbitrary module L, $\mathrm{Hom}_{\mathcal{O}_{\mathcal{F}}}(L, N(H)) = \mathrm{Hom}_{\mathbb{Q}[c_H]}(L(H), N(H))$, so that if $N(H)$ is an injective $\mathbb{Q}[c_H]$-module it is also injective over $\mathcal{O}_{\mathcal{F}}$. However, since $\mathcal{O}_{\mathcal{F}}$ is not Noetherian, it does not follow that infinite direct sums of modules of this form are injective.

LEMMA 5.3.1. *If M is a $\mathcal{O}_{\mathcal{F}}$-module with the property that $\mathcal{E}^{-1}M \cong t^{\mathcal{F}}_* \otimes U$ for some graded vector space U (for instance if $M = t^{\mathcal{F}}_* \otimes U$ or if M is an $\mathcal{F}$-finite torsion module), then for any $\mathcal{O}_{\mathcal{F}}$-module N we have $\mathrm{Ext}^s_{\mathcal{O}_{\mathcal{F}}}(M, N) = 0$ for $s \geq 2$. If in addition $N(H)$ is injective for all H then $\mathrm{Ext}^s_{\mathcal{O}_{\mathcal{F}}}(M, N) = 0$ for $s \geq 1$.*

Proof: Note that if M is $\mathcal{F}$-finite, $\mathrm{Hom}_{\mathcal{O}_{\mathcal{F}}}(M, N) = \prod_H \mathrm{Hom}_{\mathbb{Q}[c_H]}(M(H), N(H))$, so that the result holds in this case. The result holds if $M = t^{\mathcal{F}}_* \otimes U$ by 5.2.1, and the fact that $\mathcal{O}_{\mathcal{F}}$ is projective.

The general case follows by considering the morphism $M \longrightarrow \mathcal{E}^{-1}M$. Its image M'' is a submodule of $\mathcal{E}^{-1}M \cong t^{\mathcal{F}}_* \otimes U$, and so the result holds for M''; its kernel M' is an $\mathcal{F}$-finite torsion module, for which the lemma also holds. The result for M follows from the exact sequence $0 \longrightarrow M' \longrightarrow M \longrightarrow M'' \longrightarrow 0$. $\square$

This is sufficient to give us the Adams short exact sequence for arbitrary $\mathbb{T}$-spectra.

THEOREM 5.3.2. *If X is $\mathcal{F}$-contractible and Y is in $\mathbb{T}$-Spec/$\mathcal{F}$ there is a short exact sequence*

$$0 \longrightarrow \mathrm{Ext}_{\mathcal{O}_{\mathcal{F}}}(t^{\mathcal{F}}_* \otimes \pi^{\mathbb{T}}_*(\Sigma X), \pi^{\mathbb{T}}_*(Y)) \longrightarrow [X, Y]^{\mathbb{T}}_* \longrightarrow \mathrm{Hom}_{\mathcal{O}_{\mathcal{F}}}(t^{\mathcal{F}}_* \otimes \pi^{\mathbb{T}}_*(X), \pi^{\mathbb{T}}_*(Y)) \longrightarrow 0.$$

Proof: As in the construction of the Adams short exact sequence it is enough to construct a natural transformation

$$\phi : [X, Y]^{\mathbb{T}}_* \longrightarrow \mathrm{Hom}_{\mathcal{O}_{\mathcal{F}}}(t^{\mathcal{F}}_* \otimes \pi^{\mathbb{T}}_*(X), \pi^{\mathbb{T}}_*(Y))$$

and show it is an isomorphism when $\pi^{\mathbb{T}}_*(Y)$ is an $\mathcal{O}_{\mathcal{F}}$-module injective in an appropriate sense.

We define ϕ for arbitrary $\mathbb{T}$-spectra X. Observe that since $Y \simeq E\mathcal{F}_+ \wedge Y$ and $DE\mathcal{F}_+ \wedge E\mathcal{F}_+ \simeq E\mathcal{F}_+$, we have an equivalence $Y \wedge DE\mathcal{F}_+ \simeq Y$. Now define the natural transformation by taking $f : X \longrightarrow Y$, replacing it with $f \wedge 1 : X \wedge DE\mathcal{F}_+ \longrightarrow Y \wedge DE\mathcal{F}_+ \simeq Y$ and then applying homotopy: explicitly this gives a natural transformation

$$\phi : [X, Y]^{\mathbb{T}}_* \longrightarrow \mathrm{Hom}_{\mathcal{O}_{\mathcal{F}}}(\pi^{\mathbb{T}}_*(X \wedge DE\mathcal{F}_+), \pi^{\mathbb{T}}_*(Y)).$$

If $X \simeq X \wedge \tilde{E}\mathcal{F}$ then $X \simeq \bigvee_i \Sigma^{n_i}\tilde{E}\mathcal{F}$, and 5.2.3 gives an isomorphism $\pi^{\mathbb{T}}_*(X \wedge DE\mathcal{F}_+) \cong t^{\mathcal{F}}_* \otimes \pi^{\mathbb{T}}_*(\Phi^{\mathbb{T}}X)$, so that the codomain of ϕ is of the required form. Note also that in this case $[X,Y]^{\mathbb{T}} \cong \prod_i [\Sigma^{n_i}\tilde{E}\mathcal{F}, Y]^{\mathbb{T}}_*$, so that it is enough to deal with the special case $X = \tilde{E}\mathcal{F}$; we may also suppose that $\pi^{\mathbb{T}}_*(Y)$ is concentrated in even degrees.

We saw in the proof of 5.2.1 that applying $\pi^{\mathbb{T}}_*(\bullet \wedge DE\mathcal{F}_+)$ to the cofibre sequence $E\mathcal{F}_+ \longrightarrow S^0 \longrightarrow \tilde{E}\mathcal{F}$ gives a short exact sequence of $\mathcal{O}_\mathcal{F}$-modules. Hence applying ϕ we obtain a diagram

$$\begin{array}{ccccccccc} 0 & \longleftarrow & [S^0,Y]^{\mathbb{T}}_* & \longleftarrow & [\tilde{E}\mathcal{F},Y]^{\mathbb{T}}_* & \longleftarrow & [\Sigma E\mathcal{F}_+,Y]^{\mathbb{T}}_* & \longleftarrow & 0 \\ & & \cong\downarrow & & \downarrow & & \downarrow\cong & & \\ 0 & \longleftarrow & \mathrm{Hom}(\mathcal{O}_\mathcal{F}, \pi^{\mathbb{T}}_*(Y)) & \longleftarrow & \mathrm{Hom}(t^\mathcal{F}_*, \pi^{\mathbb{T}}_*(Y)) & \longleftarrow & \mathrm{Hom}(\pi^{\mathbb{T}}_*(\Sigma E\mathcal{F}_+), \pi^{\mathbb{T}}_*(Y)) & \longleftarrow & 0. \end{array}$$

The bottom lefthand zero follows from 5.3.1; the lefthand vertical is tautologically isomorphic, the righthand vertical is isomorphic by 3.1.1. The central vertical is thus an isomorphism by the 5-lemma. □

5.4. The standard model category.

The purpose of this section is to describe an abelian category $\mathcal{A} = \mathcal{A}_\mathcal{F}$ whose derived category is an algebraic model for the category of $\mathbb{T}$-spectra, and whose homological algebra provides Adams spectral sequences for both the categories. There is also a somewhat simpler abelian category $\mathcal{A}_1$ whose derived category gives an algebraic model of semifree $\mathbb{T}$-spectra, and similarly for other collections of finite subgroups.

DEFINITION 5.4.1. The objects of the *standard model* $\mathcal{A}$ are maps $N \xrightarrow{\beta} t^\mathcal{F}_* \otimes V$ of $\mathcal{O}_\mathcal{F}$-modules which become an isomorphism when the set $\mathcal{E}$ of Euler classes is inverted. Morphisms are commutative squares

$$\begin{array}{ccc} M & \xrightarrow{\theta} & N \\ \beta\downarrow & & \downarrow\gamma \\ t^\mathcal{F}_* \otimes U & \xrightarrow{1\otimes\phi} & t^\mathcal{F}_* \otimes V. \end{array}$$

We shall refer to N as the *nub* and V as the *vertex* of the object $N \longrightarrow t^\mathcal{F}_* \otimes V$. The map β is called the *basing* map.

REMARK 5.4.2. (a) We saw in 4.6.6 that $\mathcal{F}$-finite torsion modules are precisely the modules annihilated by inverting the set $\mathcal{E}$ of Euler classes. Since $t^\mathcal{F}_*$ is $\mathcal{E}$ local, the condition that $\beta : N \longrightarrow t^\mathcal{F}_* \otimes V$ becomes an isomorphism when $\mathcal{E}$ is inverted is precisely the requirement that its kernel and cokernel are $\mathcal{F}$-finite torsion modules.
(b) The role of β is to give a specific isomorphism $\mathcal{E}^{-1}N \cong t^\mathcal{F}_* \otimes V$. Thus we may view objects of $\mathcal{A}$ as given by $\mathcal{O}_\mathcal{F}$ modules together with with a particular vector subspace V of $\mathcal{E}^{-1}N$ so that the canonical extension of the inclusion $V \longrightarrow \mathcal{E}^{-1}N$ is an isomorphism. This explains why β is called a basing map.
(c) The standard model category $\mathcal{A}_\mathcal{H}$ for a set $\mathcal{H}$ of finite subgroups is obtained by replacing $\mathcal{O}_\mathcal{F}$ by $\mathcal{O}_\mathcal{H} := e_\mathcal{H}\mathcal{O}_\mathcal{F}$. The case in which $\mathcal{H}$ is finite is significantly simpler, but we shall just make explicit the case $\mathcal{H} = \{1\}$. Here the objects are $\mathbb{Q}[c_1]$-module maps $\beta : N \longrightarrow \mathbb{Q}[c_1, c_1^{-1}] \otimes V$ which become an isomorphism when c_1 is inverted. It is often helpful to

examine any algebraic construction in this case before passing to the general case $\mathcal{H} = \mathcal{F}$, and we shall adopt this policy for the more intricate bits of algebra.

Before we study the category $\mathcal{A}$ and its derived category, it will be useful to import information into $\mathcal{A}$ from categories we understand. For comparison with the model of $\mathcal{F}$-contractible spectra given by the derived category $D\mathbb{Q}$ of rational vector spaces, we let $\mathbb{Q}_*$-mod denote the category of graded rational vector spaces and define the functor

$$e : \mathbb{Q}_*\text{-mod} \longrightarrow \mathcal{A}$$

by

$$e(V) = \left(t_*^{\mathcal{F}} \otimes V \stackrel{1}{\longrightarrow} t_*^{\mathcal{F}} \otimes V\right).$$

Objects of $\mathcal{A}$ isomorphic to ones in the image of e will be called *torsion free* objects. For comparison with the model of $\mathcal{F}$-spectra we define the functor

$$f : \text{tors-}\mathcal{O}_{\mathcal{F}}^f\text{-mod} \longrightarrow \mathcal{A}$$

by

$$f(N) = (N \longrightarrow 0).$$

Objects of $\mathcal{A}$ isomorphic to ones in the image of f will be called *torsion* objects. It is easy to see that both e and f are right adjoints.

LEMMA 5.4.3. *For any object $M \longrightarrow t_*^{\mathcal{F}} \otimes U$ of $\mathcal{A}$, any graded vector space V and any $\mathcal{F}$-finite torsion $\mathcal{O}_{\mathcal{F}}$-module N we have natural isomorphisms*
(i)

$$\mathrm{Hom}_{\mathcal{A}}((M \longrightarrow t_*^{\mathcal{F}} \otimes U), e(V)) = \mathrm{Hom}_{\mathbb{Q}}(U, V).$$

(ii)

$$\mathrm{Hom}_{\mathcal{A}}((M \longrightarrow t_*^{\mathcal{F}} \otimes U), f(N)) = \mathrm{Hom}_{\mathcal{O}_{\mathcal{F}}}(M, N). \quad \square$$

COROLLARY 5.4.4. *(i) The functors e and f are full and faithful embeddings.*
(ii) For any graded vector space W and any $\mathcal{F}$-finite torsion $\mathcal{O}_{\mathcal{F}}$-module N

$$\mathrm{Hom}_{\mathcal{A}}(f(N), e(W)) = 0. \quad \square$$

5.5. Homological algebra in the standard model.

We now have the ingredients to construct injective resolutions, and thereby to prove that $\mathcal{A}$ is a split 1-dimensional abelian category.

We can import injective objects using the functors e and f, and they will prove to be sufficient for homological algebra.

LEMMA 5.5.1. *(i) The object $e(V) = (t_*^{\mathcal{F}} \otimes V \stackrel{1}{\longrightarrow} t_*^{\mathcal{F}} \otimes V)$ of $\mathcal{A}$ is injective for any graded vector space V.*
(ii) The object $f(I) = (I \longrightarrow 0)$ of $\mathcal{A}$ is injective if I is an $\mathcal{F}$-finite torsion injective.

Proof: This follows from 5.4.3 together with the obvious fact that the functors $(M \longrightarrow t_*^{\mathcal{F}} \otimes V) \longmapsto M$ and $(M \longrightarrow t_*^{\mathcal{F}} \otimes V) \longmapsto V$ are exact. □

THEOREM 5.5.2. *The category $\mathcal{A}$ is a split one dimensional graded abelian category in the sense of Definition 4.2.3.*

Proof: We must verify that $\mathcal{A}$ is abelian, that every object admits an injective resolution of length 1, and that every object X splits as a sum $X_+ \oplus X_-$ of pure pieces so that $\mathrm{Hom}(X_\delta, Y_\epsilon) = 0$ and $\mathrm{Ext}(X_\delta, Y_\epsilon) = 0$ if $\delta \neq \epsilon$.

First we make an observation about $\mathcal{F}$-finite torsion $\mathcal{O}_{\mathcal{F}}$-modules.

LEMMA 5.5.3. *If $0 \longrightarrow M' \longrightarrow M \longrightarrow M'' \longrightarrow 0$ is an exact sequence of $\mathcal{O}_{\mathcal{F}}$-modules then M is $\mathcal{F}$-finite and torsion if and only if M' and M'' are both $\mathcal{F}$-finite and torsion.* □

Now, considering the map in the square displayed in Definition 5.4.1 above, we obtain two short exact sequences:

$$\begin{array}{ccccccccc} 0 & \longrightarrow & K & \longrightarrow & M & \longrightarrow & M' & \longrightarrow & 0 \\ & & \xi\downarrow & & \beta\downarrow & & \eta\downarrow & & \\ 0 & \longrightarrow & t_*^{\mathcal{F}} \otimes U & \longrightarrow & t_*^{\mathcal{F}} \otimes V & \longrightarrow & t_*^{\mathcal{F}} \otimes V' & \longrightarrow & 0 \end{array}$$

and

$$\begin{array}{ccccccccc} 0 & \longrightarrow & M' & \longrightarrow & N & \longrightarrow & C & \longrightarrow & 0 \\ & & \eta\downarrow & & \gamma\downarrow & & \sigma\downarrow & & \\ 0 & \longrightarrow & t_*^{\mathcal{F}} \otimes V' & \longrightarrow & t_*^{\mathcal{F}} \otimes W & \longrightarrow & t_*^{\mathcal{F}} \otimes X & \longrightarrow & 0 \end{array}$$

By hypothesis, the central verticals have kernel and cokernel $\mathcal{F}$-finite and torsion. It follows from the first diagram that $ker(\xi)$ and $cok(\eta)$ are $\mathcal{F}$-finite and torsion, and from the second that $ker(\eta)$ and $cok(\sigma)$ are $\mathcal{F}$-finite and torsion. Hence $\eta : M' \longrightarrow t_*^{\mathcal{F}} \otimes V'$ is an object of $\mathcal{A}$. Now we may use the Snake Lemma and closure of $\mathcal{F}$-finite torsion spectra under extensions to deduce that $cok(\xi)$ and $ker(\sigma)$ are also $\mathcal{F}$-finite and torsion. This completes the proof that $\mathcal{A}$ is abelian.

To see that $\beta : M \longrightarrow t_*^{\mathcal{F}} \otimes V$ has a resolution of length 1, we recall that 5.5.1 allows us to import injectives from $\mathbb{Q}_*$-mod and tors-$\mathcal{O}_{\mathcal{F}}^f$-mod. Now we proceed to construct an injective resolution of length 1. First note that $ker(\beta)$ is the torsion submodule TM of M, and that there is a resolution $0 \longrightarrow TM \longrightarrow I' \longrightarrow J' \longrightarrow 0$ of TM by objects I' and J' which are injective as $\mathcal{F}$-finite torsion $\mathcal{O}_{\mathcal{F}}$-modules. Furthermore if Q is the image of β then $J'' = (t_*^{\mathcal{F}} \otimes V)/Q$ is divisible and an $\mathcal{F}$-finite torsion module, and hence injective in the same sense. In the usual way we form the diagram,

$$\begin{array}{ccccccccc}
 & & 0 & & 0 & & 0 & & \\
 & & \downarrow & & \downarrow & & \downarrow & & \\
0 & \longrightarrow & TM & \longrightarrow & M & \longrightarrow & Q & \longrightarrow & 0 \\
 & & \downarrow & & \downarrow & & \downarrow & & \\
0 & \longrightarrow & I' & \longrightarrow & I' \oplus (t^{\mathcal{F}}_* \otimes V) & \longrightarrow & t^{\mathcal{F}}_* \otimes V & \longrightarrow & 0 \\
 & & \downarrow & & \downarrow & & \downarrow & & \\
0 & \longrightarrow & J' & \longrightarrow & J' \oplus J'' & \longrightarrow & J'' & \longrightarrow & 0 \\
 & & \downarrow & & \downarrow & & \downarrow & & \\
 & & 0 & & 0 & & 0 & &
\end{array}$$

and hence the diagram

$$\begin{array}{ccccccccc}
0 & \longrightarrow & M & \longrightarrow & I' \oplus (t^{\mathcal{F}}_* \otimes V) & \longrightarrow & J' \oplus J'' & \longrightarrow & 0 \\
 & & \downarrow & & \downarrow & & \downarrow & & \\
0 & \longrightarrow & t^{\mathcal{F}}_* \otimes V & \longrightarrow & t^{\mathcal{F}}_* \otimes V & \longrightarrow & 0 & \longrightarrow & 0
\end{array}$$

This is the required resolution; injectivity of the last term was 5.5.1 (i), and injectivity of the middle term follows since it is an extension of two things known to be injective by 5.5.1.

Finally we just need to observe that any object $X = (M \longrightarrow t^{\mathcal{F}}_* \otimes V)$ splits as a sum $X_+ \oplus X_-$ where X_+ is the even graded part and X_- is the odd graded part. It is then clear that $\mathrm{Hom}(X_\delta, Y_\epsilon) = 0$ if $\delta \neq \epsilon$, and the fact that $\mathrm{Ext}(X_\delta, Y_\epsilon) = 0$ if $\delta \neq \epsilon$ follows since the resolution of Y_ϵ constructed above is entirely in parity ϵ. □

Now that we know the derived category $D\mathcal{A}$ exists by 4.1.3 we may observe that the functors e and f induce functors on derived categories.

COROLLARY 5.5.4. *The functors $e : \mathbb{Q}_*$-mod $\longrightarrow \mathcal{A}$ and f : tors-$\mathcal{O}^f_{\mathcal{F}}$-mod $\longrightarrow \mathcal{A}$ induce maps of derived categories.*

Proof: The functors e and f induce functors on categories of differential graded objects, and, by the universal property of the category of fractions, it is enough to show that the functors e and f take homology isomorphisms to homology isomorphisms. Since e and f are obviously exact functors, this is clear. □

5.6. The algebraicization of rational $\mathbb{T}$-spectra.

We have constructed a split one dimensional abelian category $\mathcal{A}$, and it has a derived category $D\mathcal{A}$ by 4.1.3; we shall show in this section that $\mathcal{A}$ is the linearization of the category of $\mathbb{T}$-spectra, and hence that $D\mathcal{A}$ is an algebraic model of the category of $\mathbb{T}$-spectra: this is the main theorem of Part I.

THEOREM 5.6.1. (Algebraicization of rational $\mathbb{T}$-spectra.) *There are equivalences of triangulated categories*

$$\mathbb{T}\text{-}Spec \simeq D\mathcal{A}.$$

Proof: Once again we aim to apply Theorem 4.3.1. It follows from the construction of Lewis-May that $\mathbb{T}$-Spec is the homotopy category of a model category. It remains to show that $\mathbb{T}$-Spec has linearization $\mathcal{A}$.

We must describe a functor $\mathbb{T}$-Spec $\longrightarrow \mathcal{A}$, and show it is essentially surjective and a linearization. For this we define

$$\pi_*^{\mathcal{A}}(X) := \pi_*^{\mathbb{T}}(X \wedge DE\mathcal{F}_+ \longrightarrow X \wedge \tilde{E}\mathcal{F} \wedge DE\mathcal{F}_+)$$

LEMMA 5.6.2. *The above definition gives a functor*

$$\pi_*^{\mathcal{A}} : \mathbb{T}\text{-}Spec \longrightarrow \mathcal{A}.$$

Proof: The definition is clearly natural, so the only point is to verify that $\pi_*^{\mathcal{A}}(X)$ is an object of $\mathcal{A}$. First we note that $\pi_*^{\mathbb{T}}(X \wedge \tilde{E}\mathcal{F} \wedge DE\mathcal{F}_+) \cong t_*^{\mathcal{F}} \otimes \pi_*^{\mathbb{T}}(\Phi^{\mathbb{T}}X)$, by 5.2.3 and we have the natural inclusion $\pi_*(\Phi^{\mathbb{T}}X) \cong \pi_*^{\mathbb{T}}(X \wedge \tilde{E}\mathcal{F}) \longrightarrow \pi_*^{\mathbb{T}}(X \wedge \tilde{E}\mathcal{F} \wedge DE\mathcal{F}_+) \cong t_*^{\mathcal{F}} \otimes \pi_*^{\mathbb{T}}(\Phi^{\mathbb{T}}X)$. This shows that the second term has the correct form. Secondly we must show that the kernel K and cokernel C of $\pi_*^{\mathbb{T}}(X \wedge DE\mathcal{F}_+) \longrightarrow \pi_*^{\mathbb{T}}(X \wedge \tilde{E}\mathcal{F} \wedge DE\mathcal{F}_+)$ are $\mathcal{F}$-finite and torsion. However the fibre of $X \wedge DE\mathcal{F}_+ \longrightarrow X \wedge \tilde{E}\mathcal{F} \wedge DE\mathcal{F}_+$ is $X \wedge E\mathcal{F}_+ \wedge DE\mathcal{F}_+ \simeq X \wedge E\mathcal{F}_+$, whose homotopy is $\mathcal{F}$-finite and torsion by 2.4.2. Since there is an extension

$$0 \longrightarrow \Sigma^{-1}C \longrightarrow \pi_*^{\mathbb{T}}(X \wedge E\mathcal{F}_+) \longrightarrow K \longrightarrow 0,$$

the result follows. □

For special types of $\mathbb{T}$-spectra we can go further, and it is useful to relate $\pi_*^{\mathcal{A}}$ to the functors we already understand; this makes it rather easy to calculate $\pi_*^{\mathcal{A}}(X)$ when X is either an $\mathcal{F}$-spectrum or $\mathcal{F}$-contractible.

LEMMA 5.6.3. *On the subcategories of $\mathcal{F}$-contractible spectra and $\mathcal{F}$-spectra the functor $\pi_*^{\mathcal{A}}$ factorizes through $\pi_*^{\mathbb{T}}$ in the sense that the following two diagrams commute:*

$$\begin{array}{ccc} \mathbb{T}\text{-}Spec/\mathbb{T} & \xrightarrow{\pi_*^{\mathbb{T}}} & D\mathbb{Q} \\ \downarrow & & \downarrow e \\ \mathbb{T}\text{-}Spec & \xrightarrow{\pi_*^{\mathcal{A}}} & D\mathcal{A} \end{array} \quad \text{and} \quad \begin{array}{ccc} \mathbb{T}\text{-}Spec/H & \xrightarrow{\pi_*^{\mathbb{T}}} & D(tors\text{-}\mathcal{O}_{\mathcal{F}}^f) \\ \downarrow & & \downarrow f \\ \mathbb{T}\text{-}Spec & \xrightarrow{\pi_*^{\mathcal{A}}} & D\mathcal{A}. \end{array}$$

Proof: For the first diagram, note that, if Y is an $\mathcal{F}$-contractible spectrum, then $Y \wedge E\mathcal{F}_+ \simeq *$, and hence the map $Y \wedge DE\mathcal{F}_+ \longrightarrow Y \wedge \tilde{E}\mathcal{F} \wedge DE\mathcal{F}_+$ is an equivalence. Now $\pi_*^{\mathbb{T}}(Y \wedge \tilde{E}\mathcal{F} \wedge DE\mathcal{F}_+) = t_*^{\mathcal{F}} \otimes \pi_*^{\mathbb{T}}(Y)$ by 5.2.3, so $\pi_*^{\mathcal{A}}(Y) = e(\pi_*^{\mathbb{T}}(Y))$ as required.

For the second diagram, we use the facts that if X is an $\mathcal{F}$-spectrum then $X \wedge \tilde{E}\mathcal{F} \simeq *$ and $X \wedge DE\mathcal{F}_+ \simeq X$, so $\pi_*^{\mathcal{A}}(X) = (\pi_*^{\mathbb{T}}(X \wedge DE\mathcal{F}_+) \longrightarrow 0) = f(\pi_*^{\mathbb{T}}(X))$. □

Next we need to observe that the Whitehead Theorem holds; this lemma was a critical point for the author in the recognition of the standard model.

LEMMA 5.6.4. (Standard model Whitehead theorem). *If $f : X \longrightarrow Y$ is a map of $\mathbb{T}$-spectra for which $\pi_*^{\mathcal{A}}(f)$ is an isomorphism, then f is an equivalence.*

Proof: Using exactness of $\pi_*^{\mathcal{A}}$ it is enough to consider the cofibre Z of f, for which $\pi_*^{\mathcal{A}}(Z) = 0$, and deduce $Z \simeq *$.

The hypothesis states that (a) $\pi_*^{\mathbb{T}}(Z \wedge DE\mathcal{F}_+) = 0$ and (b) $\pi_*^{\mathbb{T}}(Z \wedge \tilde{E}\mathcal{F} \wedge DE\mathcal{F}_+) = 0$. Now by 5.2.3, $\pi_*^{\mathbb{T}}(X \wedge \tilde{E}\mathcal{F} \wedge DE\mathcal{F}_+) = t_*^{\mathcal{F}} \otimes \pi_*(\Phi^{\mathbb{T}} X)$, so we conclude from (b) that $\Phi^{\mathbb{T}} Z \simeq *$. Thus Z is an $\mathcal{F}$-spectrum, and $Z \wedge DE\mathcal{F}_+ \simeq Z$; by 3.2.3 it follows that $Z \simeq *$. □

There is one final ingredient before we construct the Adams short exact sequence.

LEMMA 5.6.5. *Enough $\mathcal{A}$-injectives are realizable.*

Proof: The proof of Theorem 5.5.2 showed that there are enough injectives amongst those of form $f(I) = (I \longrightarrow 0)$ for an $\mathcal{F}$-finite torsion injective I and $e(V) = (1 : t_*^{\mathcal{F}} \otimes V \longrightarrow t_*^{\mathcal{F}} \otimes V)$ for a graded vector space V. Objects of the first type occur as $f(\pi_*^{\mathbb{T}}(X)) = \pi_*^{\mathcal{A}}(X)$ for $\mathcal{F}$-spectra X by 3.5.1. Those of the second type occur as $e(\pi_*^{\mathbb{T}}(Y)) = \pi_*^{\mathcal{A}}(Y)$ for $\mathcal{F}$-contractible spectra Y. □

Once we have established an Adams short exact sequence, essential surjectivity of $\pi_*^{\mathcal{A}}$ will follow by 4.2.2.

THEOREM 5.6.6. *For any rational $\mathbb{T}$-spectra X and Y, there is a natural short exact sequence*

$$0 \longrightarrow \mathrm{Ext}_{\mathcal{A}}(\pi_*^{\mathcal{A}}(\Sigma X), \pi_*^{\mathcal{A}}(Y)) \longrightarrow [X,Y]_*^{\mathbb{T}} \longrightarrow \mathrm{Hom}_{\mathcal{A}}(\pi_*^{\mathcal{A}}(X), \pi_*^{\mathcal{A}}(Y)) \longrightarrow 0,$$

and it splits unnaturally.

Proof: We are constructing an Adams spectral sequence, so we use the standard operating procedure, and resolve all spectra by injectives for which the theorem holds.

Firstly, we observe that $\pi_*^{\mathcal{A}}$ gives a natural transformation

$$\theta : [X,Y]_*^{\mathbb{T}} \longrightarrow \mathrm{Hom}_{\mathcal{A}}(\pi_*^{\mathcal{A}}(X), \pi_*^{\mathcal{A}}(Y)).$$

Next, we note that θ is an isomorphism if Y is an injective $\mathcal{F}$-spectrum or an $\mathcal{F}$-contractible spectrum.

LEMMA 5.6.7. *If $Y \simeq Y \wedge \tilde{E}\mathcal{F}$ is $\mathcal{F}$-contractible then θ is an isomorphism.*

Proof: First note that, by 5.6.3, $\pi_*^{\mathcal{A}}(Y) = e(\pi_*^{\mathbb{T}}(Y)) = e(\pi_*(\Phi^{\mathbb{T}} Y))$. Now use the commutative diagram

$$\begin{array}{ccc} [X, Y \wedge \tilde{E}\mathcal{F}]_*^{\mathbb{T}} & \longrightarrow & \mathrm{Hom}_{\mathcal{A}}(\pi_*^{\mathcal{A}}(X), e(\pi_*(\Phi^{\mathbb{T}} Y))) \\ \cong\downarrow & & \downarrow\cong \\ [\Phi^{\mathbb{T}} X, \Phi^{\mathbb{T}} Y]_* & \stackrel{\cong}{\longrightarrow} & \mathrm{Hom}_{\mathbb{Q}}(\pi_*(\Phi^{\mathbb{T}} X), \pi_*(\Phi^{\mathbb{T}} Y)). \end{array}$$

The left hand vertical is an isomorphism by obstruction theory 2.2.1, and the right hand vertical is an isomorphism by 5.4.3; the bottom isomorphism is the well known algebraicization of non-equivariant rational spectra. □

LEMMA 5.6.8. *If $Y \simeq Y \wedge E\mathcal{F}_+$ is an injective $\mathcal{F}$-spectrum then θ is an isomorphism.*

Proof: This is a little harder since there are non-zero maps from torsion free objects to torsion objects. Note first that if Y is an $\mathcal{F}$-spectrum then $\pi^{\mathcal{A}}_*(Y) = f(\pi^{\mathbb{T}}_*(Y))$ by 5.6.3. We now deal separately with various types of spectrum X.

If X is an $\mathcal{F}$-spectrum, $\pi^{\mathcal{A}}_*(X) = f(\pi^{\mathbb{T}}_*(X))$ by 5.6.3 and the lemma follows from 3.1.1, using commutativity of the diagram

$$\begin{array}{ccc} [X, Y \wedge E\mathcal{F}_+]^{\mathbb{T}}_* & \longrightarrow & \mathrm{Hom}_{\mathcal{A}}(f(\pi^{\mathbb{T}}_*(X)), f(\pi^{\mathbb{T}}_*(Y))) \\ & \searrow^{\cong} & \downarrow \cong \\ & & \mathrm{Hom}_{\mathcal{O}_{\mathcal{F}}}(\pi^{\mathbb{T}}_*(X), \pi^{\mathbb{T}}_*(Y)). \end{array}$$

The right hand vertical is an isomorphism by 5.4.3.

If X is $\mathcal{F}$-contractible $\pi^{\mathcal{A}}_*(X) = e(\pi^{\mathbb{T}}_*(X))$ by 5.6.3 and the lemma follows from 5.3.2 by commutativity of the diagram

$$\begin{array}{ccc} [X, Y \wedge E\mathcal{F}_+]^{\mathbb{T}}_* & \longrightarrow & \mathrm{Hom}_{\mathcal{A}}(e(\pi^{\mathbb{T}}_*(X)), f(\pi^{\mathbb{T}}_*(Y))) \\ & \searrow^{\cong} & \downarrow \cong \\ & & \mathrm{Hom}_{\mathcal{O}_{\mathcal{F}}}(t^{\mathcal{F}}_* \otimes \pi^{\mathbb{T}}_*(X), \pi^{\mathbb{T}}_*(Y)). \end{array}$$

The right hand vertical is an isomorphism by 5.4.3.

The general case follows by applying the 5-lemma to the diagram,

$$\begin{array}{ccccc} [X \wedge \Sigma E\mathcal{F}_+, Y]^{\mathbb{T}}_* & \longrightarrow & [X \wedge \tilde{E}\mathcal{F}, Y]^{\mathbb{T}}_* & \longrightarrow & [X, Y]^{\mathbb{T}}_* \\ \cong\downarrow & & \cong\downarrow & & \downarrow \\ \mathrm{Hom}_{\mathcal{A}}(\pi^{\mathcal{A}}_*(X \wedge \Sigma E\mathcal{F}_+), \pi^{\mathcal{A}}_*(Y)) & \longrightarrow & \mathrm{Hom}_{\mathcal{A}}(\pi^{\mathcal{A}}_*(X \wedge \tilde{E}\mathcal{F}), \pi^{\mathcal{A}}_*(Y)) & \longrightarrow & \mathrm{Hom}_{\mathcal{A}}(\pi^{\mathcal{A}}_*(X), \pi^{\mathcal{A}}_*(Y)) \end{array}$$

$$\begin{array}{ccccc} \longrightarrow & [X \wedge E\mathcal{F}_+, Y]^{\mathbb{T}}_* & \longrightarrow & [X \wedge \Sigma^{-1}\tilde{E}\mathcal{F}, Y]^{\mathbb{T}}_* \\ & \cong\downarrow & & \cong\downarrow \\ \longrightarrow & \mathrm{Hom}_{\mathcal{A}}(\pi^{\mathcal{A}}_*(X \wedge E\mathcal{F}_+), \pi^{\mathcal{A}}_*(Y)) & \longrightarrow & \mathrm{Hom}_{\mathcal{A}}(\pi^{\mathcal{A}}_*(X \wedge \Sigma^{-1}\tilde{E}\mathcal{F}), \pi^{\mathcal{A}}_*(Y)). \end{array}$$

Since $X \wedge E\mathcal{F}_+ \longrightarrow X \longrightarrow X \wedge \tilde{E}\mathcal{F}$ is a cofibre sequence, the first row is exact, and there is a long exact sequence in $\pi^{\mathcal{A}}_*$. Since $\pi^{\mathcal{A}}_*(Y)$ is injective, the second row is also exact. □

Next we observe that any $\mathbb{T}$-spectrum Y admits a convergent Adams resolution for $\pi^{\mathcal{A}}_*$ by imported injectives. More precisely, we choose a standard resolution

$$0 \longrightarrow \pi^{\mathcal{A}}_*(Y) \longrightarrow I \longrightarrow J \longrightarrow 0$$

of $\pi^{\mathcal{A}}_*(Y)$ as constructed in 5.5.2 above. Since both I and J are sums of injectives of the form $e(V)$ or $f(I)$, there are spectra $I(Y)$ and $J(Y)$ realizing them. From the injective case proved above for $I(Y)$ and $J(Y)$, we may lift the maps in the resolution to form a sequence

$$Y \longrightarrow I(Y) \longrightarrow J(Y)$$

which gives the resolution once $\pi^{\mathcal{A}}_*$ is applied. In fact it is a cofibre sequence. Indeed, since the composite is null from the injective case, the map $Y \longrightarrow I(Y)$ factors through the fibre F of $I(Y) \longrightarrow J(Y)$, and the map $Y \longrightarrow F$ is an isomorphism of $\pi_*^{\mathcal{A}}$; therefore Y is equivalent to the fibre by the Whitehead Theorem 5.6.4.

The theorem follows by applying $[X, \cdot]_*^{\mathbb{T}}$ to $Y \longrightarrow I(Y) \longrightarrow J(Y)$. □

It is worth recording here the compatibility of the models with the inclusions of categories of spectra.

COROLLARY 5.6.9. *The functors e and f induce full and faithful embeddings*

$$e : D\mathbb{Q} \longrightarrow D\mathcal{A} \text{ and } f : D(\textit{tors-}\mathcal{O}_{\mathcal{F}}^{f}) \longrightarrow D\mathcal{A}$$

of derived categories. Furthermore, these are compatible with the geometric inclusions in the sense that the diagram

$$\begin{array}{ccccc} \mathbb{T}\text{-}Spec/\mathbb{T} & \longrightarrow & \mathbb{T}\text{-}Spec & \longleftarrow & \mathbb{T}\text{-}Spec/\mathcal{F} \\ \simeq\downarrow & & \simeq\downarrow & & \simeq\downarrow \\ D\mathbb{Q} & \stackrel{e}{\longrightarrow} & D\mathcal{A} & \stackrel{f}{\longleftarrow} & D(\textit{tors-}\mathcal{O}_{\mathcal{F}}^{f}) \end{array}$$

commutes.

Proof: The fact that e and f induce full and faithful embeddings of derived categories follows from the Adams spectral sequences together with 5.4.3 and the fact that they preserve injectives.

The compatibility with geometry follows from 5.6.3 and the universal properties of categories of fractions. □

5.7. Maps between injective spectra.

We have shown that every $\mathbb{T}$-spectrum Y admits an Adams resolution

$$Y \longrightarrow I(Y) \stackrel{\delta(Y)}{\longrightarrow} J(Y),$$

and in fact $I(Y) = \bigvee_\alpha I_\alpha$ and $J(Y) = \bigvee_\beta J_\beta$ where the spectra I_α and J_β are suspensions of the standard spectra $E\langle H\rangle$ for various $H \in \mathcal{F}$ and $\tilde{E}\mathcal{F}$. Furthermore, no $\mathcal{F}$-contractible summands are required in $J(Y)$. Thus the map $\delta(Y)$ is described by its components $\delta_\alpha : I_\alpha \longrightarrow J(Y) = \bigvee_\beta J_\beta$. Each such component is thus a map $E\langle H\rangle \longrightarrow J$ or $\tilde{E}\mathcal{F} \longrightarrow J$ for an $\mathcal{F}$-free injective J. It is the purpose of the present section to show that such maps can be described as composites of inclusions of direct summands and a few maps of very special form: this gives us a rather effective way of describing the construction of an arbitrary $\mathbb{T}$-spectrum from the eminently approachable spectra $E\langle H\rangle$ and $\tilde{E}\mathcal{F}$. It is used to great effect in Chapter 24

EXAMPLE 5.7.1. *(Maps $\Sigma^i E\langle H\rangle \longrightarrow E\langle H\rangle$).* Any graded self-map of $E\langle H\rangle$ is a rational multiple of $c^n : E\langle H\rangle \longrightarrow \Sigma^{2n} E\langle H\rangle$ for some $n \geq 0$. This is immediate from 2.4.1: $[E\langle H\rangle, E\langle H\rangle]_*^{\mathbb{T}} = \mathbb{Q}[c_H]$. □

EXAMPLE 5.7.2. *(Maps $\Sigma^i \tilde{E}\langle H\rangle \longrightarrow E\langle H\rangle$).* Combining 5.3.2 with Tate duality 5.2.2 we obtain the calculation

$$[\tilde{E}\mathcal{F}, \Sigma E\mathcal{F}_+]^{\mathbb{T}}_* = t^{\mathcal{F}}_*.$$

(This can be seen more formally from the special case $F(E\mathcal{F}_+, S^0) \wedge \tilde{E}\mathcal{F} \simeq F(\tilde{E}\mathcal{F}, \Sigma E\mathcal{F}_+)$ of Warwick duality [**9**]).

In degree 0 and below, the maps factor through the quotient map $\tilde{E}\mathcal{F} \longrightarrow \Sigma E\mathcal{F}_+$, and are thus described by a map $\Sigma E\mathcal{F}_+ \longrightarrow \Sigma^{2n+1} E\mathcal{F}_+$ for some $n \geq 0$.

In positive degrees each element has a finite support. We thus choose a subgroup H containing all the supporting subgroups; the element then factors through the quotient $\tilde{E}\mathcal{F} \simeq \tilde{E}\mathcal{F} \wedge S^{-nV(H)} \longrightarrow \Sigma E\mathcal{F}_+ \wedge S^{-nV(H)}$. Thom isomorphisms 2.3.7 show that the codomain $\Sigma E\mathcal{F}_+ \wedge S^{-nV(H)}$ is a wedge of suspensions $\Sigma^{-2v(K)} E\langle K\rangle$, where $v(K) = dim(V(H)^K)$ (ie n if $K \subseteq H$ and $v(K) = 0$ otherwise). □

LEMMA 5.7.3. *If J is a wedge of suspensions of spectra $E\langle H\rangle$ for various H then*
(i) Any essential map $E\langle H\rangle \longrightarrow J$ factors as

$$E\langle H\rangle \longrightarrow \Sigma^{2n} E\langle H\rangle \longrightarrow J$$

for some n, where the first map is a scalar multiple of c^n and the second is the inclusion of a direct summand.
(ii) Any map $\tilde{E}\mathcal{F} \longrightarrow J$ factors as

$$\tilde{E}\mathcal{F} \longrightarrow \Sigma E\mathcal{F}_+ \wedge S^{-nV(H)} \simeq \bigvee_{K\in\mathcal{F}} \Sigma^{-2v(K)+1} E\langle K\rangle \longrightarrow$$
$$\bigvee_{K\in\mathcal{G}} \Sigma^{-2v(K)+1} E\langle K\rangle \longrightarrow \bigvee_{K\in\mathcal{G}} \Sigma^{-2v'(K)+1} E\langle K\rangle \longrightarrow J$$

where the first map is of the form described in 5.7.2, the second is projection onto a subwedged indexed on $\mathcal{G} \subseteq \mathcal{F}$, the third has components $c^{v(K)-v'(K)} : E\langle K\rangle \longrightarrow \Sigma^{2v(K)-2v'(K)} E\langle K\rangle$ and the last is the inclusion of a direct summand.

Proof: (i) First note that the map is described by its effect in homotopy. Next note that the image in homotopy is a quotient of $\mathbb{I}(H) = \pi^{\mathbb{T}}_*(E\langle H\rangle)$, and hence is of the form $\Sigma^{2n}\mathbb{I}(H)$. This is injective, and so the inclusion of the image is split. Since all objects concerned are injective we may realize the splitting geometrically.

(ii) The map is described by a homomorphism $\theta : t^{\mathcal{F}}_* \longrightarrow \pi^{\mathbb{T}}_*(J)$. The image of 1 is torsion, and hence $c^n\theta(1) = 0$ for some n. Hence θ factors through one of the maps $\tilde{E}\mathcal{F} \longrightarrow \Sigma^{-2n+1} E\mathcal{F}_+$ described in 5.7.2; this is clear algebraically, and thanks to injectivity may be realized geometrically. The resulting map $\Sigma^{-2n+1} E\mathcal{F}_+ \longrightarrow J$ is described by its components, each of which can be described as in Part (i). □

5.8. Algebraic cells and spheres.

In this section we make explicit the algebraic counterparts of some very important topological objects.

EXAMPLE 5.8.1. *(Algebraic basic cells.)* Let us write $L_H := \pi_*^{\mathcal{A}}(\sigma_H^0)$, and similarly for the other associated invariants.
(i) For the group $\mathbb{T}$ itself we have

$$L_{\mathbb{T}} = \left(\mathcal{O}_{\mathcal{F}} \longrightarrow t_*^{\mathcal{F}}\right);$$

this has nub $N_{\mathbb{T}} = \mathcal{O}_{\mathcal{F}}$, vertex $V_{\mathbb{T}} = \mathbb{Q}$ and torsion part $T_{\mathbb{T}} = \Sigma\mathbb{I}$.
(ii) If H is a finite subgroup, and we let $\mathbb{Q}(H)$ denote the $\mathbb{Q}[c_H]$-module $\mathbb{Q}$, regarded as a $\mathcal{O}_{\mathcal{F}}$-module in the natural way, then

$$L_H = (\Sigma\mathbb{Q}(H) \longrightarrow 0);$$

this has nub $N_H = \Sigma\mathbb{Q}(H)$, vertex $V_H = 0$ and torsion part $T_H = \Sigma\mathbb{Q}(H)$. □

In Section 12.1 we use this to calculate the homotopy Mackey functor of an arbitrary object of $\mathcal{A}$.

EXAMPLE 5.8.2. *(Algebraic spheres.)* Let $v : \mathcal{F} \longrightarrow \mathbb{Z}$ be a function with finite support, and consider the submodule

$$\mathcal{O}_{\mathcal{F}}(v) = \{x \in t_*^{\mathcal{F}} \mid c^v x \in \mathcal{O}_{\mathcal{F}}\}$$

of $t_*^{\mathcal{F}}$ consisting in degree $2n$ of functions on $\mathcal{F}$ which are only non-zero at H if $n \leq v(H)$. Define the algebraic v-sphere by

$$S^v = (\mathcal{O}_{\mathcal{F}}(v) \longrightarrow t_*^{\mathcal{F}}),$$

where the map is inclusion. Note that, since v has finite support, the map becomes an isomorphism when $\mathcal{E}$ is inverted, and hence S^v is an object of $\mathcal{A}$. We extend the notation to include suspensions in the obvious way: $S^{n+v} = \Sigma^n S^v$. □

The point of the notation is that it gives a natural extension of that familiar from topology. Recall that for any representation V with $V^{\mathbb{T}} = 0$, the Euler class is given by $\chi(V) = c^v$ where $v(H) = dim_{\mathbb{C}}(V^H)$ (see 4.6.1).

LEMMA 5.8.3. *If v is the dimension function associated to the representation V of $\mathbb{T}$ with $V^{\mathbb{T}} = 0$, then the algebraic sphere S^v models the topological sphere S^V. More generally, if in addition w is the dimension function associated to the representation W, the algebraic sphere S^{v-w} models the topological sphere S^{V-W}.*

Proof: It suffices to deal with the case $V^{\mathbb{T}} = W^{\mathbb{T}} = 0$, since integer suspensions in the algebraic and topological contexts certainly correspond.
Indeed, we may compare with the model of $\tilde{E}\mathcal{F}$. Since $\Phi^{\mathbb{T}} S^{V-W} = S^0$ there is a map $S^{V-W} \longrightarrow \tilde{E}\mathcal{F}$, unique up to multiplication by a non-zero scalar, which is an isomorphism on vertices. This map is injective on nubs: to see this, note that the mapping cone is

$S^{V-W} \wedge \Sigma E\mathcal{F}_+$, which is a $\mathcal{F}$-spectrum with all homotopy in even degrees. The nub is thus the kernel of the epimorphism

$$t_*^{\mathcal{F}} \longrightarrow \pi_*^{\mathbb{T}}(S^{V-W} \wedge \Sigma E\mathcal{F}_+).$$

Bearing in mind that there are only finitely many subgroups K so that $V^K \neq 0$, or $W^K \neq 0$, Thom isomorphisms and the identification of Euler classes identify the kernel as claimed. □

Finally, we should relate spheres to Euler classes and suspensions, building on what was done in Section 4.6. We shall need two further properties of the suspension of 4.6.7.

LEMMA 5.8.4. *(i) Multiplication by c^v gives a degree 0 isomorphism*

$$c^v : t_*^{\mathcal{F}} \xrightarrow{\cong} \Sigma^v t_*^{\mathcal{F}}.$$

(ii) We have the isomorphism

$$\Sigma^v \mathcal{O}_{\mathcal{F}} \cong \mathcal{O}_{\mathcal{F}}(v)$$

compatible with the isomorphism in (i).

Proof: Part (i) follows from the fact that $t_*^{\mathcal{F}} = \mathcal{E}^{-1}\mathcal{O}_{\mathcal{F}}$. Part (ii) is immediate from the definitions. □

This allows us to extend the definition to the standard model.

DEFINITION 5.8.5. We then define

$$\Sigma^v : \mathcal{A} \longrightarrow \mathcal{A}$$

by

$$\Sigma^v(N \longrightarrow t_*^{\mathcal{F}} \otimes V) = (\Sigma^v N \longrightarrow \Sigma^v t_*^{\mathcal{F}} \otimes V \cong t_*^{\mathcal{F}} \otimes V). \quad \square$$

This has the familiar properties of suspension.

LEMMA 5.8.6. *The generalized suspension on $\mathcal{A}$ has the properties*
(i) Σ^0 is the identity functor.
(ii) $\Sigma^v \Sigma^w = \Sigma^{v+w}$.
(iii) Σ^v is invertible.
(iv) $\Sigma^v S^0 = S^v$.
(v) $\Sigma^v e(W) \cong e(W)$.
(vi) $\Sigma^v f(M) \cong f(\Sigma^v M)$. □

It may be useful to make explicit the map $c^w : S^v \longrightarrow S^{v+w}$: it is given by

$$\begin{array}{ccc} \mathcal{O}_{\mathcal{F}}(v) & \xrightarrow{c^w} & \mathcal{O}_{\mathcal{F}}(v+w) \\ \downarrow & & \downarrow \\ t_*^{\mathcal{F}} & \longrightarrow & t_*^{\mathcal{F}}. \end{array}$$

5.9. Explicit models.

In this section we provide notation to cover our needs, and make a convenient fibrant approximation explicit: this will be useful in consideration of Mackey functor valued homotopy groups in Section 12.1.

On the one hand we want to refer to an object of $\mathcal{A}$ by a single letter like L. We shall write

$$L = \left(N_L \xrightarrow{\beta_L} t_*^{\mathcal{F}} \otimes V_L\right).$$

We call N_L the *nub* of L and V_L the *vertex* of L. The map β_L is called the *basing map*. By definition $K_L = ker(\beta_L)$ and $C_L = cok(\beta_L)$ are $\mathcal{F}$-finite torsion modules, and C_L is also divisible and hence injective. We also use the notation $k_L : K_L \longrightarrow N_L$ and $c_L : t_*^{\mathcal{F}} \otimes V_L \longrightarrow C_L$ for the inclusion and projection. Since all short exact sequences $0 \longrightarrow \Sigma^{-1}C_L \longrightarrow T \longrightarrow K_L \longrightarrow 0$ therefore split, it is reasonable to define

$$T_L := \Sigma^{-1}C_L \oplus K_L,$$

and call it the *torsion part* of L. With this notation, corresponding to the cofibre sequence

$$X \longrightarrow X \wedge \tilde{E}\mathcal{F} \xrightarrow{q_X} X \wedge \Sigma E\mathcal{F}_+$$

we have a cofibre sequence in the derived category

$$L \longrightarrow e(V_L) \xrightarrow{q_L} f(\Sigma T_L).$$

To see this, note that there is a map $L \longrightarrow e(V_L)$ which is an isomorphism on vertices, since e is right adjoint to the passage-to-vertex functor. It is clear that the kernel and cokernel are $f(K_L)$ and $f(C_L)$ respectively, and hence, since C_L is injective, the cofibre is $f(\Sigma T_L)$. Explicitly

$$(q_L : e(V_L) \longrightarrow f(\Sigma T_L)) = \begin{pmatrix} t_*^{\mathcal{F}} \otimes V_L & \xrightarrow{q'_L} & \Sigma T_L \\ \downarrow & & \downarrow \\ t_*^{\mathcal{F}} \otimes V_L & \longrightarrow & 0 \end{pmatrix}$$

If T_L is injective then q'_L is simply a homomorphism, but in general it is a morphism in a derived category of $\mathcal{O}_{\mathcal{F}}$-modules that we have not formally introduced. We shall be completely explicit about q'_L in 5.9.2 below.

Now, given a $\mathbb{T}$-spectrum X, we obtain the object $L_X = \pi_*^{\mathcal{A}}(X)$, but we would like abbreviations for its terms, so we take

$$N_X = N_{\pi_*^{\mathcal{A}}(X)} = \pi_*^{\mathbb{T}}(X \wedge DE\mathcal{F}_+)$$

and

$$V_X = V_{\pi_*^{\mathcal{A}}(X)} = \pi_*(\Phi^{\mathbb{T}} X).$$

Similarly, $\beta_X : N_X \longrightarrow t_*^{\mathcal{F}} \otimes V_X$ is the map $\pi_*^{\mathbb{T}}(X \wedge DE\mathcal{F}_+ \longrightarrow X \wedge DE\mathcal{F}_+ \wedge \tilde{E}\mathcal{F})$. Finally we let

$$T_X = T_{\pi_*^{\mathcal{A}}(X)};$$

this has the geometric significance one expects.

LEMMA 5.9.1.

$$T_X = \pi_*^{\mathbb{T}}(X \wedge E\mathcal{F}_+).$$

Proof: Since $E\mathcal{F}_+ \wedge DE\mathcal{F}_+ \simeq E\mathcal{F}_+$ we have a cofibre sequence

$$X \wedge E\mathcal{F}_+ \longrightarrow X \wedge DE\mathcal{F}_+ \longrightarrow X \wedge DE\mathcal{F}_+ \wedge \tilde{E}\mathcal{F}. \quad \square$$

It is also worth being completely explicit about some of the above constructions. We must begin by talking about injective representations of objects. We use the convention that if M is an arbitrary object then $\hat{M}$ denotes an injective object with a homology isomorphism $M \longrightarrow \hat{M}$. Of course there is usually no canonical choice of $\hat{M}$. Thus, if T is any $\mathcal{F}$-finite torsion $\mathcal{O}_{\mathcal{F}}$-module with injective resolution $0 \longrightarrow T \longrightarrow I \longrightarrow J \longrightarrow 0$, then we may take $\hat{T} = fibre(I \longrightarrow J)$. The object $e(V)$ is already injective, and $\widehat{f(T)}$ may be taken to be $f(\hat{T})$ by 5.5.1. Thus for an arbitrary object $L = (N_L \longrightarrow t_*^{\mathcal{F}} \otimes V_L)$ with zero differential we may take $\hat{L}$ to be the fibre of $q_L : e(V_L) \longrightarrow f(\Sigma \hat{T}_L)$. In what follows we shall use the notation $A \ltimes B$ to denote a dg module which is additively a direct sum of A and B, and which has B as a submodule and A as a quotient. However the sum is twisted by giving the differential a component $A \longrightarrow \Sigma B$. There is thus a short exact sequence

$$0 \longrightarrow B \longrightarrow A \ltimes B \longrightarrow A \longrightarrow 0$$

of dg modules.

LEMMA 5.9.2. *For an object L of $\mathcal{A}$ (with zero differential) an injective equivalent to L in $D\mathcal{A}$ is given by*

$$\hat{L} = \left(\begin{array}{ccc} t_*^{\mathcal{F}} \otimes V_L & \ltimes & \Sigma^{-1} C_L \oplus \hat{K}_L \\ \downarrow & & \downarrow \\ t_*^{\mathcal{F}} \otimes V_L & \oplus & 0 \end{array} \right)$$

where the differential in $(t_*^{\mathcal{F}} \otimes V_L) \ltimes (\Sigma^{-1} C_L \oplus \hat{K}_L)$ is $-q'_L : t_*^{\mathcal{F}} \otimes V_L \longrightarrow C_L \oplus \Sigma \hat{K}_L$ given explicitly by $q'_L = \{q'_C, q'_K\}$ where $q'_C = c_L$ and $q'_K \in \mathrm{Ext}(t_*^{\mathcal{F}} \otimes V_L, K_L)$ lifts the class of the extension

$$[N_L] = (0 \longrightarrow K_L \longrightarrow N_L \longrightarrow im(\beta_L) \longrightarrow 0).$$

Furthermore there is a fibre sequence

$$\hat{L} \longrightarrow e(V_L) \xrightarrow{q_L} f(\Sigma T_L)$$

where

$$(q_L : e(V_L) \longrightarrow f(\Sigma T_L)) = \left(\begin{array}{ccc} t_*^{\mathcal{F}} \otimes V_L & \xrightarrow{q'_L} & C_L \oplus \Sigma \hat{K}_L \\ \downarrow & & \downarrow \\ t_*^{\mathcal{F}} \otimes V_L & \longrightarrow & 0 \end{array} \right)$$

Proof: It is enough to construct a homology isomorphism $L \longrightarrow \hat{L}$ since the displayed $\hat{L}$ is visibly injective, and is the fibre of q_L. To simplify the picture in the reader's mind we now suppose L is even.

We shall construct a diagram

$$\begin{array}{ccc} N_L & \xrightarrow{i_L} & (t_*^{\mathcal{F}} \otimes V_L) \ltimes (\Sigma^{-1} C_L \oplus \hat{K}_L) \\ \downarrow & & \downarrow \\ t_*^{\mathcal{F}} \otimes V_L & \xrightarrow{1} & (t_*^{\mathcal{F}} \otimes V_L) \oplus 0 \end{array}$$

where $i_L = \{i'_L, 0, i'''_L\}$ with $i'_L = \beta_L$; the choice of i'''_L needs more careful description. Since the diagram commutes for any choice of i'''_L and the lower arrow is a homology isomorphism, it suffices to choose i'''_L so that i_L is a map of dg $\mathcal{O}_{\mathcal{F}}$-modules and a homology isomorphism.

For this we display the map i_L vertically and find a diagram

$$\begin{array}{ccccccccc}
0 & \longrightarrow & K_L & \longrightarrow & N_L & \longrightarrow & im(\beta_L) & \longrightarrow & 0 \\
 & & h_L \downarrow & & i_L \downarrow & & j_L \downarrow & & \\
0 & \longrightarrow & \hat{K}_L & \longrightarrow & (t_*^{\mathcal{F}} \otimes V_L) \ltimes (\Sigma^{-1} C_L \oplus \hat{K}_L) & \longrightarrow & (t_*^{\mathcal{F}} \otimes V_L) \ltimes \Sigma^{-1} C_L & \longrightarrow & 0.
\end{array}$$

We use the natural inclusion $K_L \longrightarrow \hat{K}_L$ for h_L, and j_L is the composite $im(\beta_L) \longrightarrow t_*^{\mathcal{F}} \otimes V_L \longrightarrow (t_*^{\mathcal{F}} \otimes V_L) \ltimes \Sigma^{-1} C_L$; both of these are obviously dg maps and homology isomorphisms, and the right hand square commutes for any choice of i'''_L. By the Five Lemma, it is enough to choose a map i'''_L so that i_L is a chain map and the left hand square commutes.

For this we refer to the diagram

$$\begin{array}{ccccccc}
K_L & \longrightarrow & N_L & \longrightarrow & im(i_L) & \longrightarrow & t_*^{\mathcal{F}} \otimes V_L \\
 & \searrow^{h_L} & \downarrow^{i'''_L} & & \downarrow^{\widetilde{[N_L]}} & \swarrow_{q'_K} & \\
 & & I(K_L) & \longrightarrow & J(K_L) & &
\end{array}$$

Here $\widetilde{[N_L]}$ represents the extension class $[N_L]$, and extends to q'_K by injectivity of $J(K_L)$. Now $[N_L]$ is the image of the identity map of K_L under the boundary map for the short exact sequence $0 \longrightarrow K_L \longrightarrow N_L \longrightarrow im(\beta_L) \longrightarrow 0$, and hence there is a map i'''_L in the diagram to make the square commute. Finally, from the Snake Lemma construction of the boundary map, it follows that the left hand composite is the standard inclusion $h_L : K_L \longrightarrow \hat{K}_L$ as claimed.

Since the square commutes, the map $i_L = \{i'_L, 0, i'''_L\}$ commutes with differentials. Since i'''_L extends h_L the left hand square in the map of short exact sequences commutes. □

5.10. Hausdorff modules.

It is often useful to deal with an $\mathcal{O}_{\mathcal{F}}$-module M by dealing with its summands $e_H M$. However this trick does not work for all $\mathcal{O}_{\mathcal{F}}$-modules, so we spend this section showing it is effective for all the modules that arise in the standard model.

For any $\mathcal{O}_{\mathcal{F}}$-module M, we may form the module $M^{\wedge}_{\mathcal{F}} = \prod_H e_H M$. We have chosen the notation since this can be viewed as a completion, and it is therefore natural to call a module M for which the natural map

$$M \longrightarrow M^{\wedge}_{\mathcal{F}}$$

is injective, a *Hausdorff* module. The convenient property of a Hausdorff module M is that maps into M are determined by their component maps into $e_H M$. An example of a non-Hausdorff module is the quotient $\prod_H \mathbb{Q}_H / \bigoplus_H \mathbb{Q}_H$, where $\mathcal{O}_{\mathcal{F}}$ acts on $\mathbb{Q}_H$ via the idempotent e_H. Note that, by construction, the completion functor $M \longmapsto M^{\wedge}_{\mathcal{F}}$ is exact.

LEMMA 5.10.1. *(i) All $\mathcal{F}$-finite $\mathcal{O}_{\mathcal{F}}$-modules are Hausdorff.*
(ii) The ring $\mathcal{O}_{\mathcal{F}}$ is Hausdorff as a module.
(iii) Any submodule of a Hausdorff module is Hausdorff.
(iv) The class of Hausdorff modules is closed under extensions, in the sense that if $0 \longrightarrow M' \longrightarrow M \longrightarrow M'' \longrightarrow 0$ is an exact sequence of $\mathcal{O}_{\mathcal{F}}$-modules with M' and M'' Hausdorff, then M is also Hausdorff.
(v) The class of Hausdorff modules is closed under arbitrary direct sums or direct products.

Proof: Parts (i) and (ii) are clear. Parts (iii) and (iv) follow from the Snake Lemma by exactness of completion. In Part (v), closure under direct products is obvious. For direct sums the inclusion

$$\lambda : \bigoplus_i M_i \longrightarrow \prod_i \prod_H e_H M_i$$

is injective since it is the composite of the map

$$\bigoplus_i M_i \longrightarrow \bigoplus_i \prod_H e_H M_i,$$

injective by hypothesis, and the inclusion of the sum in the product. However, λ also factors through completion, which is therefore also injective. □

COROLLARY 5.10.2. *The modules N and $t_*^{\mathcal{F}} \otimes V$ occurring in any object $N \longrightarrow t_*^{\mathcal{F}} \otimes V$ of the standard abelian category $\mathcal{A}$ are both Hausdorff.*

Proof: Since N is an extension of an $\mathcal{F}$-finite module by a submodule of $t_*^{\mathcal{F}} \otimes V$ it is sufficient, by Parts (iii) and (iv) of the lemma, to observe that $t_*^{\mathcal{F}} \otimes V$ is Hausdorff. The fact that $t_*^{\mathcal{F}}$ is Hausdorff follows from Part (iv), since $t_*^{\mathcal{F}}$ is the extension of $\mathcal{O}_{\mathcal{F}}$ over the $\mathcal{F}$-finite module $\Sigma^2\mathbb{I}$, and the fact that $t_*^{\mathcal{F}} \otimes V$ is Hausdorff now follows from Part (v). □

CHAPTER 6

The torsion model.

We begin with Section 6.1 explaining the methods we have actually used for identifying the place of various well known spectra in the standard model: the rest of the chapter places the method in a proper framework. The methods are closely tied to standard methods of equivariant topology, as described in the introduction. Whilst their success may be sufficient justification, it is desirable to explain what can and cannot be seen by these methods. We therefore introduce the geometrically natural *torsion model* in Section 6.2, investigate its homological algebra in Section 6.3 and discuss its derived category in Section 6.4. We are then ready to show algebraically in Section 6.5 that it is equivalent to the standard model. The author does not know a direct method for showing the derived category of the torsion model category is equivalent to the category of rational $\mathbb{T}$-spectra, so Section 6.6 summarizes what we do know about the relationship between the two categories.

The algebraic drawback of the torsion model is that it is of injective dimension 2, so that the natural Adams spectral sequence associated to this construction does not collapse until E_3. This means that not every homomorphism of objects in the torsion model lifts to a geometric map, and hence an object may not be be formal (ie determined by its homology). It is fortunate for the usefulness of the easy method presented here, that so many spectra of interest are formal.

6.1. Practical calculations.

Given a $\mathbb{T}$-spectrum, X we may want to identify its place in our classification. In principle this means that we must calculate $\pi_*^{\mathcal{A}}(X)$, but it is often more practical to proceed in several steps. The method outlined in this section and formalized in the rest of the chapter is used for example in Sections 12.2, 12.3, 12.4, 13.3, 13.4, 13.5 and 15.2; it is used in a more disguised form in Chapter 10 and elsewhere.

We know that X is the fibre of

$$q_X : X \wedge \tilde{E}\mathcal{F} \longrightarrow X \wedge \Sigma E\mathcal{F}_+,$$

and hence that the algebraic model $\pi_*^{\mathcal{A}}(X)$ is equivalent to the fibre of the algebraic map

$$\hat{q}_X : \pi_*^{\mathcal{A}}(X \wedge \tilde{E}\mathcal{F}) \longrightarrow \pi_*^{\mathcal{A}}(X \wedge \Sigma E\mathcal{F}_+),$$

corresponding to q_X in the derived category $D\mathcal{A}$. Furthermore we have seen in 5.6.3 that $\pi_*^{\mathcal{A}}(X \wedge \tilde{E}\mathcal{F}) = e(\pi_*(\Phi^{\mathbb{T}} X))$ and $\pi_*^{\mathcal{A}}(X \wedge \Sigma E\mathcal{F}_+) = f(\pi_*^{\mathbb{T}}(X \wedge \Sigma E\mathcal{F}_+))$.

SUMMARY 6.1.1. For any $\mathbb{T}$-spectrum X the algebraic model $\pi_*^{\mathcal{A}}(X)$ is the fibre of the map

$$\hat{q}_X : e(\pi_*(\Phi^{\mathbb{T}} X)) \longrightarrow f(\pi_*^{\mathbb{T}}(X \wedge \Sigma E\mathcal{F}_+))$$

in the derived category corresponding to $q_X : X \wedge \tilde{E}\mathcal{F} \longrightarrow X \wedge \Sigma E\mathcal{F}_+$. □

To use this method we need to identify $\Phi^{\mathbb{T}}X$, calculate $\pi_*^{\mathbb{T}}(X \wedge \Sigma E\mathcal{F}_+)$ and identify the map $q_X : X \wedge \tilde{E}\mathcal{F} \longrightarrow X \wedge \Sigma E\mathcal{F}_+$ between them, in algebraic terms. Using 5.3.2, this may be done by identifying the d- and e-invariants of $\hat{q}_X : t_*^{\mathcal{F}} \otimes \pi_*(\Phi^{\mathbb{T}}X) \longrightarrow \pi_*^{\mathbb{T}}(X \wedge \Sigma E\mathcal{F}_+)$. When $\pi_*^{\mathbb{T}}(X \wedge \Sigma E\mathcal{F}_+)$ is injective, this simply involves finding a homomorphism of $\mathcal{O}_{\mathcal{F}}$-modules.

In practice we begin by calculating $\pi_*^{\mathbb{T}}(X \wedge \Sigma E\mathcal{F}_+)$ using the spectral sequence of the skeletal filtration of $E\mathcal{F}_+$, and some ingenuity to obtain the $\mathcal{O}_{\mathcal{F}}$-module structure.

Next, we note that $\Phi^{\mathbb{T}}X$ is determined by its homotopy groups $\pi_*(\Phi^{\mathbb{T}}X) = \pi_*^{\mathbb{T}}(X \wedge \tilde{E}\mathcal{F})$. Often the most effective way of calculating this is to use the first step and the long exact homotopy sequence of $X \wedge E\mathcal{F}_+ \longrightarrow X \longrightarrow X \wedge \tilde{E}\mathcal{F}$. However the best method for calculating $\Phi^{\mathbb{T}}X$ depends on how X is given. If X is the suspension spectrum of a based space then of course $\Phi^{\mathbb{T}}X$ is the suspension spectrum of the fixed point space. In the event that X is equipped with Thom isomorphisms $X \wedge S^{V(H)} \simeq X \wedge S^2$ (for example if X is an $\mathcal{F}$-spectrum or the equivariant K-theory spectrum), then the model $\tilde{E}\mathcal{F} = \lim_{\to V^{\mathbb{T}}=0} S^V$ shows $\pi_*(\Phi^{\mathbb{T}}X) = \mathcal{E}^{-1}\pi_*^{\mathbb{T}}(X)$ where $\mathcal{E}$ is the multiplicative set of Euler classes, generated by the Euler classes of $V(H)$ for finite subgroups H. A variation of this is to use the spectral sequence of the filtration

$$S^0 \subseteq S^{V(\mathcal{F})} \subseteq S^{2V(\mathcal{F})} \subseteq \cdots \subseteq \bigcup_k S^{kV(\mathcal{F})} = S^{\infty V(\mathcal{F})} = \tilde{E}\mathcal{F}.$$

In calculating $\hat{q}_X$ we assume again that it is routine to calculate integer graded $\pi_*^{\mathbb{T}}$. In any case we will have calculated $(q_X)_* : \pi_*(\Phi^{\mathbb{T}}X) \longrightarrow \pi_*^{\mathbb{T}}(X \wedge \Sigma E\mathcal{F}_+)$ during the previous steps; since $\hat{q}_X(1 \otimes x) = (q_X)_*(x)$ this determines $\hat{q}_X$ on all elements $\tau \otimes x$ when $\tau \in t_*^{\mathcal{F}}$ is of negative degree. Many elements x of $\pi_*^{\mathbb{T}}(X \wedge \Sigma E\langle H\rangle)$ are uniquely divisible by c_H, which allows one to determine the images of elements like $c_H^{-1} \otimes x$. However this only accounts for all elements in trivial cases, and it even happens (for example with $X = S^0$) that $(q_X)_* = 0$ whilst the map q_X (and hence also $\hat{q}_X$) are highly nontrivial. In this case we need to do more work to identify the images of elements $c_H^{-k} \otimes x$. In fact the definition of c_H allows us to see that this is in effect the part of $\pi_*^{\mathbb{T}}$ twisted by a representation.

LEMMA 6.1.2. *For any integer k, the restriction*

$$\mathbb{Q}\{c_H^{-k}\} \otimes \pi_*(\Phi^{\mathbb{T}}X) \longrightarrow \Sigma^{-2k}\pi_*^{\mathbb{T}}(X \wedge \Sigma E\mathcal{F}_+)$$

of the map $\hat{q}_X$ maps into $\pi_*^{\mathbb{T}}(X \wedge \Sigma E\langle H\rangle)$ and its d-invariant is induced by

$$X \wedge \tilde{E}\mathcal{F} \simeq X \wedge \tilde{E}\mathcal{F} \wedge \sigma^{-kV(H)} \xrightarrow{q_X \wedge 1} X \wedge \Sigma E\langle H\rangle \wedge \sigma^{-kV(H)} \simeq X \wedge \Sigma^{1-2k}E\langle H\rangle.$$

Proof: This is simply naturality, applied to the square

$$\begin{array}{ccc} X \wedge \tilde{E}\mathcal{F} \wedge S^0 & \stackrel{q_X \wedge 1}{\longrightarrow} & X \wedge \Sigma E\langle H\rangle \wedge S^0 \\ \simeq\uparrow & & \uparrow c_H^k \\ X \wedge \tilde{E}\mathcal{F} \wedge \sigma^{-kV(H)} & \stackrel{q_X \wedge 1}{\longrightarrow} & X \wedge \Sigma E\langle H\rangle \wedge \sigma^{-kV(H)} \end{array}$$ □

6.2. The torsion model.

The reader will have noticed how much of our motivation and our investigation has proceeded by the following process. First we understand the $\mathcal{F}$-contractible part (i.e. the vertex), next the $\mathcal{F}$-free part (i.e. the torsion) and finally we identify the comparison map between them. It is time to formalize the process.

DEFINITION 6.2.1. The *torsion model category* is the abelian category $\mathcal{A}_t$ with objects $\mathcal{O}_\mathcal{F}$-module maps $(t^\mathcal{F}_* \otimes V \longrightarrow T)$, with T an $\mathcal{F}$-finite torsion module, and with morphisms

$$\begin{array}{ccc} t^\mathcal{F}_* \otimes U & \stackrel{1\otimes\phi}{\longrightarrow} & t^\mathcal{F}_* \otimes V \\ \downarrow & & \downarrow \\ S & \stackrel{\psi}{\longrightarrow} & T \end{array}$$

REMARK 6.2.2. The torsion model category $\mathcal{A}_{t,\mathcal{H}}$ for a set $\mathcal{H}$ of finite subgroups is defined similarly, by replacing $\mathcal{O}_\mathcal{F}$ with $\mathcal{O}_\mathcal{H} := e_\mathcal{H}\mathcal{O}_\mathcal{F}$. The case in which $\mathcal{H}$ is finite is significantly simpler, but we shall just make explicit the case $\mathcal{H} = \{1\}$. Here the objects are the $\mathbb{Q}[c_1]$-module maps $\mathbb{Q}[c_1, c_1^{-1}] \otimes V \longrightarrow T$ with T a c_1-torsion module.

Why then have we not used the torsion model category $\mathcal{A}_t$, and formed a derived category from this? The reason is that $\mathcal{A}_t$ has injective dimension 2, and this makes the model much less useful. On the one hand, it makes it hard to get a precise hold on morphisms, and on the other the objects of the abelian category (as opposed to the derived category) are not adequate to model all $\mathbb{T}$-spectra.

The purpose of this chapter is to show that $D\mathcal{A}_t$ is another model for the category of $\mathbb{T}$-spectra. This makes precise the assertion that the extra information that must be added to an object of $\mathcal{A}_t$ to specify a $\mathbb{T}$-spectrum is an extension of the cokernel of the structure map by $t^\mathcal{F}_* \otimes V$; this is exactly what is encoded by the nub in the standard algebraic model. This places our intuition about how to specify a $\mathbb{T}$-spectrum on a firm and practical basis. Furthermore, it will transpire in Part II that certain functors (the Lewis-May fixed point functor and the quotient functor) are more approachable through the torsion model. Thus the torsion model is also of theoretical value.

Before we study the category $\mathcal{A}_t$ and its derived category, it will be useful to import information into $\mathcal{A}_t$ from categories we understand. For comparison with the model of $\mathcal{F}$-contractible spectra given by the derived category $D\mathbb{Q}$ of rational vector spaces, we let $\mathbb{Q}_*$-mod denote the category of graded rational vector spaces and define the functor

$$e_t : \mathbb{Q}_*\text{-mod} \longrightarrow \mathcal{A}$$

by

$$e_t(V) = \left(t^\mathcal{F}_* \otimes V \longrightarrow 0\right).$$

Objects of $\mathcal{A}_t$ isomorphic to ones in the image of e_t will be called *torsion free* objects. We also define the functor

$$f_t : \text{tors-}\mathcal{O}^f_\mathcal{F}\text{-mod} \longrightarrow \mathcal{A}$$

by

$$f_t(N) = \left(t^\mathcal{F}_* \otimes \text{Hom}(t^\mathcal{F}_*, N) \longrightarrow N\right),$$

where the structure map is evaluation. However, unlike the case of the standard model, this is not of direct use for comparison with the model of $\mathcal{F}$-spectra, since the homotopy of $\mathcal{F}$-spectra gives objects of the quite different form $(0 \longrightarrow T)$. The justification for consideration of e_t and f_t is that they are again right adjoints.

LEMMA 6.2.3. *For any object* $t_*^{\mathcal{F}} \otimes U \longrightarrow S$ *of* $\mathcal{A}_t$*, any graded vector space* V *and any* $\mathcal{F}$*-finite torsion* $\mathcal{O}_{\mathcal{F}}$*-module* N *we have natural isomorphisms*
(i)

$$\mathrm{Hom}_{\mathcal{A}_t}((t_*^{\mathcal{F}} \otimes U \longrightarrow S), e_t(V)) = \mathrm{Hom}_{\mathbb{Q}}(U, V).$$

(ii)

$$\mathrm{Hom}_{\mathcal{A}_t}((t_*^{\mathcal{F}} \otimes U \longrightarrow S), f_t(N)) = \mathrm{Hom}_{\mathcal{O}_{\mathcal{F}}}(M, N). \quad \square$$

We warn that although the functors e_t and f_t are full and faithful embeddings. There are many non-zero morphisms $f_t(N) \longrightarrow e_t(W)$. Since $f_t(N)$ does not correspond to an $\mathcal{F}$-spectrum this should cause no disquiet.

REMARK 6.2.4. Later it will be useful to note that the torsion model category is isomorphic to the category $\mathcal{A}'_t$ of objects $V \longrightarrow \mathrm{Hom}(t_*^{\mathcal{F}}, T)$ by the usual adjunction. The point to be made is that this is also an abelian category, but epimorphisms need not be surjective. In fact a map takes the form

$$\begin{array}{ccc} U & \xrightarrow{\phi} & V \\ \downarrow & & \downarrow \\ \mathrm{Hom}(t_*^{\mathcal{F}}, S) & \xrightarrow{\mathrm{Hom}(t_*^{\mathcal{F}}, \psi)} & \mathrm{Hom}(t_*^{\mathcal{F}}, T), \end{array}$$

and its kernel and cokernel are

$$\begin{array}{ccc} \ker(\phi) & & \mathrm{cok}(\phi) \\ \downarrow & \text{and} & \downarrow \\ \mathrm{Hom}(t_*^{\mathcal{F}}, \ker(\psi)) & & \mathrm{Hom}(t_*^{\mathcal{F}}, \mathrm{cok}(\psi)). \end{array}$$

Note also that in this view the imported objects are now $e'_t(V) = (V \longrightarrow 0)$ and $f'_t(T) = (\mathrm{Hom}(t_*^{\mathcal{F}}, T) \xrightarrow{1} \mathrm{Hom}(t_*^{\mathcal{F}}, T))$ for an $\mathcal{F}$-finite torsion module T. This may perhaps make them appear more natural. We have preferred $\mathcal{A}_t$ because of its closer relationship to both the standard model and the topological motivation. $\square$

6.3. Homological algebra in the torsion model.

We can also import injective objects, and they will prove to be sufficient for homological algebra.

LEMMA 6.3.1. *(i) For any graded vector space* V *the object* $e_t(V) = (t_*^{\mathcal{F}} \otimes V \longrightarrow 0)$ *is injective in* $\mathcal{A}_t$.
(ii) For any $\mathcal{F}$*-finite torsion injective* I *the object*

$$f_t(I) = (t_*^{\mathcal{F}} \otimes \mathrm{Hom}(t_*^{\mathcal{F}}, I) \longrightarrow I),$$

whose structure map is evaluation, is injective in $\mathcal{A}_t$.

Proof: This follows from 6.2.3 together with the obvious fact that the functors $(t_*^{\mathcal{F}} \otimes U \longrightarrow S) \longmapsto U$ and $(t_*^{\mathcal{F}} \otimes U \longrightarrow S) \longmapsto S$ are exact. □

PROPOSITION 6.3.2. *The category $\mathcal{A}_t$ is abelian and of injective dimension 2.*

Proof: The fact that $\mathcal{A}_t$ is abelian follows since the category of rational vector spaces and the category of $\mathcal{F}$-finite torsion modules are both abelian.

It is plain that 6.3.1 provides enough injectives. Finally, we estimate the injective dimension as follows. For an arbitrary object $(t_*^{\mathcal{F}} \otimes V \longrightarrow T)$ we have the exact sequence

$$\begin{array}{ccccccccc} 0 & \longrightarrow & 0 & \longrightarrow & t_*^{\mathcal{F}} \otimes V & \longrightarrow & t_*^{\mathcal{F}} \otimes V & \longrightarrow & 0 \\ & & \downarrow & & \downarrow & & \downarrow & & \\ 0 & \longrightarrow & T & \longrightarrow & T & \longrightarrow & 0 & \longrightarrow & 0. \end{array}$$

Since $(t_*^{\mathcal{F}} \otimes V \longrightarrow 0)$ is injective, it suffices to show $(0 \longrightarrow T)$ has dimension ≤ 2. Suppose then that $0 \longrightarrow T \longrightarrow I \longrightarrow J \longrightarrow 0$ is an injective resolution of $\mathcal{F}$-finite torsion modules; we may then form the exact sequence

$$\begin{array}{ccccccccc} 0 & \longrightarrow & 0 & \longrightarrow & t_*^{\mathcal{F}} \otimes \mathrm{Hom}(t_*^{\mathcal{F}}, I) & \longrightarrow & t_*^{\mathcal{F}} \otimes \mathrm{Hom}(t_*^{\mathcal{F}}, I) & \longrightarrow & 0 \\ & & \downarrow & & \downarrow & & \downarrow & & \\ 0 & \longrightarrow & T & \longrightarrow & I & \longrightarrow & J & \longrightarrow & 0, \end{array}$$

and so it suffices to show $(t_*^{\mathcal{F}} \otimes \mathrm{Hom}(t_*^{\mathcal{F}}, I) \longrightarrow J)$ has injective dimension ≤ 1. For this we use the map

$$\begin{array}{ccc} t_*^{\mathcal{F}} \otimes \mathrm{Hom}(t_*^{\mathcal{F}}, I) & \longrightarrow & t_*^{\mathcal{F}} \otimes \mathrm{Hom}(t_*^{\mathcal{F}}, J) \\ \downarrow & & \downarrow \\ J & \longrightarrow & J \end{array}$$

whose kernel $(t_*^{\mathcal{F}} \otimes \mathrm{Hom}(t_*^{\mathcal{F}}, T) \longrightarrow 0)$ and cokernel $(t_*^{\mathcal{F}} \otimes \mathrm{Ext}(t_*^{\mathcal{F}}, T) \longrightarrow 0)$ are both injective.

Finally we note that the bounds can be achieved. For example

$$\mathrm{Ext}^2((t_*^{\mathcal{F}} \longrightarrow 0), (0 \longrightarrow T)) \cong \mathrm{Ext}(t_*^{\mathcal{F}}, T),$$

which need not be zero. □

6.4. The derived category of the torsion model.

We have seen in Section 4.1 that we may form the derived category $D\mathcal{A}_t$ of a category of finite injective dimension by inverting homology isomorphisms. The construction of an Adams spectral sequence for $D(\mathcal{A}_t)$ is then routine. We may define H_* on dg $\mathcal{A}_t$-objects by taking homology termwise, and this gives a functor

$$H_* : D\mathcal{A}_t \longrightarrow \mathcal{A}_t.$$

It is clear by using objects with zero differential that enough $\mathcal{A}_t$-injectives are realizable and the map

$$H_* : [X, Y] \longrightarrow \mathrm{Hom}(H_*(X), H_*(Y))$$

is isomorphic when $H_*(Y)$ is injective by 4.1.9. Now, for any object Y, we construct a sequence

$$Y \longrightarrow I_0 \longrightarrow I_1 \longrightarrow I_2$$

realizing any specified injective resolution

$$0 \longrightarrow H_*Y \longrightarrow H_*I_0 \longrightarrow H_*I_1 \longrightarrow H_*I_2 \longrightarrow 0.$$

Applying $[X, \cdot]$ we obtain a spectral sequence, which is obviously convergent since it is finite.

Let us consider the spectral sequence in more detail. Because $\mathcal{A}_t$ is only a quadratic approximation to $D\mathcal{A}_t$, it has two linked drawbacks. Firstly, two non-isomorphic objects of $D\mathcal{A}_t$ may have isomorphic homology (not all objects are 'formal' in the sense of Sullivan), and secondly, maps in $D\mathcal{A}_t$ are calculated by a spectral sequence with a non-trivial d_2 differential. In fact there is a filtration

$$0 \subseteq F_2(X,Y) \subseteq F_1(X,Y) \subseteq F_0(X,Y) = [X,Y].$$

Here $(F_1/F_2)(X,Y) = \mathrm{Ext}^1(H_*(\Sigma X), H_*(Y))$ and there is an exact sequence

$$0 \longrightarrow (F_0/F_1)(X,Y) \longrightarrow \mathrm{Hom}(H_*(X), H_*(Y)) \xrightarrow{d_2} \mathrm{Ext}^2(H_*(\Sigma X), H_*(Y)) \\ \longrightarrow F_2(X, \Sigma Y) \longrightarrow 0.$$

Thus d_2 gives the obstruction to realizing an algebraic map geometrically, and even using the splitting by parity, there may be extension problems relating $(F_0/F_1)(X,Y)$ and $F_2(X,Y)$.

We recall that an object Y of $D\mathcal{A}_t$ is said to be *formal* it is determined by its homology in the sense that $H_*(Y) \cong H_*(Y')$ implies $Y \simeq Y'$.

LEMMA 6.4.1. *If $H_*(Y)$ is of injective dimension 1 then Y is formal, and hence in particular Y is formal if the torsion part of $H_*(Y)$ is an injective $\mathcal{F}$-finite torsion $\mathcal{O}_{\mathcal{F}}$-module.*

Proof: The formality of one dimensional objects is clear, since the Adams spectral sequence collapses to let us realize an isomorphism $\phi : H_*(Y') \longrightarrow H_*(Y)$ by a map $f : Y' \longrightarrow Y$. Since $f_* = \phi$ is a homology isomorphism by construction, it is a weak equivalence.

It remains to observe that any object $(t_*^{\mathcal{F}} \otimes V \longrightarrow I)$ with I injective is of injective dimension 1. This follows since there is an obvious embedding in the injective

$$(t_*^{\mathcal{F}} \otimes V \longrightarrow 0) \oplus (t_*^{\mathcal{F}} \otimes \mathrm{Hom}(t_*^{\mathcal{F}}, I) \longrightarrow I),$$

and the cokernel clearly has zero torsion part. □

6.5. Equivalence of derived categories of standard and torsion models.

In this section we establish that the derived category of the torsion model $D\mathcal{A}_t$ is equivalent to the derived category of the standard model $D\mathcal{A}$, and hence that it too provides an algebraic model for rational $\mathbb{T}$-spectra.

THEOREM 6.5.1. *The torsion model category is equivalent to the derived category $D\mathcal{A}$ of $\mathcal{A}$, and hence we have equivalences*

$$D\mathcal{A}_t \simeq D\mathcal{A} \simeq \mathbb{T}\text{-}Spec/\mathbb{T}$$

of triangulated categories.

REMARK 6.5.2. It will be clear from the proof that the equivalence is compatible with the functors from $D\mathbb{Q}$ and $D(\text{tors-}\mathcal{O}^f_{\mathcal{F}})$ in the sense that the diagram

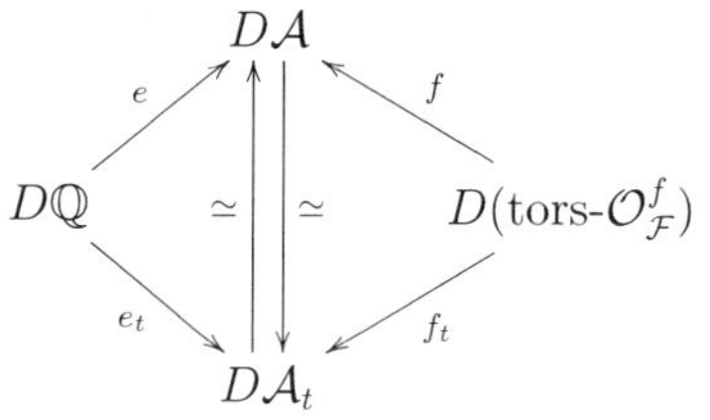

commutes up to natural isomorphism. □

Proof: We want to define functors comparing $dg\mathcal{A}_t$ and $dg\mathcal{A}$ by passage to fibre or cofibre as the case may be. Indeed we may define

$$fib : dg\mathcal{A}_t \longrightarrow dg\mathcal{A}$$

by passage to fibre, in the sense that

$$fib(t^{\mathcal{F}}_* \otimes V \stackrel{s}{\longrightarrow} \Sigma T) = (F(s) \longrightarrow t^{\mathcal{F}}_* \otimes V).$$

However if we use cofibres and define

$$cof(N \stackrel{r}{\longrightarrow} t^{\mathcal{F}}_* \otimes V) = (t^{\mathcal{F}}_* \otimes V \longrightarrow C(r))$$

the image is not in $dg\mathcal{A}_t$. Instead we consider a slight enlargement: we take $\hat{\mathcal{A}}_t$ to consist of objects $t^{\mathcal{F}}_* \otimes V \longrightarrow M$, where $\mathcal{E}^{-1}M \cong t^{\mathcal{F}}_* \otimes U$ for some graded vector space U, and then let $dgH\mathcal{A}_t$, consist of dg $\hat{\mathcal{A}}_t$-objects $t^{\mathcal{F}}_* \otimes V \longrightarrow M$ with the property that $H_*(M)$ is an $\mathcal{F}$-finite torsion module. Thus we have defined a functor

$$cof : dg\mathcal{A} \longrightarrow dgH\mathcal{A}_t.$$

Once we invert homology isomorphisms the inclusion

$$i : dg\mathcal{A}_t \longrightarrow dgH\mathcal{A}_t$$

becomes an equivalence. It is convenient to introduce a similar enlargement of $\mathcal{A}$: we take $\hat{\mathcal{A}}$ to consist of objects $N \longrightarrow t^{\mathcal{F}}_* \otimes V$, where $\mathcal{E}^{-1}N \cong t^{\mathcal{F}}_* \otimes U$ for some graded vector space

U, and then let $dgH\mathcal{A}$, consist of dg $\hat{\mathcal{A}}$-objects $N \longrightarrow t_*^{\mathcal{F}} \otimes V$ so that the homology is an isomorphism modulo $\mathcal{F}$-finite torsion modules.

LEMMA 6.5.3. *The derived categories of $dgH\mathcal{A}_t$ and $dgH\mathcal{A}$ exist and the inclusions $dg\mathcal{A}_t \longrightarrow dgH\mathcal{A}_t$ and $dg\mathcal{A} \longrightarrow dgH\mathcal{A}$ induce isomorphisms of derived categories which preserve the triangulation.*

Proof: In both cases we have the same formal situation. We have a category $\mathbb{A}$ from which we know how to form the derived category: we first form the category $dg\mathbb{A}$ of dg objects, and then invert the homology isomorphisms. To establish existence we use the class $\mathcal{I}$ of objects formed from injective objects of $\mathbb{A}$ by a finite number of cofiberings.

The category $\mathbb{A}$ is a subcategory of a larger category $\hat{\mathbb{A}}$, and we then consider the category $dgH\mathbb{A}$ of dg objects of a $\hat{\mathbb{A}}$ satisfying the condition that their homology lies in $\mathbb{A}$. We show that $D\mathbb{A}$ is also the derived category of $dgH\mathbb{A}$ by establishing that Condition 4.1.1 holds with S being the class of homology isomorphisms. No extra work is necessary for Part (ii), and Part (iii) follows by the same proof as given in Section 4.1. It therefore remains only to show that any object M of $dgH\mathbb{A}$ admits a homology isomorphism $M \longrightarrow \hat{M}$ with $\hat{M} \in \mathcal{I}$. This is sufficient to establish the existence of the derived category of $dgH\mathbb{A}$, and the fact that the inclusion $dg\mathbb{A} \longrightarrow dgH\mathbb{A}$ induces an equivalence of derived categories.

The simpler case is when $\mathbb{A} = \mathcal{A}$; we consider it first, and explain the variations necessary to deal with $\mathbb{A} = \mathcal{A}_t$. First, since $\mathcal{A}$ is one dimensional, we may take an injective resolution $0 \longrightarrow H_*M \longrightarrow I \longrightarrow J \longrightarrow 0$ of H_*M. Now let $\hat{M}$ be the fibre of $I \longrightarrow J$: it is an object of $\mathcal{I}$, so it suffices to construct a homology isomorphism $M \longrightarrow \hat{M}$.

Now for any object L of $dgH\mathcal{A}$ we have a diagram

$$\begin{array}{ccccccc} & & [L,\hat{M}] & \longrightarrow & [L,I] & \longrightarrow & [L,J] \\ & & \downarrow & & \downarrow & & \downarrow \\ 0 & \longrightarrow & \mathrm{Hom}(H_*(L),H_*(M)) & \longrightarrow & Hom(H_*(L),I) & \longrightarrow & Hom(H_*(L),J) \end{array}$$

with rows exact, where $[\cdot,\cdot]$ denotes homotopy classes. It suffices to show that the left hand vertical is surjective: then if $L = M$ we may lift the identity map. By a diagram chase it suffices to show

$$H_* : [L,I] \longrightarrow \mathrm{Hom}(H_*(L),I)$$

is an isomorphism when I is an $\mathcal{A}$-injective, and indeed it suffices to prove this for sufficiently many $\mathcal{A}$-injectives.

By 4.1.10 it suffices to show that $\mathrm{Ext}^1(L/zL,I) = 0$ and $\mathrm{Ext}^1(L/dL,I) = 0$, where $zL = ker(d)$. In view of the short exact sequence $0 \longrightarrow zL/dL \longrightarrow L/dL \longrightarrow L/zL \longrightarrow 0$, it suffices to prove the vanishing for L/zL. When $I = e(V)$ there is no difficulty, since the adjunction 5.4.3 (i) is quite general, and all rational vector spaces are injective. When $I = f(I')$, we may apply the adjunction 5.4.3 (ii). It thus remains so show $\mathrm{Ext}^1_{\mathcal{O}_{\mathcal{F}}}(P,I') = 0$ when $P = N/zN$ where N is the nub of L. By 5.3.1, the vanishing holds for $P = N$ and hence also for N/zN since d shows N/zN is isomorphic to a submodule of N.

To deal with $\mathbb{A} = \mathcal{A}_t$ we argue slightly less directly. In fact we note that, by the exact sequence $(0 \longrightarrow T) \longrightarrow (t_*^{\mathcal{F}} \otimes V \longrightarrow T) \longrightarrow (t_*^{\mathcal{F}} \otimes V \longrightarrow 0)$ it is enough to deal with objects $(t_*^{\mathcal{F}} \otimes V \longrightarrow 0)$ and $(0 \longrightarrow T)$. Those of the first type are already in the category

$\mathcal{A}_t$, so it is enough to deal with those of the second type. The category of $\mathcal{F}$-finite torsion modules is again one dimensional, we may construct a homology isomorphism $T \longrightarrow \hat{T}$ as before with $\hat{T}$ the fibre of $I \longrightarrow J$ where $0 \longrightarrow H_*T \longrightarrow I \longrightarrow J \longrightarrow 0$ is an injective resolution of $H_*(T)$; by 4.1.10 this only requires $\mathrm{Ext}^1(T/zT, I) = 0$ and $\mathrm{Ext}^1(T/dT, I) = 0$ for $\mathcal{F}$-finite torsion injectives I, and arguing from 5.3.1 as before, this condition is satisfied. Thus we have a homology isomorphism $(0 \longrightarrow T) \longrightarrow (0 \longrightarrow \hat{T})$. This completes the argument. □

Notice by the long exact sequence associated to a cofibre or fibre sequence the functors fib and cof preserve homology isomorphisms, and hence induce maps of derived categories.

We may now prove the theorem by considering the functors

$$fib' : dgH\mathcal{A}_t \longrightarrow dgH\mathcal{A} \text{ and } cof' : dgH\mathcal{A} \longrightarrow dgH\mathcal{A}_t,$$

which induce functors on derived categories for the same reasons. The standard natural transformations $cof' \circ fib' \longrightarrow 1$ and $1 \longrightarrow fib' \circ cof'$ induce homology isomorphisms. Hence fib' and cof' induce an equivalence between derived categories. □

We shall have use for one more result.

LEMMA 6.5.4. *The functor $fib : dg\mathcal{A}_t \longrightarrow dg\mathcal{A}$ takes the class $\mathcal{I}$ of fibrant objects to fibrant objects.*

Proof: It is obvious that fib takes an $\mathcal{F}$-contractible injective to an injective since

$$fib(t_*^{\mathcal{F}} \otimes V \longrightarrow 0) = (t_*^{\mathcal{F}} \otimes V \longrightarrow t_*^{\mathcal{F}} \otimes V).$$

On the other hand $fib(t_*^{\mathcal{F}} \otimes \mathrm{Hom}(t_*^{\mathcal{F}}, T) \longrightarrow T)$ is the fibre of a map

$$\begin{array}{ccc} t_*^{\mathcal{F}} \otimes \mathrm{Hom}(t_*^{\mathcal{F}}, T) & \longrightarrow & T \\ \downarrow & & \downarrow \\ t_*^{\mathcal{F}} \otimes \mathrm{Hom}(t_*^{\mathcal{F}}, T) & \longrightarrow & 0. \end{array}$$

Since fib preserves triangles the result follows. □

6.6. Relationship to topology.

Because the equivalence $\mathbb{T}$-Spec $\simeq D\mathcal{A}_t$ of 6.5.1 is so indirect, we briefly make the relationship between the two categories a little more explicit.

The conclusion is that any $\mathbb{T}$-spectrum X is modelled by a differential graded $\mathcal{A}_t$-object $T(X)$. This object is defined indirectly as a fibrant approximation (in dg $\mathcal{A}_t$) to the object

$$cof(\pi_*^{\mathcal{A}}(X)) = cof\left(\pi_*^{\mathbb{T}}(X \wedge DE\mathcal{F}_+) \longrightarrow \pi_*^{\mathbb{T}}(X \wedge DE\mathcal{F}_+ \wedge \tilde{E}\mathcal{F}) = t_*^{\mathcal{F}} \otimes \pi_*(\Phi^{\mathbb{T}} X)\right).$$

The linear approximation to this model is given by the object

$$\pi_*^{\mathcal{A}_t}(X) = \left(t_*^{\mathcal{F}} \otimes \pi_*(\Phi^{\mathbb{T}} X) = \pi_*^{\mathbb{T}}(X \wedge DE\mathcal{F}_+ \wedge \tilde{E}\mathcal{F}) \longrightarrow \pi_*^{\mathbb{T}}(X \wedge \Sigma E\mathcal{F}_+)\right),$$

of $\mathcal{A}_t$.

LEMMA 6.6.1. *The relationship between linearizations is as expected:*

$$H_*(T(X)) = \pi_*^{\mathcal{A}_t}(X).$$

Proof: Using the splitting of the standard model into even and odd parts this follows since homology takes cofibre sequences to long exact sequences. □

Transposing the warning from Section 6.4 into the topological context, we warn that $\pi_*^{\mathcal{A}_t}(X)$ does not determine X up to homotopy in general; if it does, we say X is *formal.* By 6.4.1 it follows that if $\pi_*^{\mathcal{A}_t}(X)$ is of injective dimension 0 or 1 then X is formal: for example, X is formal if $\pi_*^{\mathbb{T}}(X \wedge E\mathcal{F}_+)$ is injective.

The actual relation between $\mathcal{A}_t$ and topology is encoded in the Adams spectral sequence, which we discussed in the algebraic context in Section 6.4 above.

THEOREM 6.6.2. *For any* $\mathbb{T}$*-spectra* X *and* Y*, there is an Adams spectral sequence*

$$E_2^{*,*} = \mathrm{Ext}_{\mathcal{A}_t}^{*,*}(\pi_*^{\mathcal{A}_t}(X), \pi_*^{\mathcal{A}_t}(Y)) \Longrightarrow [X, Y]_*^{\mathbb{T}}$$

which collapses at E_3. □

Part II

Change of groups functors in algebra and topology.

CHAPTER 7

Introduction to Part II.

We begin in Section 7.1 with a summary of the models from Part I, and a broad summary of the contents of Part II. The next section briefly discusses the functors considered. Finally, in Section 7.3 we formalize a straightforward argument that is needed several times.

We remind readers that after Part I chapters are mostly independent. In particular it is not necessary to read Part II before reading about applications in Part III. First time readers should probably at least look at Chapter 13 before anything from Part II. There is further guidance at the end of Chapter 0.

7.1. General outline.

In Part I we showed that various categories of rational $\mathbb{T}$-spectra had algebraic models. Specifically we showed (4.4.1) that the category of $\mathbb{T}$-spectra over H is equivalent to the derived category of torsion $\mathbb{Q}[c_H]$-modules, (4.5.3) that the category of $\mathcal{F}$-spectra is equivalent to the derived category of $\mathcal{F}$-finite torsion $\mathcal{O}_{\mathcal{F}}$-modules, (5.6.1) that the category of semifree spectra is equivalent to the derived category of the category of $\mathbb{Q}[c]$-morphisms $N \longrightarrow \mathbb{Q}[c, c^{-1}] \otimes V$ which become isomorphisms when c is inverted and (5.6.1) that the category of all rational $\mathbb{T}$-spectra is equivalent to the derived category of the one dimensional abelian category $\mathcal{A}$ whose objects are morphisms $N \longrightarrow t_*^{\mathcal{F}} \otimes V$ with $\mathcal{F}$-finite torsion kernel and cokernel.

The main purpose of Part II is to identify the algebraic counterparts of various well known topological functors. Most interesting functors $F : \mathbb{T}\text{-Spec} \longrightarrow \mathbb{T}\text{-Spec}$ arise by passage to stable homotopy from a point set level functor f. It is therefore natural to expect that the algebraic counterpart will be the total right derived functor $Rf' : D\mathcal{A} \longrightarrow D\mathcal{A}$ of a functor $f' : \mathcal{A} \longrightarrow \mathcal{A}$, or its left derived counterpart. (Many may be content to view Rf' as defined by the formula $Rf'X = f'(\hat{X})$ where $\hat{X}$ is a fibrant approximation to X; for further discussion the reader is referred to Appendix B on Quillen closed model category structures). This expectation proves well founded in all cases we have analysed. There caveat to this statement is that at present there seems no prospect of showing that the equivalence of topological and algebraic categories arises from a string of adjunctions of Quillen closed model categories. Accordingly there is no prospect of showing that f and f' correspond before passage to derived categories, and we must proceed in an indirect way.

Thus, for each functor f the work of this part divides into two. Firstly we have to to do enough algebra to understand the behaviour of the relevant functor f', and its associated right derived functor. Secondly we have to show how to describe the functor F using only terms which we can model exactly. Of course, if we can identify FX by its homotopy (as happens in the standard model), it is relatively easy to verify the model on objects.

There is then the difficulty of indeterminacy in the modelling of morphisms, and we may have to argue rather obliquely to verify this part of the model. However, some interesting functors are hard to describe exactly, even on objects. In this case we may attempt to construct objects X from basic objects X' using sums and cofibre sequences in such a way that the value FX' of F on a basic object X' is modelled by $Rf'(M)$, and so that at each stage we can identify the maps with zero indeterminacy. It is worth emphasizing that the importance of the algebra is independent of how the equivalence of algebraic and topological constructions is proved.

Because of Part I, a great deal of the work in Part II is purely algebraic. Of course it is necessary to do a little topology from time to time if we wish to know which topological phenomena we are modelling. On the other hand, there may be readers who feel that after Part I they do not need to talk about actual $\mathbb{T}$-spectra any more, and that they can work entirely in the algebraic category. Despite the obvious dangers of this attitude, its benefits make it desirable to make it possible to read Part II without intimate acquaintance with Part I. It is thus helpful to recall the model before undertaking a full scale algebraic study.

We therefore begin by recalling that we work almost exclusively over the graded ring $\mathcal{O}_{\mathcal{F}}$, which we may view as obtained from the ring $(\mathcal{O}_{\mathcal{F}})_0 = C(\mathcal{F}, \mathbb{Q}) \cong \prod_H \mathbb{Q}$ by adjoining an indeterminate c of degree -2. We let e_H denote the idempotent of $(\mathcal{O}_{\mathcal{F}})_0$ corresponding to the projection onto the Hth factor, and we let $c_H = e_H c$ in $e_H \mathcal{O}_{\mathcal{F}} = \mathbb{Q}[c_H]$. We also need define Euler classes 4.6.1 generalizing the elements $c_1^k \in H^*(B\mathbb{T}_+)$ for $k \geq 0$. For any function $v : \mathcal{F} \longrightarrow \mathbb{Z}_{\geq 0}$ with finite support we have an associated Euler class c^v whose Hth coordinate is $c^{v(H)}$. These Euler classes are not homogeneous elements of $\mathcal{O}_{\mathcal{F}}$, but they act on a $\mathcal{O}_{\mathcal{F}}$-module M as follows. Indeed if $\phi = \{H_1, H_2, \ldots, H_n\}$ is the support of v and $e_\phi \in \mathcal{O}_{\mathcal{F}}$ is the idempotent with this support, we have

$$M = M(H_1) \oplus M(H_2) \oplus \cdots \oplus M(H_n) \oplus (1 - e_\phi)M,$$

where $M(H) = e_H M$, and c^v acts as multiplication by $c^{v(H_i)}$ on the ith factor, and as the identity on $(1 - e_\phi)M$. Thus the result of inverting c^v on M is again a graded module. We therefore consider the multiplicative set

$$\mathcal{E} = \{c^v | v : \mathcal{F} \longrightarrow \mathbb{Z}_{\geq 0} \text{ of finite support}\}$$

of Euler classes, and allow ourselves to invert it.

Next, an $\mathcal{O}_{\mathcal{F}}$-module M is said to be $\mathcal{F}$-finite if the natural inclusion is an isomorphism $\bigoplus_H e_H M \cong M$, and it is torsion if $M[c^{-1}] = 0$. It is not hard to see (4.6.6) that M is an $\mathcal{F}$-finite torsion module if and only if $\mathcal{E}^{-1}M = 0$.

The final ingredient is the twisting module $t_*^{\mathcal{F}} = \mathcal{E}^{-1}\mathcal{O}_{\mathcal{F}}$, for which we have the short exact sequence 5.2.1

$$0 \longrightarrow \mathcal{O}_{\mathcal{F}} \longrightarrow t_*^{\mathcal{F}} \longrightarrow \Sigma^2 \mathbb{I} \longrightarrow 0$$

with $\mathbb{I} = \bigoplus_H \mathbb{I}_H$, and $\Sigma^2 \mathbb{I}_H = \mathbb{Q}[c_H, c_H^{-1}]/\mathbb{Q}[c_H]$.

We now consider the category $\mathcal{A}$ whose objects are morphisms $\beta : N \longrightarrow t_*^{\mathcal{F}} \otimes V$ which become an isomorphism when Euler classes are inverted. Morphisms are required to be the identity on the tensor factor $t_*^{\mathcal{F}}$. We call N the nub, and V the vertex of the object; the map β is called the basing map. The category $\mathcal{A}$ is abelian and one dimensional, and its derived category $D\mathcal{A}$ is the standard model for rational $\mathbb{T}$-spectra (Theorem 5.6.1). Thus there is a short exact sequence for calculating morphisms in either $D\mathcal{A}$ or in $\mathbb{T}$-Spec in terms of Hom and Ext in $\mathcal{A}$, and this sequence is split. This is stated explicitly as Theorem 5.6.6.

Whilst $D\mathcal{A}$ is technically very convenient, not all geometrically interesting aspects of $\mathbb{T}$-spectra are immediately apparent. Thus, even when a construction can be performed in $D\mathcal{A}$, the result may need reinterpretation before it answers interesting questions. There also appear to be cases where $D\mathcal{A}$ is not the natural way of modelling $\mathbb{T}$-spectra. We therefore also need to consider the torsion model category $\mathcal{A}_t$ whose objects are $\mathcal{O}_{\mathcal{F}}$-maps $t_*^{\mathcal{F}} \otimes V \longrightarrow T$ with T an $\mathcal{F}$-finite torsion module. This is abelian and two dimensional and (Theorem 6.5.1) its derived category is equivalent to $D\mathcal{A}$. This completes the summary of the algebraic models.

Part II identifies algebraic models of various functors that change the ambient group of equivariance. The trick of identifying a functor by identifying its adjoint is both effective and essential, since it happens several times that only one of a pair of adjoint functors is directly accessible. We begin by considering the forgetful induction and coinduction functors, all of which are quite straightforward. The geometric $\mathbb{T}$-fixed point functors is built into our model as the vertex, and it is an easy matter to identify the model of the Lewis-May $\mathbb{T}$-fixed point functor. The most interesting functors are those which relate $\mathbb{T}$-spectra to $\mathbb{T}/K$-spectra for finite subgroups K. As would be expected, the geometric K-fixed point functor Φ^K is rather easy to describe. However both the inflation functor from $\mathbb{T}/K$ spectra to $\mathbb{T}$-spectra and its adjoint, the Lewis-May K-fixed point functor are rather subtle. It is perhaps only to be expected that quotient functors are only identified on reasonably free objects.

Although the work of Part II is of interest in its own right, it has further justification in the applications of Part III. In practice the author only undertook the analysis of a functor when forced to by the applications. The resulting clarification and simplification in the special case was most gratifying. However, since there is now a substantial amount of algebra, which may perhaps be of interest in its own right, we have collected it together in chapters of its own. This has the unfortunate effect of separating the algebra from the topology which motivated it. Therefore, to help the reader navigate Part II we will continue this introduction with more complete sketches of the analyses and motivations for the algebraic constructions: we feel justified in talking heuristically at this stage. The structure of the analyses is that we develop the necessary algebra and then explain the properties of topological constructions which relate it to the algebra. Finally the algebraic and topological functors are shown to correspond using these two parallel structures. The last section of this chapter outlines the general methods we employ for studying functors.

7.2. Modelling functors changing equivariance.

Now that we can model categories of G-spectra when G is any subquotient of the circle $\mathbb{T}$, we may attempt to understand the functors between these categories in algebraic terms.

Of course the categories of H-spectra are very simple when H is a finite subgroup, so functors involving these are rather straightforward to understand: thus we begin in Section 8.1 by modelling the induction and coinduction functors from H-spectra to $\mathbb{T}$-spectra. There is still a potential difficulty in understanding morphisms, but we recognize the algebraic functors as being left and right adjoint to the forgetful functor, which is easy to identify. It is clear in this case that the vertex plays no real part so it is unimportant whether we work with the standard or the torsion model.

More important are the fixed point functors. There are two types of fixed point functors: the geometric fixed point functors and the Lewis-May fixed point functors. Whichever we use, if we take $\mathbb{T}$-fixed points the result is a non-equivariant spectrum, and hence determined simply by its ordinary homotopy groups. These are easily deduced from the model: the geometric $\mathbb{T}$-fixed points simply give the vertex, and the Lewis-May $\mathbb{T}$-fixed points are $\Psi^{\mathbb{T}}(\cdot) = R\mathrm{Hom}(S^0, \cdot)$. We therefore concentrate on K-fixed points regarded as a functor from $\mathbb{T}$-spectra to $\mathbb{T}/K$-spectra.

The geometric K-fixed point functor is designed to pick out the isotropy parts corresponding to subgroups of $\mathbb{T}$ containing K: it is nearly obvious that this means the vertex is unchanged and that the effect on $E\mathcal{F}_+ \wedge X \simeq \bigvee_H E\langle H\rangle \wedge X$ is to throw away the summands indexed by the finite subgroups H not containing K. There is an obvious way of doing this algebraically for $\mathcal{O}_{\mathcal{F}}$-modules: one simply applies the idempotent $e \in C(\mathcal{F}, \mathbb{Q})$ with support consisting of subgroups H containing K, and identifies this set of subgroups of $\mathbb{T}$ with the subgroups of $\mathbb{T}/K$ in the obvious way. We prove in Section 8.4 that this not only gives the right functor for $\mathcal{F}$-spectra, but that simply applying e to the standard model category models the geometric fixed point functor.

The Lewis-May K-fixed point functor Ψ^K is much more complicated, and we spend Chapters 9 and 10 in studying it. It is well known that at one extreme, the geometric K-fixed point functor can be obtained from the Lewis-May fixed point functor by $\Phi^K(X) = \Psi^K(X \wedge \tilde{E}[\not\supseteq K])$. At the other extreme, the Lewis-May fixed point functor on a free spectrum is the quotient. One might hope that an understanding of these two extreme pieces of behaviour would lead to an identification of the algebraic model of the functor. Because of its intuitive appeal we present this approach in Section 10.4, but it appears only to give the answer on objects, and not as a functor.

Instead, we find it better to recognize the Lewis-May fixed point functor as the right adjoint of the inflation functor. Terminology is not quite consistent in the literature, so we must explain that if $\mathcal{U}$ is a complete $\mathbb{T}$-universe we mean the functor which takes a $\mathbb{T}/K$-spectrum X indexed on $\mathcal{U}^K$, regards it as a $\mathbb{T}$-spectrum indexed on $\mathcal{U}^K$ by pullback along the quotient map $q : \mathbb{T} \longrightarrow \mathbb{T}/K$, and then builds in representations by the relative stabilization functor corresponding to the inclusion $j : \mathcal{U}^K \longrightarrow \mathcal{U}$. The resulting spectrum is often written j_*q^*X, $q^{\#}X$ or simply j_*X. However, since we want to concentrate on the important features we write simply $X \longmapsto \mathrm{inf}_{\mathbb{T}/K}^{\mathbb{T}} X$ for this functor. The behaviour of the inflation functor is very simple on suspension spectra in the sense that $\mathrm{inf}_{\mathbb{T}/K}^{\mathbb{T}} X = X$ for a space X, but its general behaviour is rather complicated, and this is reflected algebraically. Actually, its behaviour is simple on $\mathcal{F}$-contractible spectra, in the sense that it is essentially the identity. It is also simple on $\mathcal{F}$-spectra, in that it simply increases the multiplicity of the various factors. We may view an $\mathcal{F}$-spectrum $Y \simeq \bigvee_H Y(H)$ as classified by a function $H \longmapsto N(H)$, where $N(H) = \pi_*^{\mathbb{T}}(Y(H))$ is a $\mathbb{Q}[c]$ module. Now, if X is an almost free $\mathbb{T}/K$ spectrum classified by the function $\overline{H} \longmapsto M(\overline{H})$, then $\inf_{\mathbb{T}/K}^{\mathbb{T}} X$ is classified by the function $H \longmapsto M(\overline{H})$, where $\overline{H}$ is the image of H in $\mathbb{T}/K$. Thus the summand $X(\overline{H})$, which occurs once in X, is replaced in $\inf_{\mathbb{T}/K}^{\mathbb{T}} X$ by one copy for each subgroup H of $\mathbb{T}$ whose image in $\mathbb{T}/K$ is $\overline{H}$. These two pieces are neatly spliced together by using $\inf(M) = \mathcal{O}_{\mathcal{F}} \otimes_{\overline{\mathcal{O}}_{\overline{\mathcal{F}}}} M$, where $\mathcal{O}_{\mathcal{F}}$ is an $\mathcal{O}_{\overline{\mathcal{F}}}$-module by pullback along the ring homomorphism arising from $q : \mathcal{F} \longrightarrow \overline{\mathcal{F}}$. This enables us to identify an algebraic inflation functor on the torsion model category, and then to show that it models the topological inflation functor. It then follows that the right adjoint of the geometric inflation functor models the Lewis-May fixed point functor. Following standard operating procedure, we begin in Chapter 9 by studying the algebraic inflation and its adjoints. In Section 10.1 we prove some basic properties of topological inflation and the Lewis-May fixed point functor, and use them to identify inflation on objects in Section 10.2. This prepares us to show in Section 10.3 that the algebraic and topological inflation functors correspond; it follows that the left adjoint of the algebraic inflation models the Lewis-May fixed point functor. Section 10.4 gives a more direct approach to the Lewis-May fixed point functor; it is recommended for motivation, but the approach is only successful in modelling the functor on objects. Finally we spend Section 10.5 considering the functor on the homotopy category of the standard model: the identification of a model for the Lewis-May K-fixed point functor on the standard model itself is deferred to Section 24.4, when we have the machinery of Part IV available.

The topological quotient functor is only approachable on K-free spectra, but in that case it is left adjoint to inflation. It follows that the left adjoint to the algebraic inflation functor models passage to the quotient. Since we are working rationally one might hope to extend the domain of good behaviour of the quotient functor to all $\mathcal{F}$-spectra. Although we do not do this, supporting evidence comes from the fact that the algebraic inflation functor does have a left adjoint on the category of all $\mathcal{F}$-finite torsion $\mathcal{O}_{\mathcal{F}}$-modules: we discuss these matters in Section 10.6.

7.3. Functors between split triangulated categories.

In this section we explain our general method for modelling functors. We repeatedly want to construct algebraic counterparts of topological functors between categories of spectra. In other words we have a functor $F : Spec_1 \longrightarrow Spec_2$ between two categories of spectra, and we want to find the algebraic counterpart $F' : D\mathcal{A}_1 \longrightarrow D\mathcal{A}_2$, in the sense that we complete the square

$$\begin{array}{ccc} Spec_1 & \xrightarrow{F} & Spec_2 \\ {\scriptstyle p}\downarrow\simeq & & \simeq\downarrow{\scriptstyle p} \\ D\mathcal{A}_1 & \overset{F'}{\dashrightarrow} & D\mathcal{A}_2 \end{array}$$

so that it commutes up to natural isomorphism. Typically, it is reasonably easy to guess a candidate algebraic functor $F' : D\mathcal{A}_1 \longrightarrow D\mathcal{A}_2$, and in fact F' is usually the total right derived functor of some functor $f' : \mathcal{A}_1 \longrightarrow \mathcal{A}_2$. The difficulty arises because the equivalences p were only defined indirectly and by making certain choices. Thus it is only easy to identify the homotopy of objects, and the d-invariant of morphisms. There are two grades of conclusion that we may hope to achieve. We may only be able to prove that F' is correct on objects, in the sense that if X is modelled by M then FX is modelled by $F'M$. However, in especially favourable cases, we will be able to prove that it is correct on morphisms in the sense that the above square commutes up to natural isomorphism.

Since the relevant functors F and F' preserve cofibre sequences, the idea is to show that F and F' agree on certain basic objects X', and that arbitrary objects X can be formed from these basic objects using only constructions in which every step can be modelled without indeterminacy. It is often sufficient to use mapping cones of maps $f : X_1' \longrightarrow X_2'$ in which $F(f)$ is determined by its d-invariant. If this can be done for objects we shall say that F is *object-accessible*, and if it can also be done for morphisms we shall we shall simply say that F is *accessible*.

Our only comment about object accessibility is that, if we can use a one dimensional model category $\mathcal{A}_2$, it suffices to calculate the homology of the image object, since in this case all objects are formal (in that they are determined by their homology). In the torsion model not all objects are formal, so object-accessibility is not always clear. However we will see in Part IV, where we deal with the smash product and the function spectrum, that even in the standard model, considerable work may be involved in proving object accessibility.

Since it is often possible to work with one dimensional models, it is therefore valuable to record the minimal data that needs to be checked to ensure the diagram commutes up to natural isomorphism. In practice we compare the two composites $F_1 = pF$ and $F_2 = F'p$ from the top left to the bottom right.

THEOREM 7.3.1. *Given a split linear triangulated category Spec and a one dimensional split abelian category $\mathcal{A}$, two functors $F_1, F_2 : Spec \longrightarrow D\mathcal{A}$ are naturally isomorphic provided the following conditions are satisfied.*
(1) Both functors preserve (a) triangles, (b) injectives, and (c) pure parity objects
(2) The functors agree up to isomorphism (a) on enough injective objects and (b) on maps into injective objects.

We remark that in the motivating case it is enough to check Condition 1 for the functors F and F' since the functors p are equivalences of triangulated categories, and used to define the notions of parity and injectivity.

Proof: First note that for any object Y we have $F_1(Y) \simeq F_2(Y)$. Indeed we may choose an injective resolution $Y \longrightarrow I \xrightarrow{d} J$, and thus a triangle $F_iY \longrightarrow F_iI \xrightarrow{F_id} F_iJ$. Since F_iJ is injective by (1)(b), it follows that the morphism $F_id : F_iI \longrightarrow F_iJ$ is determined by its d-invariant. Once the objects are identified using (2)(a) the two maps F_1d and F_2d are homotopic by (2)(b). Hence $F_1Y \simeq F_2Y$ as required, and we write simply FY.

By property (1)(c) it is enough to check that the two maps $[X, Y] \longrightarrow [FX, FY]$ agree when X and Y are of even parity. In this case the Adams short exact sequence for $[X, Y]$ is split by parity. Now suppose we have an Adams resolution $Y \longrightarrow I \longrightarrow J$ with image $FY \longrightarrow FI \longrightarrow FJ$. Suppose first that $f : X \longrightarrow Y$ is of even degree, and consider the composite $X \longrightarrow Y \longrightarrow I$. Since FI is injective by (1)(b), the two images of the composite $FX \longrightarrow FY \longrightarrow FI$ are equal. However the indeterminacy in the map $FX \longrightarrow FY$ with such a composite is the image of the group $[FX, \Sigma^{-1}FJ]$, which is zero since FX and FJ are of even parity and FJ is injective. Since there is no indeterminacy $F_1 f = F_2 f$.

Next, if $f : X \longrightarrow Y$ is of odd degree, it factors as $X \longrightarrow \Sigma^{-1}J \longrightarrow Y$, and the image of this is the composite $FX \longrightarrow \Sigma^{-1}FJ \longrightarrow FY$. The first map is a map into an injective, and so F_1 and F_2 agree; the second map is part of the triangle $FY \longrightarrow FI \longrightarrow FJ$. □

CHAPTER 8

Induction, coinduction and geometric fixed points.

This chapter treats the simplest change of groups functors. In Section 8.1 we give algebraic models of the forgetful functor from $\mathbb{T}$-spectra to H-spectra, and its adjoints, the induction and coinduction functors. In Section 8.2 we identify the algebraic counterpart of the Lewis-May $\mathbb{T}$-fixed point functor, leaving the more complicated Lewis-May K-fixed point functor until Chapters 9 and 10. Finally, in Section 8.3 we motivate and define an algebraic counterpart of the geometric K-fixed point functor, and in Section 8.4 we prove it is indeed a model.

8.1. Forgetful, induction and coinduction functors.

It is the purpose of this section to describe the algebraic counterparts of the forgetful functor $U : \mathbb{T}\text{-Spec} \longrightarrow H\text{-Spec}$, the induction functor $ind : H\text{-Spec} \longrightarrow \mathbb{T}\text{-Spec}$, $X \longmapsto \mathbb{T} \ltimes_H X$, and the coinduction functor $coind : H\text{-Spec} \longrightarrow \mathbb{T}\text{-Spec}$, $X \longmapsto F_H[\mathbb{T}, X)$. The induction and coinduction functors are left and right adjoint to the forgetful functor respectively, so that if X is a $\mathbb{T}$-spectrum and Y an H-spectrum

$$[\mathbb{T} \ltimes_H Y, X]^{\mathbb{T}} = [Y, UX]^H \text{ and } [UX, Y]^H = [X, F_H[\mathbb{T}, Y)]^{\mathbb{T}}.$$

We must begin by describing the category H-Spec. First, we recall that an arbitrary rational H-spectrum splits as a product of Eilenberg-MacLane spectra [**14**]; thus H-Spec is equivalent to the derived category of the 0-dimensional abelian category of rational H-Mackey functors.

For any finite group H, the rational Mackey functors are easy to describe, and are sums of the functors arising from representations of the Weyl groups $N_H(K)/K$ for $K \subseteq H$. In our case H is abelian, so we need only explain that a module V for the quotient H/K gives rise to a Mackey functor $R^H_K(V)$ defined by

$$R^H_K(V)(L) = \begin{cases} V^L & \text{if } L \supseteq K \\ 0 & \text{if } L \not\supseteq K \end{cases}$$

The restriction maps are given by inclusions of fixed point sets, and the transfer maps are given by coset sums. We let $\mathfrak{M}_H$ denote the category of all rational Mackey functors, and $\mathfrak{M}_H^{triv}$ denote the full subcategory of Mackey functors with trivial Weyl group action (i.e. sums of functors of form $R^H_K(\mathbb{Q})$. Thus

$$H\text{-Spec} \simeq D(\mathfrak{M}_H).$$

Because the category of rational Mackey functors is of global dimension 0, and the free functors are realizable, all spectra are generalized Eilenberg-MacLane spectra and it is easy to see that the equivalence is given by $X \longmapsto \underline{\pi}^H_*(X)$. Therefore the algebraic counterpart of the forgetful functor $\mathbb{T}$-Spec $\longrightarrow$ H-Spec is the functor $D\mathcal{A} \longrightarrow D(\mathfrak{M}_H)$ given by the the same condition. However, it is perhaps clearer to compose with the equivalence

$$D\mathfrak{M}_H \xrightarrow{\simeq} \prod_K \mathbb{Q}H/K - \text{mod}$$

whose Kth component is $M \longmapsto e_K M(K)$.

SUMMARY 8.1.1. The forgetful map $\mathbb{T}$-Spec $\longrightarrow$ H-Spec corresponds to the functor $D\mathcal{A} \longrightarrow \prod_K D(\mathbb{Q}H/K\text{-mod})$ with Kth component $M \longmapsto [L_K, M]$, where $L_K = (\Sigma\mathbb{Q}(H) \longrightarrow 0)$ is the algebraic counterpart to the basic cell σ_K^0 (see 5.8.1). □

We have two methods open to us to identify the algebraic induction and coinduction functors. We could calculate the behaviour of the geometric functors, but instead we shall simply guess the answers and prove they have the requisite adjointness properties. Actually, we 'guess' the functors by a calculation on geometric objects. Because H-spectra split as a wedge of Eilenberg-MacLane spectra, and the induction functor commutes with wedges, the induction will be determined by its values on Eilenberg-MacLane spectra. Finally, it is useful to use the fact [**19**, II.6.2] that $F_H[\mathbb{T}, X) \simeq \Sigma^{-1}\mathbb{T} \ltimes_H X$. Now we simply observe that any induced spectrum $\mathbb{T} \ltimes_H X$ is an $\mathcal{F}$-spectrum, and is thus determined by its homotopy groups: $\pi_*^{\mathbb{T}}(\mathbb{T} \ltimes_H Y) = \pi_*^{\mathbb{T}}(\Sigma F_H[\mathbb{T}, Y)) = \pi_*^H(\Sigma Y)$.

LEMMA 8.1.2. *If the H-Mackey functor M corresponds to $V = (V_K)_K$ with V_K a $\mathbb{Q}H/K$-module (ie $M = \bigoplus_K R_K^H V_K$), then*

$$M(H) = \bigoplus_K (V_K)^H. \quad \square$$

DEFINITION 8.1.3. If V_K is a graded $\mathbb{Q}H/K$-module for each $K \subseteq H$, then we define induction and coinduction on $V = (V_K)$ by

$$\text{ind}(V) = \Sigma(\bigoplus_K (V_K)^H \longrightarrow 0)$$

and

$$\text{coind}(V) = (\bigoplus_K (V_K)^H \longrightarrow 0),$$

where the Chern class c acts as zero in both cases.

Evidently these are both exact functors $\prod_K \mathbb{Q}H/K$-mod $\longrightarrow \mathcal{A}$, and hence induce functors on the derived category.

PROPOSITION 8.1.4. *These induction and coinduction functors are left and right adjoint to the forgetful functor $\mathcal{A} \longrightarrow \prod_K \mathbb{Q}H/K$-mod, and the same holds at the level of derived categories.*

Before sketching the proof we record the desired corollary.

COROLLARY 8.1.5. *The algebraic and topological induction and coinduction functors correspond in the sense that the diagrams*

$$\begin{array}{ccc} H\text{-}Spec & \stackrel{\mathbb{T}\ltimes_H}{\longrightarrow} & \mathbb{T}\text{-}Spec \\ \simeq\downarrow & & \downarrow\simeq \\ \prod_K \mathbb{Q}H/K\text{-}mod & \stackrel{\text{ind}}{\longrightarrow} & D\mathcal{A} \end{array} \quad \text{and} \quad \begin{array}{ccc} H\text{-}Spec & \stackrel{F_H[\mathbb{T},\cdot)}{\longrightarrow} & \mathbb{T}\text{-}Spec \\ \simeq\downarrow & & \downarrow\simeq \\ \prod_K \mathbb{Q}H/K\text{-}mod & \stackrel{\text{coind}}{\longrightarrow} & D\mathcal{A} \end{array}$$

are commutative up to natural isomorphism. □

Proof of 8.1.4: As usual we must record the units and counits of the adjunctions: we leave the verification of the triangular identities to the reader. We must first recall the natural exact sequence

$$0 \longrightarrow e_K M/ce_K M \longrightarrow [L_K, M] \longrightarrow \Sigma^{-1}ann(c, e_K M) \longrightarrow 0.$$

To describe the counit $\text{ind}(UM) \longrightarrow M$ suppose $M = (N \longrightarrow t_*^{\mathcal{F}} \otimes W)$. It is enough to describe the map of nubs since the domain is c-torsion and hence maps to zero in $t_*^{\mathcal{F}} \otimes W$, so the counit is necessarily consistent on vertices. The counit has Kth component the evident map

$$\Sigma([L_K, M]) \longrightarrow \Sigma(\Sigma^{-1}ann(c, e_K N)) \longrightarrow e_K N.$$

The unit $V \longrightarrow U\text{ind}V$ has as its Kth component the composite

$$V_K \longrightarrow (V_K)^H \longrightarrow [L_K, \text{ind}V] \cong (V_K)^H \oplus \Sigma(V_K)^H$$

where the first map is given by coset averages, and the second uses the fact that the short exact sequence for $[L_K, \text{ind}V]$ splits naturally in V.

Next, to describe the unit $M \longrightarrow \text{coind}(UM)$ it is obviously enough to discuss nubs, since the codomain has zero vertex. Its Kth component is

$$e_K N \longrightarrow e_K N/ce_K N \longrightarrow [L_K, M].$$

The counit $U\text{coind}(V) \longrightarrow V$ has Kth component the composite

$$[L_K, \text{coind}(V)] \longrightarrow (V_K)^H \longrightarrow V_K. \quad \square$$

8.2. The Lewis-May $\mathbb{T}$-fixed point functor.

The other elementary case is given by the adjoint pair

$$[\text{inf}_1^{\mathbb{T}} X, Y]^{\mathbb{T}} = [X, \Psi^{\mathbb{T}} Y],$$

where $\text{inf}_1^{\mathbb{T}} : \text{Spec} \longrightarrow \mathbb{T}\text{-Spec}$ is the inflation functor building in non-trivial representations, and $\Psi^{\mathbb{T}} : \mathbb{T}\text{-Spec} \longrightarrow \text{Spec}$ is the Lewis-May $\mathbb{T}$-fixed point functor. The more complicated Lewis-May K-fixed point functor is treated in Chapters 9 and 10.

By definition, inflation preserves suspension spectra, so that for any space X we have $\text{inf}_1^{\mathbb{T}} X = X$. Now an arbitrary spectrum X is rationally a wedge of spheres, and inflation preserves wedges so $\text{inf}_1^G X = S^0[\pi_*(X)]$. It is therefore straightforward to identify its algebraic model, and that of its right adjoint, the Lewis-May fixed point functor $\Psi^{\mathbb{T}}$.

PROPOSITION 8.2.1. *(a) The algebraic functor* $S^0\otimes : \mathbb{Q}_*\text{-mod} \longrightarrow \mathcal{A}$ *models the geometric completion functor in the sense that*

$$\begin{array}{ccc} \mathit{Spec} & \xrightarrow{\inf_1^{\mathbb{T}}} & \mathbb{T}\text{-}\mathit{Spec} \\ \simeq\downarrow & & \downarrow\simeq \\ D\mathbb{Q} & \xrightarrow{S^0\otimes} & D\mathcal{A} \end{array}$$

commutes up to natural isomorphism. In short, we have

$$\inf_1^{\mathbb{T}} V = S^0 \otimes V$$

at the level of derived categories.
(b) The algebraic functor $\Psi^{\mathbb{T}} = \mathrm{Hom}(S^0, \cdot) : \mathcal{A} \longrightarrow \mathbb{Q}_*\text{-mod}$ *models the geometric Lewis-May* $\mathbb{T}$*-fixed point functor in the sense that*

$$\begin{array}{ccc} \mathbb{T}\text{-}\mathit{Spec} & \xrightarrow{\Psi^{\mathbb{T}}} & \mathit{Spec} \\ \simeq\downarrow & & \downarrow\simeq \\ D\mathcal{A} & \xrightarrow{R\mathrm{Hom}(S^0,\cdot)} & D\mathbb{Q} \end{array}$$

commutes up to natural isomorphism. In short,

$$R\Psi^{\mathbb{T}} C = R\mathrm{Hom}(S^0, C)$$

models the Lewis-May $\mathbb{T}$*-fixed point functor on the homotopy category.*

REMARK 8.2.2. (i) Conventions common in topology would allow us to abbreviate $R\Psi^{\mathbb{T}}$ to $\Psi^{\mathbb{T}}$, provided it is clear the functor on derived categories is intended.
(ii) It is possible to give an analogous treatment for the K-fixed point functor, but the Hom functor involved is more sophisticated. We defer this to Section 24.4 of Part IV.

Proof: In topology we have $\inf_1^T X = S^0[\pi_*(X)]$. Since tensor product with V is exact, the algebraic inflation functor

$$V \longmapsto S^0 \otimes V$$

gives a model of the inflation functor.

LEMMA 8.2.3. *At the level of abelian categories, the right adjoint of the algebraic inflation functor is*

$$X \longmapsto \mathrm{Hom}(S^0, X)$$

where Hom *refers to morphisms in the standard model.* □

Since the algebraic inflation functor is exact, the adjunction at the level of abelian categories passes to derived categories as claimed. □

8.3. An algebraic model for geometric fixed points.

In this section and the next, we give an analysis of the geometric K-fixed point functor, regarded as a functor from $\mathbb{T}$-spectra to $\mathbb{T}/K$-spectra for a finite subgroup K. The functor $\Phi^{\mathbb{T}}$ is corresponds to taking the vertex, and is an integral part of our analysis. It therefore requires no further comment. This section motivates the definition of an algebraic model, and Section 8.4 proves it is indeed a model.

To minimize confusion in the coming discussion we let $\overline{\mathbb{T}} = \mathbb{T}/K$ and use bars to indicate reference to the ambient group $\overline{\mathbb{T}}$; when necessary, we denote the quotient map by $q : \mathbb{T} \longrightarrow \overline{\mathbb{T}}$. For example $\overline{\mathcal{F}}$ is the family of finite subgroups of $\overline{\mathbb{T}}$, and $\overline{H}$ is the image in $\overline{\mathbb{T}}$ of the subgroup H of $\mathbb{T}$. Note that the $\overline{\mathbb{T}}$-space $E\overline{\mathcal{F}}_+$ may be regarded as a $\mathbb{T}$-space, and as such it is $E\mathcal{F}_+$. Nonetheless, we shall use use the notation which best indicates which group is acting.

In one sense, the functor Φ^K is obvious from our construction. Indeed, we analyse X using the cofibre sequence

$$X \longrightarrow X \wedge \tilde{E}\mathcal{F} \longrightarrow X \wedge \Sigma E\mathcal{F}_+,$$

and if we apply Φ^K we obtain

$$\Phi^K X \longrightarrow \Phi^K(X \wedge \tilde{E}\mathcal{F}) \longrightarrow \Phi^K(X \wedge \Sigma E\mathcal{F}_+).$$

Now we use the fact that $X \wedge \tilde{E}\mathcal{F} = (\Phi^{\mathbb{T}} X) \wedge \tilde{E}\mathcal{F}$ so that $\Phi^K(X \wedge \tilde{E}\mathcal{F}) = (\Phi^{\mathbb{T}} X) \wedge \tilde{E}\overline{\mathcal{F}}$. On the other hand, we have conducted our analysis using the stable splitting

$$E\mathcal{F}_+ \simeq \bigvee_H E\langle H \rangle$$

where $E\langle H \rangle$ is the space which is the cofibre of the universal map $E[\subset H]_+ \longrightarrow E[\subseteq H]_+$. Since Φ^K is the extension of the K-fixed point space functor the basic fact is the following.

LEMMA 8.3.1. *For any finite subgroups H and K of $\mathbb{T}$, we have the equivalence*

$$(E\langle H \rangle)^K \simeq \begin{cases} E\langle \overline{H} \rangle & \text{if} \quad H \supseteq K \\ * & \text{if} \quad H \not\supseteq K \end{cases}$$

of based $\overline{\mathbb{T}}$-spaces.

Proof: The characterization of universal spaces by their fixed point spaces gives $(E\mathcal{H})^K = E\overline{\mathcal{H}}$ for any family $\mathcal{H}$ of subgroups of $\mathbb{T}$, where $\overline{\mathcal{H}} = \{\overline{H} \mid H \in \mathcal{H}\}$. □

Restated in the stable language, this states that, whenever $K \subseteq H$,

$$\Phi^K E\langle H \rangle \simeq E\langle \overline{H} \rangle.$$

The natural guess is now that the invariants for $\Phi^K(X \wedge E\mathcal{F}_+)$ are obtained from the sequence of spectra $X(H) = X \wedge E\langle H \rangle$ and their characteristic $\mathbb{Q}[c_H]$-modules $\pi_*^{\mathbb{T}}(X(H))$ by ignoring those terms with $H \not\supseteq K$. This turns out to be correct. To make sense of it, we need to observe the there is a natural identification of the rings of operations $\mathbb{Q}[c_H]$ and $\mathbb{Q}[c_{\overline{H}}]$.

The following observation will be used several times.

LEMMA 8.3.2. *If $K \subseteq H$ and Y is any $\mathbb{T}$-spectrum over H then Φ^K gives an isomorphism*

$$\Phi^K : [X,Y]^{\mathbb{T}}_* \xrightarrow{\cong} [\Phi^K X, \Phi^K Y]^{\overline{\mathbb{T}}}_*.$$

Proof: Since

$$[\Phi^K X, \Phi^K Y]^{\overline{\mathbb{T}}}_* = [X, \tilde{E}[\not\supseteq K] \wedge Y]^{\mathbb{T}}_*,$$

and we have the cofibre sequence

$$E[\not\supseteq K]_+ \longrightarrow S^0 \longrightarrow \tilde{E}[\not\supseteq K],$$

it is enough to show $E[\not\supseteq K]_+ \wedge Y \simeq *$. However $Y \simeq Y \wedge E\langle H \rangle$, and, by considering fixed points, $E[\not\supseteq K]_+ \wedge E\langle H \rangle$ is a contractible space. □

Using the fact that $\Phi^K E\langle H \rangle \simeq E\langle \overline{H} \rangle$, we obtain an identification of rings of operations.

COROLLARY 8.3.3. *Whenever $K \subseteq H$, Φ^K induces an isomorphism*

$$\mathbb{Q}[c_H] = [E\langle H \rangle, E\langle H \rangle]^{\mathbb{T}}_* \xrightarrow{\cong} [E\langle \overline{H} \rangle, E\langle \overline{H} \rangle]^{\overline{\mathbb{T}}}_* = \mathbb{Q}[c_{\overline{H}}]. \quad □$$

Furthermore, we can now confirm intuition.

COROLLARY 8.3.4. *If $K \subseteq H$ then Φ^K induces an isomorphism*

$$\pi^{\overline{\mathbb{T}}}_*((\Phi^K X)(\overline{H})) = \pi^{\mathbb{T}}_*(X(H))$$

of $\mathbb{Q}[c_H]$-modules.

Proof: We calculate

$$(\Phi^K X)(\overline{H}) = (\Phi^K X) \wedge E\langle \overline{H} \rangle = \Phi^K (X \wedge E\langle H \rangle) \simeq \Phi^K (X(H)). \quad □$$

This motivates the definition of the algebraic analogue. We consider the cofibre sequence

$$M \longrightarrow e(V_M) \xrightarrow{\hat{q}_M} f(\Sigma T_M)$$

and apply Φ^K to obtain

$$\Phi^K M \longrightarrow \Phi^K e(V_M) \xrightarrow{\Phi^K \hat{q}_M} \Phi^K f(\Sigma T_M).$$

Thus, it is enough to identify Φ^K on $e(V_M)$ and $f(T_M)$. Since $X(\mathbb{T}) = \Phi^{\mathbb{T}} X \wedge \tilde{E}\mathcal{F}$ and $\Phi^{\mathbb{T}} X = \Phi^{\overline{\mathbb{T}}} \Phi^K X$, we take $\Phi^K e(V_M) = \overline{e}(V_M)$; in other words, vertices are identical $V_{\Phi^K M} = V_M$. On the other hand, we take $\Phi^K f(T_M) = \overline{f}(\overline{T_M})$, where $T_{\Phi^K M} = \overline{T_M}$ is obtained by deleting the factors not containing K. More succinctly, we let e denote the idempotent of $\mathcal{O}_{\mathcal{F}}$ with support the set of subgroups containing K. Now view $\overline{\mathcal{O}}_{\overline{\mathcal{F}}}$ as $e\mathcal{O}_{\mathcal{F}}$ and take $\overline{T_M} = eT_M$. Also, once we have observed that $t^{\overline{\mathcal{F}}}_* \cong et^{\mathcal{F}}_*$ it is natural to take $\hat{q}_{\Phi^K M} = \overline{\hat{q}_M}$ to be $e\hat{q}_M$.

DEFINITION 8.3.5. Let $e \in \mathcal{O}_{\mathcal{F}}$ be the idempotent whose support is the set of finite subgroups containing K. Using the identifications $\overline{\mathcal{O}}_{\overline{\mathcal{F}}} \cong e\mathcal{O}_{\mathcal{F}}$ and $t^{\overline{\mathcal{F}}}_* \cong et^{\mathcal{F}}_*$, we define the functor

$$\Phi^K : \mathcal{A} \longrightarrow \overline{\mathcal{A}}$$

by

$$\Phi^K(N_M \xrightarrow{r_M} t_*^{\mathcal{F}} \otimes V_M) := (eN_M \xrightarrow{er_M} et_*^{\mathcal{F}} \otimes V_M),$$

or more briefly $\Phi^K M := eM$. □

8.4. Analysis of geometric fixed points.

We may now set about showing the algebraic functor of 8.3.5 has the desired properties.

THEOREM 8.4.1. *The algebraic functor Φ^K induces a functor*

$$\Phi^K : D\mathcal{A} \longrightarrow D\overline{\mathcal{A}}$$

so that the diagram

$$\begin{array}{ccc} \mathbb{T}\text{-}Spec & \xrightarrow{\Phi^K} & \overline{\mathbb{T}}\text{-}Spec \\ \simeq\downarrow & & \downarrow\simeq \\ D\mathcal{A} & \xrightarrow{\Phi^K} & D\overline{\mathcal{A}} \end{array}$$

commutes up to natural isomorphism.

Before we prove this, we should verify that the identifications

$$\overline{\mathcal{O}}_{\overline{\mathcal{F}}} \cong e\mathcal{O}_{\mathcal{F}} \text{ and } t_*^{\overline{\mathcal{F}}} \cong et_*^{\mathcal{F}}$$

correspond to suitable geometric statements. However, some care is necessary at this point: several naive expectations are false.

First, note that we may view a $\overline{\mathbb{T}}$-space Y as a $\mathbb{T}$-space, and hence we may form the dual $D_{\mathbb{T}}Y$ of Y as a $\mathbb{T}$-spectrum; we may also form the dual $\overline{\mathbb{T}}$-spectrum $D_{\overline{\mathbb{T}}}Y$ and view it as a $\mathbb{T}$-spectrum by building in representations. (The inflation functor functor building in representations is often written j_* where $j : \mathcal{U}^K \longrightarrow \mathcal{U}$ is the inclusion of universes, but in this section we shall follow the convention for suspension spectra and omit the notation j_*, which will always be made clear from the context. When notation is required for emphasis we write $\inf D_{\overline{\mathbb{T}}}Y$.) Furthermore, by regarding the $\overline{\mathbb{T}}$-map $D_{\overline{\mathbb{T}}}Y \wedge Y \longrightarrow S^0$ as a $\mathbb{T}$-map, we obtain a comparison map

$$\mu : D_{\overline{\mathbb{T}}}Y \longrightarrow D_{\mathbb{T}}Y$$

of $\mathbb{T}$-spectra. We warn that this is definitely not an equivalence for $Y = E\overline{\mathcal{F}}_+$; the easiest way to see this is to observe that the $\mathbb{T}$-equivariant homotopy groups of $\tilde{E}[\not\supseteq K] \wedge D_{\overline{\mathbb{T}}}E\overline{\mathcal{F}}_+$ are zero in positive degrees, whilst those of $\tilde{E}[\not\supseteq K] \wedge D_{\mathbb{T}}E\overline{\mathcal{F}}_+$ are not. This means that

$$\Phi^K DE\mathcal{F}_+ \not\simeq DE\overline{\mathcal{F}}_+.$$

However, the following positive result is what we need.

LEMMA 8.4.2. *The map*

$$\mu : D_{\overline{\mathbb{T}}}E\overline{\mathcal{F}}_+ \longrightarrow D_{\mathbb{T}}E\mathcal{F}_+$$

is an $\mathcal{F}$-equivalence.

Proof: Apply the forgetful functor to the construction of μ; since $E\mathcal{F}_+$ is H-equivariantly S^0 for all finite subgroups H, the lemma follows from the space level construction of the Spanier-Whitehead dual of a finite $\overline{\mathbb{T}}$-complex. $\square$

We may now find the geometric basis for the algebraic definition of Φ^K on nubs.

LEMMA 8.4.3. *There is a natural equivalence $eDE\mathcal{F}_+ \simeq D_{\overline{\mathbb{T}}}(E\overline{\mathcal{F}}_+) \wedge \tilde{E}[\not\supseteq K]$, of $\mathbb{T}$-spectra and hence*

$$X \wedge eD_{\mathbb{T}}E\mathcal{F}_+ \simeq \Phi^K X \wedge D_{\overline{\mathbb{T}}}E\overline{\mathcal{F}}_+ \wedge \tilde{E}[\not\supseteq K],$$

and

$$e\pi_*^{\mathbb{T}}(X \wedge D_{\mathbb{T}}E\mathcal{F}_+) \cong \pi_*^{\overline{\mathbb{T}}}(\Phi^K X \wedge D_{\overline{\mathbb{T}}}E\overline{\mathcal{F}}_+).$$

Proof: First observe that $E\mathcal{F}_+ = eE\mathcal{F}_+ \vee (1-e)E\mathcal{F}_+$, and that $(1-e)E\mathcal{F}_+ \simeq E[\not\supseteq K]_+$ whilst $eE\mathcal{F}_+ \simeq E\mathcal{F}_+ \wedge \tilde{E}[\not\supseteq K]$. Thus $E[\not\supseteq K]_+ \wedge eD_{\mathbb{T}}E\mathcal{F}_+ \simeq (1-e)E\mathcal{F}_+ \wedge eD_{\mathbb{T}}E\mathcal{F}_+ \simeq *$, so that $eD_{\mathbb{T}}E\mathcal{F}_+ \simeq \tilde{E}[\not\supseteq K] \wedge eD_{\mathbb{T}}E\mathcal{F}_+$.

We may therefore consider the natural map ν which is the composite

$$D_{\overline{\mathbb{T}}}E\overline{\mathcal{F}}_+ \wedge \tilde{E}[\not\supseteq K] \xrightarrow{\mu\wedge 1} D_{\mathbb{T}}E\mathcal{F}_+ \wedge \tilde{E}[\not\supseteq K] \longrightarrow eD_{\mathbb{T}}E\mathcal{F}_+ \wedge \tilde{E}[\not\supseteq K] \simeq eD_{\mathbb{T}}E\mathcal{F}_+.$$

Since μ is an $\mathcal{F}$-equivalence, it follows from the definition of e that ν is an $\mathcal{F}$-equivalence.

For the rest, we want to show that the map $\tilde{E}[\not\supseteq K] \wedge D_{\overline{\mathbb{T}}}E\overline{\mathcal{F}}_+ \longrightarrow eD_{\mathbb{T}}E\mathcal{F}_+$ induces a bijection of $[S^0, \cdot]_*^{\mathbb{T}}$. First note that a diagram

$$\begin{array}{ccc} & S^0 & \\ \swarrow & & \searrow \\ \tilde{E}[\not\supseteq K] \wedge D_{\overline{\mathbb{T}}}E\overline{\mathcal{F}}_+ & \longrightarrow & eD_{\mathbb{T}}E\mathcal{F}_+ \end{array}$$

corresponds under adjunction to a diagram

$$\begin{array}{ccc} & eE\mathcal{F}_+ & \\ \swarrow & & \searrow \\ \tilde{E}[\not\supseteq K] \wedge D_{\overline{\mathbb{T}}}E\overline{\mathcal{F}}_+ \wedge eE\mathcal{F}_+ & \longrightarrow & S^0. \end{array}$$

Now observe that

$$\tilde{E}[\not\supseteq K] \wedge D_{\overline{\mathbb{T}}}E\overline{\mathcal{F}}_+ \wedge eE\mathcal{F}_+ \simeq \tilde{E}[\not\supseteq K] \wedge eE\mathcal{F}_+ \simeq eE\mathcal{F}_+,$$

and the horizontal is induced by the collapse map $E\mathcal{F}_+ \longrightarrow S^0$. The cofibre of the horizontal thus becomes $(1-e)E\mathcal{F}_+$ when smashed with $E\mathcal{F}_+$, and hence the horizontal induces a bijection in $[eE\mathcal{F}_+, \cdot]_*^{\mathbb{T}}$ as required. $\square$

Proof of 8.4.1: For the first statement, we need only observe that the above definition gives an exact functor on $\mathcal{A}$; it then induces a functor on the category of dg $\mathcal{A}$-objects, and preserves homology isomorphisms. The existence of the functor then follows from the universal property of a category of fractions.

To see the algebraic Φ^K is compatible with the topological one, we apply the functor comparison theorem 7.3.1. Note first that Condition 1 holds. Indeed, it is obvious that both functors preserve cofibre sequences. It is also obvious that the algebraic functor preserves both parity and injectives; the analogous fact for the topological functor follows from the agreement of the functors on objects used to motivate the definition. This gives (2)(a), and finally we must check (2)(b), that if X and Y are each either $\mathcal{F}$-contractible or injective $\mathcal{F}$-spectra, then for any map $f : X \longrightarrow Y$ we have $\Phi^K(\pi_*^{\mathcal{A}}(f)) = \pi_*^{\overline{\mathcal{A}}}(\Phi^K f)$.

If Y is $\mathcal{F}$-contractible, we use the diagram

$$\begin{array}{ccc} [X,Y]_*^{\mathbb{T}} & \stackrel{\cong}{\longrightarrow} & \mathrm{Hom}(\pi_*(\Phi^{\mathbb{T}}X), \pi_*(\Phi^{\mathbb{T}}Y)) \\ \Phi^K \downarrow\cong & & \downarrow= \\ [\Phi^K X, \Phi^K Y]_*^{\overline{\mathbb{T}}} & \stackrel{\cong}{\longrightarrow} & \mathrm{Hom}(\pi_*(\Phi^{\overline{\mathbb{T}}}\Phi^K X), \pi_*(\Phi^{\overline{\mathbb{T}}}\Phi^K Y)), \end{array}$$

which commutes since $\Phi^{\overline{\mathbb{T}}}\Phi^K = \Phi^{\mathbb{T}}$. If Y is $\mathcal{F}$-free and injective we use the diagram

$$\begin{array}{ccc} [X,Y]_*^{\mathbb{T}} & \stackrel{\cong}{\longrightarrow} & \mathrm{Hom}(\pi_*^{\mathbb{T}}(X \wedge DE\mathcal{F}_+), \pi_*^{\mathbb{T}}(Y)) \\ \Phi^K \downarrow & & \downarrow \Phi^K \\ [\Phi^K X, \Phi^K Y]_*^{\overline{\mathbb{T}}} & \stackrel{\cong}{\longrightarrow} & \mathrm{Hom}(\pi_*^{\overline{\mathbb{T}}}(\Phi^K X \wedge DE\overline{\mathcal{F}}_+), \pi_*^{\overline{\mathbb{T}}}(\Phi^K Y)). \end{array}$$

We must explain why the diagram commutes. First note that, since Y is an $\mathcal{F}$-spectrum, e is defined as a self-map of Y and $eY \simeq Y \wedge \tilde{E}[\not\supseteq K]$ so that $[\Phi^K X, \Phi^K Y]_*^{\overline{\mathbb{T}}} = [X, eY]_*^{\mathbb{T}}$. The diagram would therefore commute if the right hand vertical were replaced by

$$\mathrm{Hom}(\pi_*^{\mathbb{T}}(X \wedge DE\mathcal{F}_+), \pi_*^{\mathbb{T}}(Y)) \longrightarrow \mathrm{Hom}(\pi_*^{\mathbb{T}}(X \wedge DE\mathcal{F}_+), e\pi_*^{\mathbb{T}}(Y)).$$

It therefore suffices to prove that multiplication by e on nubs corresponds to the geometric construction, which was 8.4.3 above. □

It seems worth recording one consequence of the above discussion.

LEMMA 8.4.4. *If $Y \longrightarrow I(Y) \longrightarrow J(Y)$ is an Adams resolution, then it remains so after taking geometric fixed points.* □

CHAPTER 9

Algebraic inflation and deflation.

Given a finite subgroup $K \subseteq \mathbb{T}$, with $\overline{\mathbb{T}} = \mathbb{T}/K$ we have the adjoint pair

$$\overline{\mathbb{T}}\text{-Spec} \underset{\Psi^K}{\overset{\inf}{\rightleftarrows}} \mathbb{T}\text{-Spec}$$

(with left adjoint on top), where Ψ^K denotes the Lewis-May K-fixed point functor, and inf is the inflation functor. This chapter constructs an adjoint pair of functors

$$\overline{\mathcal{A}}_t \underset{\Psi^K}{\overset{\inf}{\rightleftarrows}} \mathcal{A}_t$$

which pass to derived categories to give algebraic models of the topological functors. The fact that the algebraic functors do model the topological ones is proved in Chapter 10. Some readers may find the algebra natural in itself, but others will want to look in Section 10.4 for topological motivation. Finally, for those willing to invest in the machinery of Part IV, there is a fully satisfying reinterpretation in Section 24.4.

In Section 9.1 we define the inflation and deflation functors, and show they have the requisite adjointness properties on the category of $\mathcal{O}_{\mathcal{F}}$-modules, and in Section 9.2 this is extended to the torsion model category.

9.1. Algebraic inflation and deflation of $\mathcal{O}_{\mathcal{F}}$-modules.

Given a quotient homomorphism $G \longrightarrow G/N$ one obtains various change of group results, and we generically refer to functors from G/N-equivariant structure to G-equivariant structure as *inflation*, and to functors in the reverse direction as *deflation*. This section is devoted to a particular algebraic case of this which will be important in the analysis of Lewis-May fixed points.

We suppose given a finite subgroup K of $\mathbb{T}$ and follow the conventions of Section 8.3 on notation by using bars to indicate when the ambient group is the quotient $\overline{\mathbb{T}} = \mathbb{T}/K$. First, note that the quotient map $q : \mathbb{T} \longrightarrow \mathbb{T}/K = \overline{\mathbb{T}}$ induces a map of subgroups, and in particular a surjective map $q_* : \mathcal{F} \longrightarrow \overline{\mathcal{F}}$. Since any subgroup mapping to a fixed subgroup $\overline{H}$ of $\overline{\mathbb{T}}$ lies inside $q^{-1}(\overline{H})$, it is clear that the fibres of q_* are finite.

Accordingly, since $\mathcal{O}_{\mathcal{F}} = C(\mathcal{F}, \mathbb{Q}[c])$ the map q_* induces a map

$$q^* : \overline{\mathcal{O}}_{\overline{\mathcal{F}}} \longrightarrow \mathcal{O}_{\mathcal{F}},$$

and since the fibres of q_* are finite this extends to

$$q^* : t_*^{\overline{\mathcal{F}}} \longrightarrow t_*^{\mathcal{F}}.$$

Explicitly, q^* is the diagonal map on each factor in the sense that $q^*(e_{\overline{H}}) = \Sigma_{\overline{L}=\overline{H}} e_L$. Pullback along q^* induces a deflation functor

$$\mathrm{def} : \mathcal{O}_{\mathcal{F}}\text{-mod} \longrightarrow \overline{\mathcal{O}}_{\overline{\mathcal{F}}}\text{-mod}.$$

If we think of an $\mathcal{F}$-finite module M as corresponding to a sequence of modules $e_L M$, then the effect of deflation is to collect together all the summands $e_L M$ with $\overline{L} = \overline{H}$ and make this the summand $e_{\overline{H}}\mathrm{def}M$. After this section we shall often omit notation for deflation, since the underlying set is unchanged.

We also want to obtain an inflation functor in the reverse direction.

DEFINITION 9.1.1. The inflation functor

$$\inf : \overline{\mathcal{O}}_{\overline{\mathcal{F}}}\text{-mod} \longrightarrow \mathcal{O}_{\mathcal{F}}\text{-mod},$$

is defined by

$$\inf(M) = \mathcal{O}_{\mathcal{F}} \otimes_{\overline{\mathcal{O}}_{\overline{\mathcal{F}}}} M,$$

where $\mathcal{O}_{\mathcal{F}}$ is an $\overline{\mathcal{O}}_{\overline{\mathcal{F}}}$-module via q^*.

The word inflation is also suggestive of the construction, at least on $\overline{\mathcal{F}}$-finite torsion modules.

LEMMA 9.1.2. *(i) The inflation functor is exact.*
(ii) Inflation takes $\overline{\mathcal{F}}$-finite torsion modules to $\mathcal{F}$-finite torsion modules. If $\overline{N}$ is an $\overline{\mathcal{F}}$-finite torsion module then

$$\inf(\overline{N}) = \bigoplus_H e_{\overline{H}}\overline{N},$$

so the functor simply gives each summand $e_{\overline{H}}\overline{N}$ the multiplicity of the fibre of $q_*^{-1}(\overline{H})$.

Proof: Both parts follow from an analysis of $\mathcal{O}_{\mathcal{F}}$ as an $\overline{\mathcal{O}}_{\overline{\mathcal{F}}}$-module. Consider the subset

$$\overline{\phi}_i = \{\overline{H} \subseteq \overline{\mathbb{T}} \mid |(q^*)^{-1}(\overline{H})| = i\}$$

of $\overline{\mathcal{F}}$, and note that only finitely many of these sets are non-empty. Now observe that $e_{\overline{\phi}_i}\mathcal{O}_{\mathcal{F}}$ is a sum of i copies of $e_{\overline{\phi}_i}\overline{\mathcal{O}}_{\overline{\mathcal{F}}}$.

Part (i) follows since this shows that $\mathcal{O}_{\mathcal{F}}$ is a projective $\overline{\mathcal{O}}_{\overline{\mathcal{F}}}$-module. Part (ii) follows by tensoring the decomposition of $\mathcal{O}_{\mathcal{F}}$ with the module $\overline{N}$. □

Now let $\overline{\mathcal{E}}$ denote the set of Euler classes in $\overline{\mathcal{O}}_{\overline{\mathcal{F}}}$, as in 4.6.

LEMMA 9.1.3. *The image of $\overline{\mathcal{E}}$ under the map $q^* : \overline{\mathcal{O}}_{\overline{\mathcal{F}}} \longrightarrow \mathcal{O}_{\mathcal{F}}$ generates the same saturated multiplicative set as $\mathcal{E}$. Thus for any $\mathcal{O}_{\mathcal{F}}$-module M*

$$\overline{\mathcal{E}}^{-1}M \cong \mathcal{E}^{-1}M.$$

Proof: If $\overline{v} : \overline{\mathcal{F}} \longrightarrow \mathbb{Z}_{\geq 0}$ is a function with finite support, then so is $\overline{v}q_* : \mathcal{F} \longrightarrow \mathbb{Z}_{\geq 0}$, and $q^*c^{\overline{v}} = c^{\overline{v}q_*}$. Furthermore, any function $v : \mathcal{F} \longrightarrow \mathbb{Z}_{\geq 0}$ is dominated by one of the form $\overline{v}q_*$. □

COROLLARY 9.1.4. *(i) For any $\overline{\mathcal{O}}_{\overline{\mathcal{F}}}$-module $\overline{N}$*

$$\mathrm{inf}(\overline{\mathcal{E}}^{-1}\overline{N}) = \mathcal{E}^{-1}\mathrm{inf}(\overline{N}).$$

(ii) For any graded vector space V

$$\mathrm{inf}(t_*^{\overline{\mathcal{F}}} \otimes V) = t_*^{\mathcal{F}} \otimes V.$$

Proof: Part (ii) is a special case of Part (i) since $t_*^{\overline{\mathcal{F}}} = \overline{\mathcal{E}}^{-1}\overline{\mathcal{O}}_{\overline{\mathcal{F}}}$, and Part (i) is immediate by the associativity of the tensor product. □

We may now deduce the required adjunction.

PROPOSITION 9.1.5. *The functors* inf *and* def *are both left and right adjoint to each other. Thus if M is an $\mathcal{O}_{\mathcal{F}}$-module and $\overline{N}$ is an $\overline{\mathcal{O}}_{\overline{\mathcal{F}}}$-module we have isomorphisms*

$$\mathrm{Hom}(\mathrm{def}M, \overline{N}) = \mathrm{Hom}(M, \mathrm{inf}\overline{N}) \text{ and } \mathrm{Hom}(\mathrm{inf}\overline{N}, M) = \mathrm{Hom}(\overline{N}, \mathrm{def}M).$$

Proof: The second adjunction is the statement

$$\mathrm{Hom}_{\mathcal{O}_{\mathcal{F}}}(\mathcal{O}_{\mathcal{F}} \otimes_{\overline{\mathcal{O}}_{\overline{\mathcal{F}}}} \overline{N}, M) = \mathrm{Hom}_{\overline{\mathcal{O}}_{\overline{\mathcal{F}}}}(\overline{N}, q_* M).$$

For the first adjunction, observe that by the decomposition in 9.1.2 we have

$$\mathrm{Hom}_{\overline{\mathcal{O}}_{\overline{\mathcal{F}}}}(\mathcal{O}_{\mathcal{F}}, \overline{N}) \cong \mathcal{O}_{\mathcal{F}} \otimes_{\overline{\mathcal{O}}_{\overline{\mathcal{F}}}} \overline{N}.$$

The adjunction then follows from the standard adjunction

$$\mathrm{Hom}_{\mathcal{O}_{\mathcal{F}}}(M, \mathrm{Hom}_{\overline{\mathcal{O}}_{\overline{\mathcal{F}}}}(\mathcal{O}_{\mathcal{F}}, \overline{N})) = \mathrm{Hom}_{\overline{\mathcal{O}}_{\overline{\mathcal{F}}}}(q_* M, \overline{N}).$$

□

9.2. Inflation and its right adjoint on the torsion model category.

The use we make of the results of the previous section is in constructing functors to correspond to the inflation functor $\overline{\mathbb{T}}$-Spec $\longrightarrow$ $\mathbb{T}$-Spec building in representations, and its right adjoint. This right adjoint is the Lewis-May fixed point functor [**19**, II.4.4]. It is difficult to identify the Lewis-May fixed point functor directly, since it does not satisfy all the hypotheses of Theorem 7.3.1; instead we shall identify the algebraic inflation functor, which does satisfy the relevant conditions, and then identify the Lewis-May fixed point functor as its adjoint. To define the Lewis-May fixed point functor using the standard abelian models $\mathcal{A}$ and $\overline{\mathcal{A}}$, we need more sophisticated algebra introduced in Part IV. For the present we define it using the torsion models $\mathcal{A}_t$ and $\overline{\mathcal{A}}_t$.

DEFINITION 9.2.1. (i) The inflation functor

$$\mathrm{inf} : \overline{\mathcal{A}}_t \longrightarrow \mathcal{A}_t$$

is the termwise extension of the functor 9.1.1 on Hausdorff modules:

$$\mathrm{inf}(t_*^{\overline{\mathcal{F}}} \otimes V \longrightarrow T) = (t_*^{\mathcal{F}} \otimes V = \mathrm{inf}(t_*^{\overline{\mathcal{F}}} \otimes V) \longrightarrow \mathrm{inf}T).$$

We have seen in 9.1.2 this is an exact functor.
(ii) The algebraic Lewis-May fixed point functor

$$\Psi^K : \mathcal{A}_t \longrightarrow \overline{\mathcal{A}}_t$$

is defined on an object $M = (t_*^{\mathcal{F}} \otimes V \longrightarrow T)$ by taking

$$\Psi^K M = (t_*^{\overline{\mathcal{F}}} \otimes V \xrightarrow{q^* \otimes 1} t_*^{\mathcal{F}} \otimes V \longrightarrow T),$$

where $t_*^{\mathcal{F}}$ and T are regarded as $\overline{\mathcal{O}}_{\overline{\mathcal{F}}}$-modules by pullback along q^*. This definition extends in an obvious way to morphisms, and is visibly exact.

The reason for naming this after the Lewis-May fixed point functor will appear in Section 10.1 below, although the following proposition may seem sufficient explanation in view of [**19**, II.4.4].

PROPOSITION 9.2.2. *The algebraic Lewis-May fixed point functor*

$$\Psi^K : \mathcal{A}_t \longrightarrow \overline{\mathcal{A}}_t$$

is right adjoint to the inflation functor

$$\mathrm{inf} : \overline{\mathcal{A}}_t \longrightarrow \mathcal{A}.$$

Thus we have natural isomorphisms

$$\mathrm{Hom}(\mathrm{inf}\overline{N}, M) = \mathrm{Hom}(\overline{N}, \Psi^K M)$$

for any object $\overline{N}$ of $\overline{\mathcal{A}}_t$ and any object M of $\mathcal{A}_t$.

Proof: As usual it suffices to construct the unit

$$\eta' : \overline{N} \longrightarrow \Psi^K \mathrm{inf}\overline{N}$$

and the counit

$$\epsilon' : \mathrm{inf}\Psi^K M \longrightarrow M$$

and verify the triangular identities. Before we begin, notice that the map $q^* : t_*^{\overline{\mathcal{F}}} \longrightarrow t_*^{\mathcal{F}} = \mathrm{def} \circ \mathrm{inf}(t_*^{\overline{\mathcal{F}}})$ is the unit η of the second adjunction of 9.1.5.

For the unit, suppose $\overline{N} = (t_*^{\overline{\mathcal{F}}} \otimes V \xrightarrow{\overline{s}} \overline{T})$ and define η' by the diagram

$$\begin{array}{ccc}
t_*^{\overline{\mathcal{F}}} \otimes V & \xrightarrow{1} & t_*^{\overline{\mathcal{F}}} \otimes V \\
 & & \downarrow{\scriptstyle q^* \otimes 1} \\
\downarrow{\scriptstyle \overline{s}} & & t_*^{\mathcal{F}} \otimes V \\
 & & \downarrow{\scriptstyle \mathrm{inf}(\overline{s})} \\
\overline{T} & \xrightarrow{\eta} & \mathrm{def} \circ \mathrm{inf}(\overline{T})
\end{array}$$

where the lower horizontal is the unit of the adjunction of 9.1.5. The commutativity of the square states $\eta \circ \overline{s} = \mathrm{inf}(\overline{s}) \circ (q^* \otimes 1)$, which follows from the triangular identity of 9.1.5

together with naturality of η. For the counit suppose $M = (t_*^{\mathcal{F}} \otimes V \stackrel{s}{\longrightarrow} T)$ and define ϵ' by the diagram

$$\begin{array}{ccc} t_*^{\mathcal{F}} \otimes V & \stackrel{1}{\longrightarrow} & t_*^{\mathcal{F}} \otimes V \\ {\scriptstyle \inf(q^*)} \downarrow & & \\ \inf \circ \operatorname{def}(t_*^{\mathcal{F}}) \otimes V & & \downarrow {\scriptstyle s} \\ \downarrow & & \\ \inf \circ \operatorname{def}(T) & \stackrel{\epsilon}{\longrightarrow} & T \end{array}$$

where the lower horizontal is the counit of the adjunction of 9.1.5. The commutativity of the square states $s = \inf(s) \circ (\inf(q^*) \otimes 1)$, which follows from the triangular identity of 9.1.5 together with the naturality of ϵ. The triangular identities are immediate from those of 9.1.5. □

We remark that we cannot use a similar construction for a left adjoint of inf because of the role of q^*. We would need to know that the unit $t_*^{\mathcal{F}} \longrightarrow \inf \circ \operatorname{def}\, t_*^{\mathcal{F}}$ of the first adjunction of 9.1.5 was the inflation of some map $t_*^{\mathcal{F}} \longrightarrow \operatorname{def}\, t_*^{\mathcal{F}}$, which is not the case. In fact, inf does not have a left adjoint, even in the homotopy category, since it does not preserve arbitrary products.

EXAMPLE 9.2.3. *(Inflation does not preserve products.)* The claim is the algebraic counterpart of the fact that the natural map

$$\nu : \inf(\prod_n \Sigma^{-n} E\overline{\mathbb{T}}_+) \longrightarrow \prod_n (\inf \Sigma^{-n} E\overline{\mathbb{T}}_+)$$

is not an equivalence, where n runs through the natural numbers. This follows from the fact (see 10.1.2 below) that, when Y is a $\overline{\mathbb{T}}$-spectrum, $\pi_*^{\overline{\mathbb{T}}}(Y) = \pi_*^{\mathbb{T}}(Y \wedge \Psi^K S^0)$. Indeed, the Lewis-May fixed point spectrum $\Psi^K S^0$ is a wedge of S^0 and terms including $E\overline{\mathbb{T}}_+$. Thus if ν were an equivalence, then the natural map

$$\nu' : E\overline{\mathbb{T}}_+ \wedge \prod_n \Sigma^{-n} E\overline{\mathbb{T}}_+ \longrightarrow \prod_n E\overline{\mathbb{T}}_+ \wedge \Sigma^{-n} E\overline{\mathbb{T}}_+$$

would also be an equivalence. However, ν' is not an equivalence since its fibre is

$$\tilde{E}\overline{\mathbb{T}} \wedge \prod_n \Sigma^{-n} E\overline{\mathbb{T}}_+ .$$

The homotopy groups of this fibre are obtained from $\prod_n \Sigma^{1-n} \mathbb{I}$ by inverting c_1, and are thus nonzero. □

However we shall see later that inflation does have a left adjoint on suitable subcategories.

We end this section by briefly discussing the induced functors on derived categories. The point is that both inf : $\overline{\mathcal{A}}_t \longrightarrow \mathcal{A}_t$ and $\Psi^K : \mathcal{A}_t \longrightarrow \overline{\mathcal{A}}_t$ are exact functors, and hence the functors on dg categories that they induce preserve homology isomorphisms. They therefore induce functors inf : $D\overline{\mathcal{A}}_t \longrightarrow D\mathcal{A}_t$ and def : $D\mathcal{A}_t \longrightarrow D\overline{\mathcal{A}}_t$ on derived categories in the usual way. It is reasonable to use the notation inf and def, as we have done, because of the exactness of the functors on abelian categories.

It is also clear that inflation is defined as a functor $\overline{\mathcal{A}} \longrightarrow \mathcal{A}$ by applying inf termwise, and it is still exact by 9.1.2. It therefore induces a functor $D\overline{\mathcal{A}} \longrightarrow D\mathcal{A}$ on derived categories. It will be valuable to know that this corresponds with the original functor under the equivalence between standard and torsion model categories; this is obvious by the fibre construction of the equivalences $fib : dg\mathcal{A}_t \xrightarrow{\simeq} dg\mathcal{A}$ and $fib : dg\overline{\mathcal{A}}_t \xrightarrow{\simeq} dg\overline{\mathcal{A}}$ from 6.5.1.

LEMMA 9.2.4. *The diagram*

$$\begin{array}{ccc} D\overline{\mathcal{A}} & \xrightarrow{\text{inf}} & D\mathcal{A} \\ \simeq\downarrow & & \downarrow\simeq \\ D\overline{\mathcal{A}}_t & \xrightarrow{\text{inf}} & D\mathcal{A}_t \end{array}$$

commutes up to natural isomorphism. □

We shall identify the Lewis-May fixed point functor on the standard model in Section 24.4, but for homotopy level calculations we give an identification on objects in Section 10.4.

CHAPTER 10

Inflation, Lewis-May fixed points and quotients.

We begin in Section 10.1 by recording the formal properties of the inflation and Lewis-May fixed point functors in topology, and in Section 10.2 we deduce the facts we need later in the chapter. There are then two natural routes through the chapter: to simply obtain a model for the Lewis-May fixed point functor, continue with Section 10.3. There we apply the Functor Comparison Theorem 7.3.1 to deduce that the algebraic inflation functor models the topological one, and conclude that their right adjoints, the Lewis-May fixed point functors, also correspond. However this route is rather indirect, and some readers will prefer a more concrete approach. We therefore give a more direct analysis of the Lewis-May fixed point functor in Section 10.4, by considering the $\mathcal{F}$-free and $\mathcal{F}$-contractible pieces separately. In the same spirit, Section 10.5 makes the functor explicit in the derived category of the standard model, but does not approach morphisms at all. However, it does illustrate the behaviour of the functor, and it gives a splitting theorem.

The chapter ends with the very short Section 10.6, drawing attention to the fact that we have also modelled the quotient on sufficiently free spectra, since it is left adjoint to inflation.

10.1. The topological inflation and Lewis-May fixed point functors.

We continue to use the notation introduced in Section 8.3, with K being a finite subgroup of $\mathbb{T}$, and $q : \mathbb{T} \longrightarrow \overline{\mathbb{T}}$ the quotient map. We have explained in Section 7.2 that the inflation functor $\inf = \inf_{\overline{\mathbb{T}}}^{\mathbb{T}} : \overline{\mathbb{T}}\text{-Spec} \longrightarrow \mathbb{T}\text{-Spec}$ regarding a $\overline{\mathbb{T}}$-spectrum as $\mathbb{T}$-spectrum is the composite of the pull-back along the quotient $q : \mathbb{T} \longrightarrow \overline{\mathbb{T}}$ and the functor building in representations which are not K-fixed.

The purpose of this section is to provide the various facts we shall need about the topological inflation and Lewis-May fixed point functors. This will give us enough information to apply the Functor Comparison Theorem 7.3.1 to the algebraic and geometric inflation functors. Although we are first concerned with inflation, there are various points where it seems convenient to use properties of Lewis-May fixed points, so we discuss the two functors together. Because of its important place in the discussion, we include notation for the inflation functor throughout this section.

The fundamental fact [**19**, II.4.4] is that for any $\overline{\mathbb{T}}$-spectrum Y, there is an adjunction

$$[\inf_{\overline{\mathbb{T}}}^{\mathbb{T}} Y, X]_*^{\mathbb{T}} = [Y, \Psi^K X]_*^{\overline{\mathbb{T}}}.$$

The crudest implication, which we shall often use, is that, since S^0 is an inflation, $\pi_*^{\mathbb{T}}(X) = \pi_*^{\overline{\mathbb{T}}}(\Psi^K X)$. We shall also need to understand the fixed point functor on an inflated spectrum. Thus if we take $X = \inf Y$, the $\mathbb{T}$-equivariant identity map on the left corresponds to the unit $\eta : Y \longrightarrow \Psi^K \text{inf } Y$ of the adjunction. Accordingly, for any $\mathbb{T}$-spectrum Y we may consider the composite

$$\Psi^K(X) \wedge Y \longrightarrow \Psi^K(X) \wedge \Psi^K(\text{inf } Y) \longrightarrow \Psi^K(X \wedge \text{inf } Y).$$

The following curious lemma is the key to the analysis.

LEMMA 10.1.1. *If Y is a $\overline{\mathbb{T}}$-spectrum then the natural map*

$$\Psi^K(X) \wedge Y \longrightarrow \Psi^K(X \wedge \text{inf } Y)$$

is an equivalence of $\overline{\mathbb{T}}$-spectra.

Proof: We have a natural map which commutes with direct limits and cofibre sequences in the variable Y, so it is enough to verify it is an equivalence when Y is a cell, $\overline{\mathbb{T}}/\overline{H}_+$ for some subgroup $\overline{H} \subseteq \overline{\mathbb{T}}$. However, the lemma for any finite $\overline{\mathbb{T}}$-complex Y is clear by playing with adjunctions, since the $\overline{\mathbb{T}}$-dual of Y regarded as a $\mathbb{T}$-spectrum is the $\mathbb{T}$-dual. □

The particular case $X = S^0$ is often useful:

COROLLARY 10.1.2. *For any $\overline{\mathbb{T}}$-spectrum Y we have a natural equivalence*

$$\Psi^K(S^0) \wedge Y \xrightarrow{\simeq} \Psi^K \text{inf } Y,$$

where $\Psi^K S^0$ denotes the Lewis-May fixed point spectrum. □

This becomes most useful in conjunction with tom Dieck's calculation of $\Psi^K S^0$, which we recall in Example 10.5.6 below.

Now we turn to the behaviour of inflation, and begin by considering how basic cells behave. Recall that there are only finitely many subgroups L with $\overline{L} = \overline{H}$.

LEMMA 10.1.3. *If we inflate the basic $\overline{\mathbb{T}}$-cell $\sigma^0_{\overline{H}}$ to a $\mathbb{T}$-spectrum we find an equivalence*

$$\text{inf}\sigma^0_{\overline{H}} \simeq \bigvee_{\overline{L}=\overline{H}} \sigma^0_L$$

of $\mathbb{T}$-spectra.

Proof: Let $H = q^{-1}(\overline{H})$ and consider the quotient map $q : H \longrightarrow \overline{H}$. This induces a map $q^* : A(\overline{H}) \longrightarrow A(H)$, and for any $x \in A(\overline{H})$, whenever $L \subseteq H$, we have $\phi_L(q^*(x)) = \phi_{\overline{L}}(x)$. Thus in particular

$$q^*(e_{\overline{H}}) = \Sigma_{\overline{L}=\overline{H}} e_L.$$

The lemma follows. □

Since the the $\overline{\mathbb{T}}$-space $E\overline{\mathcal{F}}_+$ is $E\mathcal{F}_+$ when regarded as a $\mathbb{T}$-space (i.e. $\text{inf}(E\overline{\mathcal{F}}_+) \simeq E\mathcal{F}_+$), we immediately deduce the required consequence.

COROLLARY 10.1.4. *If $E\langle\overline{H}\rangle$ is regarded as a $\mathbb{T}$-spectrum we have an equivalence*

$$E\langle\overline{H}\rangle \simeq \bigvee_{\overline{L}=\overline{H}} E\langle L\rangle$$

of $\mathbb{T}$-spectra. □

It follows from this and the calculation of the self maps of $E\langle H\rangle$, that the inflation map

$$\mathbb{Q}[c_{\overline{H}}] = [E\langle\overline{H}\rangle, E\langle\overline{H}\rangle]_*^{\overline{\mathbb{T}}} \longrightarrow [E\langle\overline{H}\rangle, E\langle\overline{H}\rangle]_*^{\mathbb{T}} = \prod_{\overline{L}=\overline{H}} \mathbb{Q}[c_L]$$

is the diagonal inclusion $c_{\overline{H}} \longmapsto (c_L)_L$.

COROLLARY 10.1.5. *The inflation functor*

$$\overline{\mathcal{O}}_{\overline{\mathcal{F}}} = [E\overline{\mathcal{F}}_+, E\overline{\mathcal{F}}_+]_*^{\overline{\mathbb{T}}} \longrightarrow [E\mathcal{F}_+, E\mathcal{F}_+]_*^{\mathbb{T}} = \mathcal{O}_{\mathcal{F}}$$

induces the ring homomorphism q^*, where $q : \mathcal{F} \longrightarrow \overline{\mathcal{F}}$ is induced by the quotient map $\mathbb{T} \longrightarrow \overline{\mathbb{T}}$ (see Section 9.1). □

10.2. Inflation on objects.

From the facts summarized in the previous section it is quite easy to deduce the effect of inflation on objects, and that it agrees with the algebraic inflation as defined in 9.2.1. This will be used in the next section as an ingredient in showing the model is functorial.

Since inf is exact on $\mathcal{F}$-finite modules we deduce the following.

COROLLARY 10.2.1. *For an $\overline{\mathcal{F}}$-spectrum X*

$$\pi_*^{\mathbb{T}}(\mathrm{inf} X) = \mathrm{inf}\ \pi_*^{\overline{\mathbb{T}}}(X).$$

Proof: We apply inflation to an Adams resolution of X. Now use the fact that 10.1.4 also identifies inflations of maps between $\mathcal{F}$-injectives with their algebraic counterparts. We will give a detailed proof of a generalization of this immediately below. □

We would like to make the analogous statement for arbitrary spectra. The remaining obstacle is the identification of maps between injectives. The key to understanding this is a calculation.

First note that we have a projection map $\mathrm{inf} E\langle\overline{H}\rangle \longrightarrow E\langle H\rangle$, with adjoint $E\langle\overline{H}\rangle \longrightarrow \Psi^K E\langle H\rangle$. Similarly the injection map $E\langle H\rangle \longrightarrow \mathrm{inf} E\langle\overline{H}\rangle$ can be used to form

$$E\langle H\rangle \wedge \mathrm{inf} D_{\overline{\mathbb{T}}} E\langle\overline{H}\rangle \longrightarrow \mathrm{inf} E\langle\overline{H}\rangle \wedge \mathrm{inf} D_{\overline{\mathbb{T}}} E\langle\overline{H}\rangle \simeq \mathrm{inf}(E\langle\overline{H}\rangle \wedge D_{\overline{\mathbb{T}}} E\langle\overline{H}\rangle) \longrightarrow S^0$$

with adjoint $\mathrm{inf} D_{\overline{\mathbb{T}}} E\langle\overline{H}\rangle \longrightarrow D_{\mathbb{T}} E\langle H\rangle$; the adjoint of this is a map $D_{\overline{\mathbb{T}}} E\langle\overline{H}\rangle \longrightarrow \Psi^K(D_{\mathbb{T}} E\langle H\rangle)$.

LEMMA 10.2.2. *The natural maps described above give equivalences of $\overline{\mathbb{T}}$-spectra*

$$E\langle\overline{H}\rangle \xrightarrow{\simeq} \Psi^K E\langle H\rangle \text{ and } DE\langle\overline{H}\rangle \xrightarrow{\simeq} \Psi^K(DE\langle H\rangle)$$

Proof: Both facts follows from 8.3.2. The second begins with the calculation

$$[T, \Psi^K(D_{\mathbb{T}}E\langle H\rangle)]^{\overline{\mathbb{T}}} = [\inf(T), D_{\mathbb{T}}E\langle H\rangle]^{\mathbb{T}} = [\inf(T)\wedge E\langle H\rangle, S^0]^{\mathbb{T}} = [\inf(T)\wedge E\langle H\rangle, E\langle H\rangle]^{\mathbb{T}},$$

and continues by saying this is the same as

$$[T, D_{\overline{\mathbb{T}}}E\langle \overline{H}\rangle]^{\overline{\mathbb{T}}} = [T \wedge E\langle \overline{H}\rangle, S^0]^{\overline{\mathbb{T}}} = [T \wedge E\langle \overline{H}\rangle, E\langle \overline{H}\rangle]^{\overline{\mathbb{T}}}.$$

Under passage to geometric fixed points this is an isomorphism by 8.3.2. Since $E\langle H\rangle$ is a space over K, geometric fixed points coincides with Lewis-May fixed points. The reader should now check that this map coincides with that induced by the map described above. The first equivalence is slightly easier. □

COROLLARY 10.2.3. *Under the identification of $\mathbb{Q}[c_H]$ and $\mathbb{Q}[c_{\overline{H}}]$, for any map $f : X \longrightarrow Y$ of $\overline{\mathbb{T}}$-spectra with Y an $\overline{\mathcal{F}}$-spectrum, the map*

$$\inf(X) \wedge D_{\mathbb{T}}E\langle H\rangle \longrightarrow \inf(Y) \wedge D_{\mathbb{T}}E\langle H\rangle \simeq \inf(Y) \wedge E\langle H\rangle$$

induces the same map in $\pi_*^{\mathbb{T}}$ as

$$X \wedge D_{\overline{\mathbb{T}}}E\langle \overline{H}\rangle \longrightarrow Y \wedge D_{\overline{\mathbb{T}}}E\langle \overline{H}\rangle \simeq Y \wedge E\langle \overline{H}\rangle$$

in $\pi_*^{\overline{\mathbb{T}}}$.

Proof: We use the fact that $\pi_*^{\mathbb{T}}(Z) = \pi_*^{\overline{\mathbb{T}}}(\Psi^K Z)$ and $\Psi^K(\inf(Y) \wedge Z) \simeq Y \wedge \Psi^K Z$. We now obtain a diagram

$$\begin{array}{ccccc} X \wedge \Psi^K(D_{\mathbb{T}}E\langle H\rangle) & \longrightarrow & Y \wedge \Psi^K(E\langle H\rangle) \wedge \Psi^K(D_{\mathbb{T}}E\langle H\rangle) & \longrightarrow & Y \wedge \Psi^K(E\langle H\rangle) \\ \uparrow & & \uparrow & & \uparrow \\ X \wedge D_{\overline{\mathbb{T}}}E\langle \overline{H}\rangle & \longrightarrow & Y \wedge E\langle \overline{H}\rangle \wedge D_{\overline{\mathbb{T}}}E\langle \overline{H}\rangle & \longrightarrow & Y \wedge E\langle \overline{H}\rangle \end{array}$$

in which the first and last verticals are equivalences by 10.2.2 above. The commutativity of the left square is naturality of the equivalence $\Psi^K(\inf(Y) \wedge Z) \simeq Y \wedge \Psi^K Z$. For the right hand square, we first explain that the upper horizontal is the composite of

$$\Psi^K(E\langle H\rangle) \wedge \Psi^K(D_{\mathbb{T}}E\langle H\rangle) \longrightarrow \Psi^K(E\langle H\rangle \wedge D_{\mathbb{T}}E\langle H\rangle)$$

and the K fixed points of evaluation to S^0, which lifts uniquely to $E\langle H\rangle$. Commutativity now follows, since the square with Y omitted is the adjoint of the diagram

$$\begin{array}{ccccc} \inf\left(\Psi^K(E\langle H\rangle) \wedge \Psi^K(D_{\mathbb{T}}E\langle H\rangle)\right) & \longrightarrow & E\langle H\rangle \wedge D_{\mathbb{T}}E\langle H\rangle & \longrightarrow & E\langle H\rangle \\ \uparrow & & & & \uparrow \\ \inf(E\langle \overline{H}\rangle) \wedge D_{\mathbb{T}}E\langle H\rangle & & \longrightarrow & & \inf(E\langle \overline{H}\rangle) \end{array}$$

whose commutativity is clear from the definitions. □

PROPOSITION 10.2.4. *For an arbitrary $\overline{\mathbb{T}}$-spectrum X,*

$$\pi_*^{\mathbb{T}}(\inf(X) \wedge D_{\mathbb{T}}E\mathcal{F}_+) = \inf\left(\pi_*^{\overline{\mathbb{T}}}(X \wedge D_{\overline{\mathbb{T}}}E\overline{\mathcal{F}}_+)\right).$$

Hence also we have

$$\pi_*^{\mathcal{A}}(\inf(X)) = \inf\left(\pi_*^{\overline{\mathcal{A}}}(X)\right).$$

Proof: We have already seen this is true if X is injective. The result follows once we know that topological inflation corresponds to algebraic inflation for maps between two injective spectra. This is clear if both injectives are $\mathcal{F}$-spectra, so the remaining case is covered by considering a map

$$f : \tilde{E}\overline{\mathcal{F}} \longrightarrow \Sigma^n E\overline{\mathcal{F}}_+.$$

Since we are considering maps into a Hausdorff module it suffices to consider its idempotent parts. In other words we need to show that the map

$$\tilde{E}\mathcal{F} \wedge D_{\mathbb{T}}E\langle H\rangle \simeq \tilde{E}\overline{\mathcal{F}} \wedge e_H D_{\mathbb{T}} E\mathcal{F}_+ \longrightarrow \Sigma^n E\overline{\mathcal{F}}_+ \wedge e_H D_{\mathbb{T}} E\mathcal{F}_+ \simeq \Sigma^n E\langle H\rangle$$

induces the same as

$$\tilde{E}\overline{\mathcal{F}} \wedge D_{\overline{\mathbb{T}}}E\langle \overline{H}\rangle \simeq \tilde{E}\overline{\mathcal{F}} \wedge e_{\overline{H}} D_{\overline{\mathbb{T}}} E\overline{\mathcal{F}}_+ \longrightarrow \Sigma^n E\overline{\mathcal{F}}_+ \wedge e_{\overline{H}} D_{\overline{\mathbb{T}}} E\overline{\mathcal{F}}_+ \simeq \Sigma^n E\langle \overline{H}\rangle.$$

This is a special case of 10.2.3. □

We could at this point give a direct approach to the Lewis-May fixed point functor on objects. To emphasize the logical structure of the argument, we defer the direct approach to Section 10.4, but it is still recommended for motivation.

10.3. Correspondence of Algebraic and geometric inflation functors.

In this section we show that the algebraic inflation and Lewis-May fixed point functors defined in 9.2.1 model their topological counterparts.

THEOREM 10.3.1. *The algebraic inflation functor of 9.2.1 induces a functor*

$$\inf : D\mathcal{A}_t \longrightarrow D\overline{\mathcal{A}}_t$$

so that the diagram

$$\begin{array}{ccc} \overline{\mathbb{T}}\text{-}Spec & \stackrel{\inf}{\longrightarrow} & \mathbb{T}\text{-}Spec \\ \simeq\downarrow & & \downarrow\simeq \\ D\overline{\mathcal{A}}_t & \stackrel{\inf}{\longrightarrow} & D\mathcal{A}_t \end{array}$$

commutes.

Proof: The existence of the functor on derived categories is immediate from the fact the algebraic functor $\overline{\mathcal{A}}_t \longrightarrow \mathcal{A}_t$ of Section 9.1 is exact.

We want to apply the Functor Comparison Theorem 7.3.1 to the two functors $\overline{\mathbb{T}}$-Spec $\longrightarrow$ $D\mathcal{A}$. We are therefore implicitly using the equivalence $D\mathcal{A}_t \stackrel{\simeq}{\longrightarrow} D\mathcal{A}$ to identify $D\mathcal{A}_t$ and $D\mathcal{A}$.

We work through Condition 1, starting with the topological functor $\overline{\mathbb{T}}$-Spec $\longrightarrow$ $\mathbb{T}$-Spec. It certainly preserves triangles. Enough topological injectives are obtained from wedges of suspensions of $\tilde{E}\overline{\mathcal{F}}$ and the various spectra $E\langle \overline{H}\rangle$. The inflation of $\tilde{E}\overline{\mathcal{F}}$ is $\tilde{E}\mathcal{F}$ and, by

10.1.4, the inflation of $E\langle\overline{H}\rangle$ is $\bigvee_L E\langle L\rangle$ where the wedge is over the subgroups L with $\overline{L}=\overline{H}$. The fact that it preserves objects of pure parity is immediate from 10.2.4.

The algebraic functor $D\overline{\mathcal{A}}_t \longrightarrow D\mathcal{A}_t$ preserves triangles by construction. It preserves injectives by 9.1.2. The fact that it preserves objects of pure parity is immediate from exactness of inflation as manifested in 9.2.4.

We now turn to Condition 2 of 7.3.1. The definition of the algebraic inflation functor was designed to agree with the topological one on enough injectives, as one sees from 10.2.4. Finally it remains to prove that our functors agree for maps into injectives, and we recall that 9.1.2 showed the algebraic inflation functor on the standard model was also termwise application of inflation. Suppose that X and Y are $\overline{\mathbb{T}}$-spectra. First consider the case when Y is $\overline{\mathcal{F}}$-contractible. We use the diagram

$$\begin{array}{ccccc} [X,Y]^{\mathbb{T}}_* & \stackrel{\cong}{\longrightarrow} & \mathrm{Hom}(\pi^{\mathcal{A}}_*(X),\pi^{\mathcal{A}}_*(Y)) & \cong & \mathrm{Hom}(\pi_*(\Phi^{\mathbb{T}}X),\pi_*(\Phi^{\mathbb{T}}Y)) \\ \mathrm{inf}\uparrow\cong & & \uparrow\mathrm{inf} & & \uparrow= \\ [X,Y]^{\overline{\mathbb{T}}}_* & \stackrel{\cong}{\longrightarrow} & \mathrm{Hom}(\pi^{\overline{\mathcal{A}}}_*(X),\pi^{\overline{\mathcal{A}}}_*(Y)) & \cong & \mathrm{Hom}(\pi_*(\Phi^{\overline{\mathbb{T}}}X),\pi_*(\Phi^{\mathbb{T}}Y)). \end{array}$$

The diagram commutes since the horizontals factor through passage to total geometric fixed points. Both algebraic and topological inflation functors are the identity on vertices.

Next we suppose that Y is an injective $\overline{\mathcal{F}}$-spectrum, and use the diagram

$$\begin{array}{ccccc} [X,Y]^{\mathbb{T}}_* & \stackrel{\cong}{\longrightarrow} & \mathrm{Hom}(\pi^{\mathcal{A}}_*(X),\pi^{\mathcal{A}}_*(Y)) & \cong & \mathrm{Hom}(\pi^{\mathbb{T}}_*(X\wedge DE\mathcal{F}_+),\pi^{\mathbb{T}}_*(Y)) \\ \mathrm{inf}\uparrow & & \uparrow\mathrm{inf} & & \uparrow \\ [X,Y]^{\overline{\mathbb{T}}}_* & \stackrel{\cong}{\longrightarrow} & \mathrm{Hom}(\pi^{\overline{\mathcal{A}}}_*(X),\pi^{\overline{\mathcal{A}}}_*(Y)) & \cong & \mathrm{Hom}(\pi^{\overline{\mathbb{T}}}_*(X\wedge DE\overline{\mathcal{F}}_+),\pi^{\overline{\mathbb{T}}}_*(Y)). \end{array}$$

We must explain why the diagram commutes. In other words we must show that if $\pi^{\overline{\mathcal{A}}}_*(X)=(\overline{N}\longrightarrow t^{\overline{\mathcal{F}}}_*\otimes V)$ and the map $f:X\longrightarrow Y$ corresponds to $\theta:\overline{N}\longrightarrow\pi^{\overline{\mathbb{T}}}_*(Y)$, so that $\theta=\pi^{\overline{\mathbb{T}}}_*(X\wedge DE\overline{\mathcal{F}}_+\longrightarrow Y\wedge DE\overline{\mathcal{F}}_+\simeq Y)$, then $\mathrm{inf}(\theta)$ is $\pi^{\mathbb{T}}_*(X\wedge DE\mathcal{F}_+\longrightarrow Y\wedge DE\mathcal{F}_+\simeq Y)$. A map into a Hausdorff $\mathcal{O}_{\mathcal{F}}$-module (such as $\pi^{\mathbb{T}}_*(Y)$) is determined by its idempotent parts. It is thus sufficient to show that $\pi^{\mathbb{T}}_*(e_H(X\wedge DE\mathcal{F}_+)\longrightarrow e_HY)$ is the same as $\pi^{\overline{\mathbb{T}}}_*(e_{\overline{H}}(X\wedge DE\overline{\mathcal{F}}_+)\longrightarrow e_{\overline{H}}Y)$. Since $e_HX\wedge DE\mathcal{F}_+\simeq X\wedge e_HDE\mathcal{F}_+\simeq X\wedge DE\langle H\rangle$ and similarly $e_{\overline{H}}X\wedge DE\overline{\mathcal{F}}_+\simeq X\wedge DE\langle\overline{H}\rangle$, this follows from 10.2.3. □

It is now formal to deduce the correspondence between the algebraic and topological Lewis-May fixed point functors.

THEOREM 10.3.2. *The algebraic Lewis-May K-fixed point functor Ψ^K of 9.2.1 induces a functor*

$$\Psi^K : D\mathcal{A}_t \longrightarrow D\overline{\mathcal{A}}_t$$

so that the diagram

$$\begin{array}{ccc} \mathbb{T}\text{-}Spec & \stackrel{\Psi^K}{\longrightarrow} & \overline{\mathbb{T}}\text{-}Spec \\ \simeq\downarrow & & \downarrow\simeq \\ D\mathcal{A}_t & \stackrel{\Psi^K}{\longrightarrow} & D\overline{\mathcal{A}}_t \end{array}$$

commutes.

Proof: The existence of the functor on derived categories is immediate from the fact that Ψ^K is an exact functor $\mathcal{A}_t \longrightarrow \overline{\mathcal{A}}_t$.

We have seen that the algebraic Lewis-May fixed point functor is right adjoint to inflation at the level of abelian categories. The unit and counit of the adjunction induce maps of derived categories, and the triangular identities continue to hold. Since the topological Lewis-May fixed point functor is right adjoint to inflation by [**19**, II.4.4], and since we have seen in 10.3.1 that the left adjoints correspond, the right adjoints also correspond. □

10.4. A direct approach to the Lewis-May fixed point functor.

In the previous section we completed the proof that the algebraic Lewis-May K-fixed point functor corresponds to the geometric one, and for logical purposes all further discussion could take place in the algebraic models. However, we think it is important to make the link between the algebraic and geometric forms of the functor more direct, and that is the purpose of this section. This can be viewed as belated motivation for the definition of the algebraic Lewis-May fixed point functor of 9.2.1, and can be used to give an alternative proof of Theorem 10.3.2 for objects. However, we defer a functorial construction of the Lewis-May fixed point functor on the standard model until Section 24.4, when we may apply the more sophisticated algebraic techniques of Part IV.

The construction of the model is rather natural if viewed correctly. The point is that (in the language of Appendix A) our decomposition $X(\mathcal{F}) \simeq \bigvee_H X(H)$ follows the decomposition of a Mackey functor as an $h\mathcal{SB}$-module. Thus

$$\pi_*^H(X) = \bigoplus_{L \subseteq H} \pi_*^H(X(L)).$$

It follows that when $H \supseteq K$ we may calculate $\pi_*^{\overline{H}}(\Psi^K X) = \pi_*^H(X)$. We see that, whatever the subgroup $\overline{H}$, $\pi_*^{\overline{H}}(\Psi^K X)$ contains the summands for $X(L)$ for all $L \subseteq K$. The end result is that when we decompose $(\Psi^K X)(\mathcal{F})$ as a wedge of spectra $(\Psi^K X)(\overline{H})$, the term $(\Psi^K X)(\overline{1})$ contains information equivalent to $\bigvee_{L \subseteq K} X(L)$, and in general the term $(\Psi^K X)(\overline{H})$ contains information equivalent to $X(L)$ for all L with $\overline{L} = \overline{H}$.

Once again we need to give a context in which to make this meaningful. The basic idea is that there is a natural way of identifying $\mathbb{Q}[c_H]$ and $\mathbb{Q}[c_{\overline{H}}]$. Indeed, we gave one identification in 8.3.3 in terms of geometric fixed points; for our present purpose we must show the Φ^K fixed point map $\mathbb{Q}[c_H] \longrightarrow \mathbb{Q}[c_{\overline{H}}]$ coincides with passage to Lewis-May fixed points.

LEMMA 10.4.1. *If H is a finite subgroup containing K, then passage to Lewis-May K-fixed points defines the isomorphism*

$$\mathbb{Q}[c_H] = [E\langle H\rangle, E\langle H\rangle]_*^{\mathbb{T}} \xrightarrow{\cong} [E\langle\overline{H}\rangle, E\langle\overline{H}\rangle]_*^{\mathbb{T}} = \mathbb{Q}[c_{\overline{H}}]$$

also described as passage to geometric fixed points in 8.3.3.

Proof: Since $E\langle H\rangle$ is concentrated over K in the sense that $E\langle H\rangle \simeq E\langle H\rangle \wedge \tilde{E}[\not\supseteq K]$, Lewis-May fixed points coincide with Φ^K fixed points. □

Note that by 10.1.5 the inflation map

$$\mathbb{Q}[c_{\overline{H}}] = [E\langle\overline{H}\rangle, E\langle\overline{H}\rangle]^{\overline{\mathbb{T}}}_* \longrightarrow [E\langle\overline{H}\rangle, E\langle\overline{H}\rangle]^{\mathbb{T}}_* = \prod_{\overline{L}=\overline{H}} \mathbb{Q}[c_L]$$

is the diagonal inclusion $c_{\overline{H}} \longmapsto (c_L)_L$. This gives sense to the following consequence.

COROLLARY 10.4.2. *If H is a finite subgroup of $\mathbb{T}$ containing K then*

$$\pi^{\overline{\mathbb{T}}}_*((\Psi^K X)(\overline{H})) \cong \bigoplus_{\overline{L}=\overline{H}} \pi^{\mathbb{T}}_*(X(L)). \quad \square$$

The curious Lemma 10.1.1 also lets us deduce the $\mathcal{F}$-contractible part.

COROLLARY 10.4.3. *For any $\mathbb{T}$-spectrum X we have an equivalence*

$$\Phi^{\overline{\mathbb{T}}}(\Psi^K X) \simeq \Phi^{\mathbb{T}} X.$$

In particular

$$\pi^{\overline{\mathbb{T}}}_*(\Psi^K(X) \wedge \tilde{E}\overline{\mathcal{F}}) \cong \pi^{\mathbb{T}}_*(X \wedge \tilde{E}\mathcal{F}).$$

Proof: Letting $Y = \tilde{E}\overline{\mathcal{F}}$ in 10.1.1, we find that $\Psi^K(X) \wedge \tilde{E}\overline{\mathcal{F}} \simeq \Psi^K(X \wedge \tilde{E}\overline{\mathcal{F}})$. Since $\tilde{E}\overline{\mathcal{F}}$ regarded as a $\mathbb{T}$-space is $\tilde{E}\mathcal{F}$, the result follows. $\square$

The result may seem peculiar to begin with, so we outline an alternative argument applying when suitable Euler classes are defined. We note that $\pi^{\overline{\mathbb{T}}}_*(\Psi^K(X) \wedge \tilde{E}\overline{\mathcal{F}})$ is the localization of $\pi^{\overline{\mathbb{T}}}_*(\Psi^K X) = \pi^{\mathbb{T}}_*(X)$ so as to invert Euler classes of all representations $\overline{V}$ of $\overline{\mathbb{T}}$ with $\overline{V}^{\overline{\mathbb{T}}} = 0$. On the other hand $\pi^{\mathbb{T}}_*(X \wedge \tilde{E}\mathcal{F})$ is the localization so as to invert all representations V of $\mathbb{T}$ with $V^{\mathbb{T}} = 0$. However, when V is one dimensional, the Euler class of V divides the Euler class of $V^{\otimes n}$ and if K is of order n then $V^{\otimes n}$ is a representation of $\overline{\mathbb{T}}$. The two localizations are therefore equal.

We now complete the alternative approach to Lewis-May fixed points by checking directly that the algebraic construction is correct on objects.

LEMMA 10.4.4. *The algebraic Lewis-May fixed point functor 9.2.1 is compatible with the geometric one on objects in the sense that*

$$\Psi^K(p(X)) \simeq \overline{p}(\Psi^K X)$$

for any $\mathbb{T}$-spectrum X.

Proof: The statement is true when X is $\mathcal{F}$-contractible by 10.4.3 and when X is $\mathcal{F}$-free by 10.4.2.

However, because the definitions on torsion and vertices are so different, the core of the matter is that the structure maps are correct. For this we consider the diagram

$$\begin{array}{ccccc}
\Psi^K(X)\wedge DE\overline{\mathcal{F}}_+\wedge \tilde{E}\overline{\mathcal{F}} & \longrightarrow & \Psi^K(X)\wedge DE\overline{\mathcal{F}}_+\wedge \Sigma E\overline{\mathcal{F}}_+ & \stackrel{\simeq}{\longleftarrow} & \Psi^K(X)\wedge \Sigma E\overline{\mathcal{F}}_+ \\
\simeq\downarrow & & \simeq\downarrow & & \\
\Psi^K(X\wedge DE\overline{\mathcal{F}}_+)\wedge \tilde{E}\overline{\mathcal{F}} & \longrightarrow & \Psi^K(X\wedge DE\overline{\mathcal{F}}_+)\wedge \Sigma E\overline{\mathcal{F}}_+ & & \\
\downarrow & & \downarrow & & \downarrow \\
\Psi^K(X\wedge DE\mathcal{F}_+)\wedge \tilde{E}\overline{\mathcal{F}} & \longrightarrow & \Psi^K(X\wedge DE\mathcal{F}_+)\wedge \Sigma E\overline{\mathcal{F}}_+ & & \\
\simeq\downarrow & & \simeq\downarrow & & \\
\Psi^K(X\wedge DE\mathcal{F}_+\wedge \tilde{E}\mathcal{F}) & \longrightarrow & \Psi^K(X\wedge DE\mathcal{F}_+\wedge \Sigma E\mathcal{F}_+) & \stackrel{\simeq}{\longleftarrow} & \Psi^K(X\wedge \Sigma E\mathcal{F}_+)
\end{array}$$

in which the equivalences come from 10.1.1. Since the top horizontal is q and the bottom horizontal is $\overline{q}$, and the right hand vertical is an equivalence, it remains to remark that the central vertical induces inf in homotopy.

This in turn follows from the fact that $DE\overline{\mathcal{F}}_+ \longrightarrow \Psi^K(DE\mathcal{F}_+)$ induces $q^*: \mathcal{O}_{\overline{\mathcal{F}}} \longrightarrow \mathcal{O}_{\mathcal{F}}$, as is clear from the fact that it is two adjunctions from the $\mathbb{T}$-map $E\mathcal{F}_+ \simeq E\mathcal{F}_+\wedge DE\overline{\mathcal{F}}_+ \longrightarrow S^0$. □

10.5. The homotopy type of Lewis-May fixed points.

It is useful to identify the behaviour of the fixed point functor on objects, using the fact that objects of $D\mathcal{A}$ correspond to objects of $\mathcal{A}$ in the standard model. This is entirely based on a homotopy level analysis, and we defer discussion of a functorial construction of the Lewis-May fixed point functor on the standard model until Section 24.4.

We suppose given an object $M = (N \stackrel{\beta}{\longrightarrow} t_*^{\mathcal{F}} \otimes V)$ of the standard model $\mathcal{A}$. Letting I denote the image of β we consider the modules associated to M by the exact sequences

$$0 \longrightarrow K \longrightarrow N \longrightarrow I \longrightarrow 0 \text{ and } 0 \longrightarrow I \longrightarrow t_*^{\mathcal{F}} \otimes V \longrightarrow C \longrightarrow 0;$$

the torsion part is then given by $T = K \oplus \Sigma^{-1}C$. The definition is based on the fact that the torsion part of the Lewis-May fixed point object $\Psi^K M$ is simply T regarded as an $\mathcal{O}_{\overline{\mathcal{F}}}$-module by pullback along q^*, and its vertex is the same as that for M.

We shall make a definition of an object $\Psi_c^K M$ of $\overline{\mathcal{A}}$, which will be of the homotopy type of $\Psi^K M$; the notation for the associated modules will be indicated systematically by the subscript c. Thus $\Psi_c^K M$ has torsion module $T_c = q^*T$, and vertex $V_c = V$. From now on we regard all $\mathcal{O}_{\mathcal{F}}$-modules as $\mathcal{O}_{\overline{\mathcal{F}}}$-modules, and omit the notation q^*. We therefore need to define an $\mathcal{O}_{\overline{\mathcal{F}}}$-homomorphism $s_c : t_*^{\overline{\mathcal{F}}} \otimes V \longrightarrow \Sigma T$; it is natural to use the composite

$$t_*^{\overline{\mathcal{F}}} \otimes V \stackrel{\inf\otimes 1}{\longrightarrow} t_*^{\mathcal{F}} \otimes V \stackrel{\beta}{\longrightarrow} \Sigma T.$$

This gives a map of short exact sequences

$$\begin{array}{ccccccccc}
0 & \longrightarrow & t_*^{\overline{\mathcal{F}}} \otimes V & \longrightarrow & t_*^{\mathcal{F}} \otimes V & \longrightarrow & t_*^{\mathcal{F}}/t_*^{\overline{\mathcal{F}}} \otimes V & \longrightarrow & 0 \\
 & & \downarrow & & \downarrow & & \downarrow & & \\
0 & \longrightarrow & C & \longrightarrow & C & \longrightarrow & 0 & \longrightarrow & 0
\end{array}.$$

In particular, we note that although $T_c = T$, the contributions from the kernel and cokernel of the basing map are quite different: C_c is bigger than C and K_c is correspondingly smaller than K.

Now consider the map $\inf \otimes 1 : t_*^{\mathcal{F}} \otimes V \longrightarrow t_*^{\overline{\mathcal{F}}} \otimes V$; it is injective and we view it as an inclusion, so that we may take $I_c = (t_*^{\overline{\mathcal{F}}} \otimes V) \cap I$. Thus we have an exact sequence

$$0 \longrightarrow I_c \longrightarrow I \longrightarrow (t_*^{\mathcal{F}}/t_*^{\overline{\mathcal{F}}}) \otimes V \longrightarrow D_c \longrightarrow 0.$$

Now let $N' = \beta^{-1}(I_c)$, so that we have an exact sequence

$$0 \longrightarrow K \longrightarrow N' \longrightarrow I_c \longrightarrow 0.$$

DEFINITION 10.5.1. We define the crude Lewis-May fixed point object of an object $M = (N \stackrel{\beta}{\longrightarrow} t_*^{\mathcal{F}} \otimes V)$ of $\mathcal{A}$ by taking $\Psi_c^K M = (N_c \stackrel{\beta_c}{\longrightarrow} t_*^{\overline{\mathcal{F}}} \otimes V)$, where $N_c = \Sigma^{-1} D_c \oplus N'$ and β_c is the composite $N_c = \Sigma^{-1} D_c \oplus N' \longrightarrow N' \longrightarrow I_c \longrightarrow t_*^{\overline{\mathcal{F}}} \otimes V$.

REMARK 10.5.2. With the above definition we have a splitting

$$\Psi_c^K M \cong f(\Sigma^{-1} D_c) \oplus \Lambda^K M,$$

with D_c injective. Note in particular that, even if M is even, $\Psi_c^K M$ will not usually be even. The fact that the splitting is not canonical is the reason this construction is not sufficient to construct the algebraic Lewis-May fixed points as a functor. □

We should begin by observing that $\Psi_c^K M$ is actually an object of $\overline{\mathcal{A}}$.

LEMMA 10.5.3. *The kernel and cokernel of the map β_c just defined are both $\overline{\mathcal{F}}$-finite and torsion, and hence $\Psi_c^K M$ is an object of $\overline{\mathcal{A}}$.*

Proof: By construction we see that, since C_c is injective, $C = C_c \oplus D_c$ and also $K_c = K \oplus \Sigma^{-1} D_c$. This establishes that K_c and C_c are $\overline{\mathcal{F}}$-finite and torsion. □

It now makes sense to claim that the Ψ_c^K construction describes Lewis-May fixed points.

PROPOSITION 10.5.4. *The Ψ_c^K construction gives the Lewis-May fixed point functor on objects in the sense that if we identify objects of $D\mathcal{A}$ up to isomorphism with those of $\mathcal{A}$, then*

$$\Psi^K M \simeq \Psi_c^K M.$$

Proof: The object $N_c \stackrel{\beta_c}{\longrightarrow} t_*^{\overline{\mathcal{F}}} \otimes V$ is determined by the image of β_c, and its extension class in $\mathrm{Ext}(im(\beta_c), ker(\beta_c)) = \mathrm{Ext}(I_c, K \oplus \Sigma^{-1} D_c)$. It follows from the construction that the image of β_c is the kernel of the composite $t_*^{\overline{\mathcal{F}}} \otimes V \longrightarrow t_*^{\mathcal{F}} \otimes V \longrightarrow \Sigma T$, as required. Since D_c is injective, the extension class of N_c is the pullback of that of N in $\mathrm{Ext}(I, K)$ by construction. □

Combining 10.3.2, 10.5.2 and 10.5.4 we obtain a topological splitting result.

COROLLARY 10.5.5. *There is an unnatural splitting*

$$\Psi^K X \simeq \Sigma^{-1}\Delta^K X \vee \Lambda^K X$$

with $\Delta^K X$ an injective $\overline{\mathcal{F}}$-spectrum. □

EXAMPLE 10.5.6. Consider the algebraic 0-sphere, $L_{\mathbb{T}} = (\mathcal{O}_{\mathcal{F}} \longrightarrow t_*^{\mathcal{F}})$ of 5.8.1. We find $C = \Sigma^2\mathbb{I}$, $K = 0$ and $I = \mathcal{O}_{\mathcal{F}}$. Thus $I_c = \overline{\mathcal{O}}_{\overline{\mathcal{F}}}$, $C_c = \Sigma^2\overline{\mathbb{I}}$, and $K_c = \Sigma I/I_c$. More explicitly, I/I_c is a sum of injectives $\mathbb{I}(\overline{H})$, with multiplicity one less than the number of subgroups L with $\overline{L} = \overline{H}$. For example, if K is of order 2 there is a single summand $\mathbb{I}(\overline{H})$ whenever $\overline{H}$ is of odd order.

This should be compared with the equivalence of [**19**, V.11.1], which gives

$$\Psi^K S^0 \simeq \bigvee_{\Lambda \subseteq K} E[\cap K/\Lambda = 1]_+/(K/\Lambda),$$

where the quotient of the universal K/Λ-free $\mathbb{T}/\Lambda$-space, $E[\cap K/\Lambda = 1]_+$ by K/Λ is viewed as a $\overline{\mathbb{T}}$-space using the isomorphism $\overline{\mathbb{T}} = \mathbb{T}/K \cong (\mathbb{T}/\Lambda)/(K/\Lambda)$. Note first that the summand with $\Lambda = K$ is S^0, and that all other terms are $\overline{\mathcal{F}}$-spaces.

It seems worth making the consistency with the algebraic description explicit. Let $K^* = K/\Lambda$, and, more generally, use asterisks to denote reference to $\mathbb{T}/\Lambda$. We have observed that if $\Lambda = K$ the summand is S^0, so now suppose Λ is a proper subgroup of K. Since K^* is finite, the quotient of a K^*-free universal space is again rationally a universal space. Indeed

$$E[\cap K^* = 1] = \bigvee_{L^* \cap K^* = 1} E\langle L^* \rangle,$$

and when $L^* \cap K^* = 1$, $E\langle L^* \rangle / K^* = E\langle \overline{L} \rangle$. Thus

$$E[\cap K^* = 1]_+/K^* \simeq \bigvee_{L^* \cap K^* = 1} E\langle \overline{L} \rangle.$$

Thus, a subgroup L gives rise to a summand $E\langle \overline{L} \rangle$ for each subgroup $\Lambda \subset K$ with $(|L^*|, |K^*|) = 1$; this happens exactly once for each subgroup not containing K. The multiplicity of the summand $E\langle \overline{L} \rangle$ is thus $|q_*^{-1}(\overline{L})| - 1$. □

The following observation is also useful.

LEMMA 10.5.7. *If $Y \longrightarrow I(Y) \longrightarrow J(Y)$ is an Adams resolution, then it remains so after taking Lewis-May fixed points.*

Proof: We have seen that passage to Lewis-May fixed points preserves $\mathcal{F}$-contractible objects and $\mathcal{F}$-free objects. By the classification of torsion $\mathcal{F}$-finite injectives we see that the fixed point spectrum of an $\mathcal{F}$-free injective is also injective.

Finally we must observe that

$$0 \longrightarrow \pi_*^{\overline{A}}(\Psi^K Y) \longrightarrow \pi_*^{\overline{A}}(\Psi^K I(Y)) \longrightarrow \pi_*^{\overline{A}}(\Psi^K J(Y)) \longrightarrow 0$$

is exact. This is a diagram chase using the fact that the first map is an isomorphism of vertices, and that, because the torsion sequence is exact, the sequences of kernels and of cokernels are also both exact. □

Finally we remark that it is not hard to prove directly that the d invariants of the algebraic and topological Lewis-May fixed points of a map agree. This falls short of a complete direct analysis because Lewis-May fixed points do not preserve pure parity objects.

10.6. Quotient functors.

It is notorious that the quotient functor is only well behaved on sufficiently free spectra. This phenomenon has already shown itself in the algebra. Topologically, even even the homotopy groups of quotients of arbitrary spectra are quite inaccessible.

In fact the properties of the deflation functor show that $\mathrm{def} : D(\text{tors-}\mathcal{O}^f_{\mathcal{F}}) \longrightarrow D(\text{tors-}\overline{\mathcal{O}}^f_{\overline{\mathcal{F}}})$ is left adjoint to $\mathrm{inf} : D(\text{tors-}\overline{\mathcal{O}}^f_{\overline{\mathcal{F}}}) \longrightarrow D(\text{tors-}\mathcal{O}^f_{\mathcal{F}})$. Hence, whenever the topological quotient is left adjoint to the topological inflation functor, the algebraic deflation models the effect on homotopy groups of the topological quotient. Applying [**19**, I.3.8 and II.2.8] we immediately deduce the required result.

PROPOSITION 10.6.1. *The algebraic K-deflation functor corresponds to the topological quotient by K on the category of K-free spectra.* □

It seems reasonable to refer to the topological functor $/K$ defined by the diagram

$$\begin{array}{ccc} \mathbb{T}\text{-Spec}/\mathcal{F} & \xrightarrow{/K} & \mathbb{T}\text{-Spec} \;/\; \overline{\mathcal{F}} \\ \simeq\downarrow & & \downarrow\simeq \\ D(\text{tors-}\mathcal{O}^f_{\mathcal{F}}) & \xrightarrow{\mathrm{def}} & D(\text{tors-}\overline{\mathcal{O}}^f_{\overline{\mathcal{F}}}) \end{array}$$

as the quotient by K. We have seen that it agrees with the quotient by K on K-free spectra, and it extends this functor to arbitrary rational $\mathcal{F}$-spectra. It would be interesting to know if there is a topological definition of this functor.

Part III

Applications.

CHAPTER 11

Introduction to Part III.

11.1. General Outline.

The material in Part III is a collection of applications of the general theory developed in Parts I and II. Accordingly, the chapters are completely independent of each other. Since most of the work has already been incorporated in the general framework, the sections are rather short, and uncluttered by technicalities.

We begin by considering the homotopy Mackey functor and Eilenberg-MacLane spectra. In the context of compact Lie groups of positive dimension there are two other variants of Eilenberg-MacLane spectra: co-Eilenberg-MacLane spectra representing ordinary *ho*mology, and Brown-Comenetz spectra representing the Mackey duals of homotopy. Since one is used to these coinciding with Eilenberg-MacLane spectra rationally for finite groups of equivariance, it seems worth giving a detailed analysis to show how different they are in the present case.

Next comes Chapter 13, which consists of five independent sections answering the main questions motivating this study. Perhaps the most obvious problem of all, in the light of Haeberly's example, is to understand the behaviour of the Atiyah-Hirzebruch spectral sequence. We show that it collapses at E_2 for all $\mathcal{F}$-spaces if and only if $\pi^{\mathbb{T}}_*(K \wedge E\mathcal{F}_+)$ is injective over $\mathcal{O}_{\mathcal{F}}$. In general, the differentials encode the Adams short exact sequence, but there are differentials of arbitrary length. For arbitrary spaces, the main message is that the Atiyah-Hirzebruch spectral sequence is not a natural tool. Alternatively, we can always use a cellular decomposition of the domain to try to understand maps. In our case, the graded orbit category can be made quite explicit, and we can therefore do homological algebra over the category of additive functors on this category. We may then construct a spectral sequence, whose E_2 term is calculable and given by homological algebra over the graded orbit category. It is also obviously convergent, but it does not seem a practical tool in general, because it is usually a half-plane spectral sequence. The moral is that we should not decompose spectra by Eilenberg-MacLane spectra or by cells: injectives in the standard model are much more effective.

Another basic construction of spectra is by taking the suspension spectrum of a space. Equivariantly it is known that suspension spectra have certain special properties, such as tom Dieck splitting, which are not enjoyed by all spectra. In particular, the inclusion of the H-fixed points can be regarded as a map of $\mathbb{T}$-spaces, and this has implications for the model. However, since our model has no record of purely unstable information, we cannot hope for a precise characterization of suspension spectra.

The special case of K-theory is interesting because it has Bott periodicity, and hence Euler classes of its own. The representing spectrum also turns out to be formal in the torsion model, so that the K-theory of any spectrum depends only on its homology in the torsion model. On the other hand, our analysis is really only the beginning of a study melding the formalisms of the present work with the geometric information of K-theory: in particular it would be interesting to understand the Chern character in more detail and to relate our results to the work of Brylinski and collaborators [**3**].

The rational Segal conjecture is now an elementary example: in the torsion model $DE\mathbb{T}_+$ is formal, and represented by the natural map

$$t_*^{\mathcal{F}} \otimes t_*^{\mathcal{F}} \longrightarrow t_*^{\mathcal{F}} \longrightarrow t_*^{\mathcal{F}}/\mathcal{O}_{\mathcal{F}} = \Sigma^2\mathbb{I} \longrightarrow \mathbb{Q}[c, c^{-1}]/\mathbb{Q}[c].$$

In Part IV we will identify the model for function spectra, thereby giving an alternative approach to functional duals in general.

In Chapter 14 we consider the well known $\mathbb{T}$-equivariant cohomology theory given by cyclic cohomology. This is very simple rationally; more generally, rational Tate spectra are also rather simple, but we may make certain intriguing algebraic connections. Finally we are able to identify the *integral* Tate spectrum $t(K\mathbb{Z})$ of *integral* complex K-theory $K\mathbb{Z}$. This is of interest because $t(K\mathbb{Z})$ is known to be H-equivariantly rational for all finite subgroups H. In fact, we identify the spectra $t(K\mathbb{Z}) \wedge \tilde{E}\mathcal{F}$ and $t(K\mathbb{Z}) \wedge \Sigma E\mathcal{F}_+$ integrally, together with the map of which $t(K\mathbb{Z})$ is the fibre. The first is obtained from K-theory with suitable coefficients by inflating and smashing with $\tilde{E}\mathcal{F}$. The second is in fact rational, and it is therefore determined by the injective Euler-torsion $\mathcal{O}_{\mathcal{F}}$-module $\pi_*^{\mathbb{T}}(t(K\mathbb{Z}) \wedge \Sigma E\mathcal{F}_+)$, which we calculate.

The final Chapter 15 is more substantial. We turn to examples gaining their importancc from algebraic K-theory. Bökstedt, Hsiang and Madsen define the topological cyclic cohomology of a ring or a space [**2**]. It is obtained by performing various constructions on the topological Hochschild homology spectrum and can be used as a close approximation to the completed algebraic K-theory of suitable rings. Hesselholt and Madsen [**18**] identify the structure required of a $\mathbb{T}$-spectrum X for one to be able to construct $TC(X)$. Spectra X with this structure are called *cyclotomic* spectra. The motivation for the notion of a cyclotomic spectrum comes from the free loop space $\Lambda X = map(\mathbb{T}, X)$ on a $\mathbb{T}$-fixed space X. This has the property that if we take K-fixed points we obtain the $\mathbb{T}/K$-space $map(\mathbb{T}/K, X)$, and if we identify the circle $\mathbb{T}$ with the circle $\mathbb{T}/K$ by the $|K|$th root isomorphism we recover ΛX. For spectra, one also needs to worry about the indexing universe, but a cyclotomic spectrum is basically one whose geometric fixed point spectrum $\Phi^K X$, regarded as a $\mathbb{T}$-spectrum, is the original $\mathbb{T}$-spectrum X. After the suspension spectrum of a free loop space, the principal example comes from the topological Hochschild homology of $THH(F)$ of a functor F with smash products. Given such a cyclotomic spectrum X, one may construct the topological cyclic spectrum $TC(X)$ of Bökstedt-Hsiang-Madsen [**2**], which is a non-equivariant spectrum. An intermediate construction of some interest is the $\mathbb{T}$-spectrum $TR(X)$. Although these constructions are principally of interest profinitely, it is instructive to identify the cyclotomic spectra in our model and follow the constructions through. We identify the rational cyclotomic spectra in the torsion model: they are the spectra X so that the function $[N] : \mathcal{F} \longrightarrow$ torsion $\mathbb{Q}[c]$-modules modelling $E\mathcal{F}_+ \wedge X$ is constant, and so that the structure map $t_*^{\mathcal{F}} \otimes V \longrightarrow \Sigma N$ commutes with any translation of the finite subgroups. It therefore factors through $t_*^{\mathcal{F}} \otimes V \longrightarrow t_*^{\mathcal{F}}/\mathcal{O}_{\mathcal{F}} \otimes V$ and the map $t_*^{\mathcal{F}}/\mathcal{O}_{\mathcal{F}} \otimes V \longrightarrow \Sigma N$ is a direct sum of copies of the single map $\mathbb{Q}[c, c^{-1}]/\mathbb{Q}[c] \otimes V \longrightarrow \Sigma[N](1)$. Furthermore, we may recover Goodwillie's theorem [**7**] that for any cyclotomic spectrum X we have $TC(X) = X^{h\mathbb{T}}$: topological cyclic cohomology coincides with cyclic cohomology in the rational setting.

11.2. Prospects and problems.

The main theoretical problem is to show that the equivalence of the category of $\mathbb{T}$-spectra and the derived category of the standard model can be obtained from a chain of equivalences arising from adjoint pairs of functors on underlying Quillen model categories. This would inevitably be linked with a better understanding of the meaning of the standard model.

One of the most interesting prospective applications is that of understanding rational $\mathbb{T}$-equivariant elliptic cohomology. Constructions have recently been given by Grojnowski [**16**] and by Ginzburg-Kapranov-Vasserot [**6**], and the cohomology of any $\mathbb{T}$-space is a sheaf over an elliptic curve. One can ask if these theories are represented. This would involve considering sheaves of $\mathbb{T}$-spectra over an elliptic curve, and it would seem a sensible first step to consider sheaves of objects of $\mathcal{A}$.

Both in this case and that of K-theory, there is the task of relating the general model to the geometry of the cohomology theory: in practice this will involve concentration on the Chern character, and comparison with the work of Brylinski [**3**]. There are a number of other classes of spectra which we do not understand as well as we should like, such as suspension spectra, free loop spaces, THH, ...

In the present work we have concentrated entirely on the circle group $\mathbb{T}$. Although it is unlikely to be possible to give so complete a picture as we have done for $\mathbb{T}$-spectra, we hope to consider other small groups in due course. The continuous quaternion and dihedral groups are prime candidates, both by virtue of their simplicity and the prospects for applications. A model for rational stable equivariant homotopy theory in these cases is given in [**12**], but further structure and examples need to be made explicit. However, consideration of the case of Mackey functors [**10**] shows that it is necessary to replace the underlying algebra of $\mathcal{O}_{\mathcal{F}}$-modules by that of sheaves over spaces of subgroups, since the topology on the space of subgroups can no longer be ignored. For groups of rank greater than 1, the injective dimension of the relevant algebraic categories will be greater than 1. There is therefore no prospect of obtaining splittings for formal reasons, and models must be based on a more complete geometric understanding than we have used here.

CHAPTER 12

Homotopy Mackey functors and related constructions.

This chapter investigates a number of constructions associated with a Mackey functor, and the reader is advised to glance at Appendix A before reading further.

A cohomology theory $k_G^*(\cdot)$ satisfying the dimension axiom is characterized by the Mackey functor $M : G/H \longmapsto k_G^0(G/H_+)$ describing its values on homogeneous spaces. It is represented by the Eilenberg-MacLane spectrum HM. A homology theory $l_*^G(\cdot)$ satisfying the dimension axiom is characterized by the coMackey functor $N : G/H \longmapsto l_0^G(G/H_+)$ describing its values on homogeneous spaces. It is represented by the co-Eilenberg-MacLane spectrum JN. Finally, given an injective Mackey functor I we can form the Brown-Comenetz cohomology theory by taking $hI_{\mathbb{T}}^n(X) = \mathrm{Hom}(\underline{\pi}_n^{\mathbb{T}}(X), I)$, and this is represented by a spectrum hI. When a Mackey functor M admits an injective resolution of length 1, we can also define the spectrum hM. For finite groups the orbit category is self-dual, so that there is a natural way of identifying Mackey functors and co-Mackey functors so that a Mackey functor M gives rise to three representing spectra, HM, JM and hM. It is a deeply ingrained fact that these three spectra coincide. However, it relies on the facts that G/H is a 0-dimensional manifold, and that $H^*(BH; \mathbb{Q})$ is concentrated in degree 0. Both these facts fail for positive dimensional compact Lie groups. We shall identify the Eilenberg-MacLane, co-Eilenberg-MacLane and Brown-Comenetz spectra for the circle group, and it is apparent that they are completely different in general. However it is interesting to observe that all three classes of spectra are formal in the torsion model: in fact their torsion parts are injective.

First, we consider ordinary *cohomology*, and the usual Eilenberg-MacLane spectrum. An Eilenberg-MacLane spectrum is recognized by the fact that its homotopy groups are only nonzero in a single degree. It is characterized by the homotopy groups in that degree, regarded as a Mackey functor. We therefore begin in Section 12.1 by making explicit how to recover the Mackey functor homotopy groups from the standard model. By way of illustration we deduce a functorial construction of the model of an Eilenberg-MacLane spectrum from a Mackey functor. In Section 12.2 we take the opposite approach and begin with ordinary cohomology and deduce a model, and provide decompositions of Eilenberg-MacLane spectra which are interesting from a topological point of view. We follow this by Section 12.3 which gives an analogous approach to the analysis of representing objects for ordinary *homology*. We have not made the algebraic approach to co-Eilenberg-MacLane spectra explicit.

We find Brown-Comenetz spectra to be particularly interesting, and devote Section 12.4 to them. One reason for interest is that by definition these spectra are well suited to the construction of Adams spectral sequences, and the author's first approach to the algebraic modelling of $\mathbb{T}$-spectra therefore used them. However the extreme complexity of the Brown-Comenetz spectra seems to make the resulting Adams spectral sequence impractical except in certain special cases. On the other hand the spectra do form a natural class of unbounded spectra, and they illuminate the rational Segal conjecture: they therefore provide an interesting test case for the present methods.

12.1. The homotopy Mackey functor on $\mathcal{A}$.

We have the information to calculate the homotopy Mackey functor of any object of $D\mathcal{A}$. We shall make use of the notation described in Section 5.9, and we recall from Example 5.8.1 that the algebraic basic cells are $L_H = (\Sigma\mathbb{Q}(H) \longrightarrow 0)$ and $L_{\mathbb{T}} = (\mathcal{O}_{\mathcal{F}} \longrightarrow t_*^{\mathcal{F}})$. We use the obvious injective form of the algebraic fixed cell $\hat{L}_{\mathbb{T}} = (t_*^{\mathcal{F}} \ltimes \Sigma\mathbb{I} \longrightarrow t_*^{\mathcal{F}})$ where the the semidirect product $t_*^{\mathcal{F}} \ltimes \Sigma\mathbb{I}$ is additively the direct sum, but with differential given by the augmentation $t_*^{\mathcal{F}} \longrightarrow \Sigma^2\mathbb{I}$. We proceed in steps, using the notation $\pi_*^{\mathbb{T}}(M) = [L_{\mathbb{T}}, M]$ and $e\pi_*^H(M) = [L_H, M]$ for an object M of $D\mathcal{A}$. The symbol abbreviates the more descriptive $e_H\pi_*^H$, the point being that L_H is the algebraic counterpart of the basic cell σ_H^0 rather than G/H_+. Since $M \simeq H_*(M)$ in $D\mathcal{A}$ we may work with objects with zero differential.

LEMMA 12.1.1. *Suppose $M = (N \xrightarrow{\beta} t_*^{\mathcal{F}} \otimes V)$ is an object with zero differential and torsion part $T_M = ker(\beta) \oplus \Sigma^{-1}cok(\beta)$. Let s_M be the composite*

$$V_M \longrightarrow t_*^{\mathcal{F}} \otimes V_M \xrightarrow{d(q_M)} \Sigma T_M,$$

then there is a natural isomorphism

$$\pi_*^{\mathbb{T}}(M) \cong \Sigma^{-1}cok(s_M) \oplus ker(s_M).$$

Proof: This is straightforward from the triangle $M \longrightarrow e(V_M) \longrightarrow f(\Sigma T_M)$. Indeed $[L_{\mathbb{T}}, e(V_M)] = [V_{\mathbb{T}}, V_M] = \mathrm{Hom}(\mathbb{Q}, V_M) = V_M$, and because $N_{\mathbb{T}} = \mathcal{O}_{\mathcal{F}}$ is projective we have $[L_{\mathbb{T}}, f(\Sigma T_M)] = [N_{\mathbb{T}}, \Sigma T_M] = \mathrm{Hom}(\mathcal{O}_{\mathcal{F}}, \Sigma T_M) = \Sigma T_M$. The map is easily identified.

The splitting is constructed as follows. If $v \in ker(s_M)$ then a map $L_{\mathbb{T}} \longrightarrow M$ is explicitly represented by the homomorphism $L_{\mathbb{T}} \longrightarrow \hat{M}$ displayed in the diagram

$$\begin{array}{ccc} \mathcal{O}_{\mathcal{F}} & \xrightarrow{\{1\otimes v, 0\}} & (t_*^{\mathcal{F}} \otimes V_M) \ltimes \hat{T}_M \\ \downarrow & & \downarrow \\ t_*^{\mathcal{F}} & \xrightarrow{1\otimes v} & t_*^{\mathcal{F}} \otimes V_M. \end{array} \quad \square$$

The homotopy groups $\pi_*^H(M)$ are a little easier, since L_H is torsion.

LEMMA 12.1.2. *Suppose $M = (N \xrightarrow{\beta} t_*^{\mathcal{F}} \otimes V)$ is an object with zero differential and torsion part $T_M = ker(\beta) \oplus \Sigma^{-1}cok(\beta)$. There is a short exact sequence*

$$0 \longrightarrow T_M(H)/c_H \longrightarrow e\pi_*^H(M) \longrightarrow \Sigma^{-1}ann(c_H, T_M(H)) \longrightarrow 0,$$

where $T_M(H) = e_H T_M$ is regarded as a module over $\mathbb{Q}[c_H] \cong e_H\mathcal{O}_{\mathcal{F}}$ in the obvious way. The sequence splits unnaturally.

Proof: This follows from the Adams short exact sequence 3.1.1 by calculating

$$\mathrm{Hom}(\mathbb{Q}(H), T_M) = ann(c_H, T_M(H)) \text{ and } \mathrm{Ext}(\mathbb{Q}(H), T_M) = \Sigma^2 T_M(H)/c_H. \quad \square$$

Before continuing, let us see what this implies about objects with homotopy groups only nonzero in a single degree: the Eilenberg-MacLane objects.

EXAMPLE 12.1.3. If M is an Eilenberg-MacLane object in the standard model with zero differential and homotopy groups $\pi_*^{\mathbb{T}}(M) = W(\mathbb{T})$ and $e\pi_*^H(M) = W(H)$, concentrated in degree zero, then M has vertex

$$V_M = W(\mathbb{T}) \oplus \bigoplus_H \Sigma^2\mathbb{I}(H) \otimes W(H)$$

and torsion part

$$\Sigma T_M = \bigoplus_H \Sigma^2\mathbb{I}(H) \otimes W(H).$$

The map $s_M : V_M \longrightarrow \Sigma T_M$ is surjective.

Proof: The object M is the sum of its even and odd degree parts. Since any non-zero object of with zero differential has non-zero homotopy one of these is zero. It is easy to adapt the following discussion to prove that $M = 0$ if M is odd. We therefore suppose M is even.

Now if $W(\mathbb{T})$ is a graded vector space concentrated in degree 0, $e(W(\mathbb{T}))$ is an Eilenberg-MacLane object with homotopy in degree 0, and the theorem holds with $M = e(W(\mathbb{T}))$, and $W(H) = 0$ for finite subgroups H. If the natural map $M \longrightarrow e(W(\mathbb{T}))$ is not an isomorphism, then $T_M \neq 0$, and we examine the possibilities. First, we claim that T_M is injective and in odd degrees. Indeed, we have a decomposition $T_M = T_{ev} \oplus T_{od}$ into even and odd parts, and we apply 12.1.2 to analyse each. If $T_{ev} \neq 0$ then $\Sigma^{-1}ann(c_H, T_{ev}(H)) \neq 0$ for some H, since T_{ev} is torsion: this would contribute odd homotopy groups to $e\pi_*^H(M)$, so we conclude $T_M = T_{od}$. For the same reason, an element of $T_{od}(H)$ annihilated by c_H in degree $n+1$ contributes homotopy to M in degree n, and we conclude that all c_H-torsion is in degree 1. Since the odd degree homotopy of M is zero, $T_{od}(H)/c_H = 0$ for all H. Hence $T_M(H) = T_{od}(H)$ is divisible and hence injective. By 3.4.1 we conclude that $T_M(H) = \Sigma\mathbb{I}(H) \otimes W(H)$, where $W(H)$ is concentrated in degree zero.

Now consider $s_M : V_M \longrightarrow \Sigma T_M = \Sigma^2\mathbb{I}(H) \otimes W(H)$. The cokernel gives homotopy in odd degrees, so that s_M is onto. Since $\pi_*^{\mathbb{T}}(M)$ is concentrated in degree 0, s_M is an isomorphism, except perhaps in degree 0. □

To determine the entire Mackey functor we need to identify the restriction map $\pi_*^{\mathbb{T}}(M) \longrightarrow e\pi_*^H(M)$ induced by the generator $p_H : L_H \longrightarrow L_{\mathbb{T}}$ of $[L_H, L_{\mathbb{T}}] \cong \mathbb{Q}$. We must begin by making explicit that p_H is represented by the homomorphism

$$\left(\hat{p}_H : L_H \longrightarrow \hat{L}_{\mathbb{T}}\right) = \left(\begin{array}{ccc} \Sigma\mathbb{Q}(H) & \stackrel{\{0,\iota_H\}}{\longrightarrow} & t_*^{\mathcal{F}} \ltimes \Sigma\mathbb{I} \\ \downarrow & & \downarrow \\ 0 & \longrightarrow & t_*^{\mathcal{F}} \end{array}\right),$$

where $\iota_H : \mathbb{Q}(H) \longrightarrow \mathbb{I}$ is the inclusion of the degree 0 part of $e_H\mathbb{I}$. This is readily verified from the definitions.

PROPOSITION 12.1.4. *The restriction map* $\pi_*^{\mathbb{T}}(M) \longrightarrow e\pi_*^H(M)$ *is the direct sum of the natural quotient map*

$$res' : \Sigma^{-1} cok(s_M) \longrightarrow T_M(H)/c_H$$

given by $res'(\bar{t}) = \bar{\bar{t}}$ *and the map*

$$res'' : ker(s_M) \longrightarrow \Sigma^{-1} ann(c_H, T_M(H))$$

given by $res''(v) = \beta_M(c_H^{-1} \otimes v)$.

Proof: The proof is by unravelling definitions. Suppose given a homotopy class $\theta : L_{\mathbb{T}} \longrightarrow M$. This can be represented by a homomorphism $L_{\mathbb{T}} \longrightarrow \hat{M}$ of the same name where hats to denote fibrant approximation as usual. We then find a homomorphism $\hat{\theta}$ so that the diagram

$$\begin{array}{ccccc} & & L_{\mathbb{T}} & \xrightarrow{\theta} & \hat{M} \\ & & \downarrow & & \downarrow = \\ L_H & \xrightarrow{\hat{p}_H} & \hat{L}_{\mathbb{T}} & \xrightarrow{\hat{\theta}} & \hat{M} \end{array}$$

commutes. Then $res(\theta)$ is determined by $\hat{\theta} p_H(1) = \hat{\theta}(0, \iota_H(1)) = \hat{\theta}(d(c_H^{-1}, 0)) = d\hat{\theta}(c_H^{-1}, 0)$. Now $\hat{\theta}$ takes the form

$$\begin{array}{ccc} t_*^{\mathcal{F}} \ltimes \Sigma\mathbb{I} & \xrightarrow{\hat{\theta}'} & (t_*^{\mathcal{F}} \otimes V_M) \ltimes \hat{T}_M \\ \downarrow & & \downarrow \\ t_*^{\mathcal{F}} & \xrightarrow{\theta''} & t_*^{\mathcal{F}} \otimes V_M. \end{array}$$

Next observe that if the map of nubs is $\theta' = \{\theta'_V, \theta'_T\} : \mathcal{O}_{\mathcal{F}} \longrightarrow (t_*^{\mathcal{F}} \otimes V_M) \ltimes \hat{T}_M$ then the extension of θ' over $t_*^{\mathcal{F}}$ determines $\hat{\theta}'$ since the differential $t_*^{\mathcal{F}} \longrightarrow \Sigma\mathbb{I}$ is surjective, and the map into the $t_*^{\mathcal{F}} \otimes V_M$ factor must be θ''. It remains to specify an extension $\tilde{\theta} : t_*^{\mathcal{F}} \longrightarrow \hat{T}_M$ of θ'_T and verify that the result commutes with differentials. Fortunately we need only do this when θ is of two particularly simple forms.

For res' we suppose θ corresponds to $\bar{t} \in cok(s_M : V_M \longrightarrow \Sigma T_M)$ for some element $t \in T_M$ (which may be of even or odd degree). In this case θ factors through $f(T_M) \longrightarrow M$. Now suppose $\hat{T}_M = fibre(I \longrightarrow J)$, and view t as an element of I. We may then extend the map $\theta : \mathcal{O}_{\mathcal{F}} \longrightarrow I$ to $\tilde{\theta} : t_*^{\mathcal{F}} \longrightarrow I$, and any such $\tilde{\theta}$ gives rise to a map $\hat{L}_{\mathbb{T}} \longrightarrow f(\hat{T}_M)$ and hence to $\hat{\theta}$ as the composite $\hat{L}_{\mathbb{T}} \longrightarrow f(\hat{T}_M) \longrightarrow \hat{M}$. Therefore $res'(\bar{t})$ is represented by $d\tilde{\theta}(c_H^{-1}) \in J$, which is an element dy where $y \in I$ satisfies $c_H y = t$; the result follows by considering the resolution beginning $0 \longrightarrow T_M(H) \longrightarrow (T_M(H)/c_H) \otimes \mathbb{I}(H)$.

For res'' we use the explicit splitting given in 12.1.1 above. Suppose that θ corresponds to $v \in V_M$, so that $\theta'_V = 1 \otimes v$, $\theta'_T = 0$ and $\theta'' = 1 \otimes v$. We choose $\tilde{\theta} = 0$, and it is immediate that this defines a map $\hat{\theta}'$ commuting with differentials. Thus $\hat{\theta}'(c_H^{-1}) = (c_H^{-1} \otimes v, 0)$; the result follows. □

This allows us to complete the discussion of Eilenberg-MacLane objects.

EXAMPLE 12.1.5. We continue with the notation of Example 12.1.3. First observe that since s_M is surjective, $res' = 0$. Thus for $v \in \pi_0^{\mathbb{T}}(M) = W(\mathbb{T})$ we have $res(v) = \beta_M(c_H^{-1} \otimes v)$. The map $q'_M : t_*^{\mathcal{F}} \otimes V_M \longrightarrow \Sigma T_M$ may thus be chosen to have components

$$q'_{\mathbb{T}} : t_*^{\mathcal{F}} \otimes W(\mathbb{T}) \longrightarrow \bigoplus_H \Sigma^2 \mathbb{I}(H) \otimes W(H)$$

and

$$q'_H : t_*^{\mathcal{F}} \otimes \Sigma^2 \mathbb{I}(H) \otimes W(H) \longrightarrow \bigoplus_H \Sigma^2 \mathbb{I}(H) \otimes W(H)$$

described as follows. First, $q'_{\mathbb{T}}$ is the composite of projection

$$t_*^{\mathcal{F}} \otimes W(\mathbb{T}) \longrightarrow \Sigma^2 \mathbb{I} \otimes W(\mathbb{T}) \cong \bigoplus_H \Sigma^2 \mathbb{I}(H) \otimes W(\mathbb{T})$$

and the map which is restriction on each factor. Next q'_H is the composite of projection

$$t_*^{\mathcal{F}} \otimes \Sigma^2 \mathbb{I}(H) \otimes W(H) \longrightarrow e_H t_*^{\mathcal{F}} \otimes \Sigma^2 \mathbb{I}(H) \otimes W(H) \cong \mathbb{Q}[c_H, c_H^{-1}] \otimes \Sigma^2 \mathbb{I}(H) \otimes W(H)$$

and the map $c_H^k \otimes c_H^l \otimes w \longmapsto c_H^{k+l} \otimes w$, where c_H^{k+l} is interpreted as zero if $k + l \geq 0$. □

12.2. Eilenberg-MacLane spectra.

In the previous section we worked entirely in the algebraic model to identify objects with homotopy only in one degree. In the present section we start from Eilenberg-MacLane spectra as is more natural from a topological point of view. The identification of the model of a topological Eilenberg-MacLane spectrum gives an alternative to the algebraic deduction, but the point of view is sufficiently different for this second method to be worth including.

We shall use notation and terminology for Mackey functors established in Appendix A, which the reader should refer to as necessary. In particular we recall that a Mackey functor M corresponds to a collection of vector spaces $M^e(H)$, one for each finite subgroup H and a vector space $M^e(\mathbb{T})$ with restriction maps $M^e(\mathbb{T}) \longrightarrow M^e(H)$.

The easy example is the Eilenberg-MacLane spectrum associated to the Mackey functor $L[U]$ concentrated over $\mathbb{T}$, for which we obviously have $H(L[U]) \simeq \tilde{E}\mathcal{F} \wedge S^0[U]$. This deals with all $\tilde{\mathcal{F}}$-injectives. On the other hand, the basic $\mathcal{F}$-injective is the functor R_K concentrated over $\mathbb{T}$ and K. An arbitrary $\mathcal{F}$-injective is a product of injectives which are sums of the basic injective R_K, so the same is true of the corresponding Eilenberg-MacLane spectra. Therefore the following examples describe the structure of Eilenberg-MacLane spectra for all projective and injective spectra.

EXAMPLE 12.2.1. (i) For the indecomposable projective Mackey functor $P_{\mathbb{T}}$ (the 'constant at $\mathbb{Q}$' functor) we have a cofibre sequence

$$\Sigma \tilde{E}\mathcal{F} \wedge S^0[\mathbb{Q}\mathcal{F}] \longrightarrow S^0 \longrightarrow HP_{\mathbb{T}}.$$

(ii) For any finite subgroup K with corresponding indecomposable projective Mackey functor P_K concentrated over K we have a cofibre sequence

$$\Sigma\tilde{E}\mathcal{F} \longrightarrow E\langle K\rangle \longrightarrow HP_K.$$

(iii) For any finite subgroup K with corresponding injective Mackey functor R_K we have cofibre sequences

$$\Sigma HR_K \longrightarrow \sigma_K^0 \longrightarrow HP_K$$

and

$$HP_K \longrightarrow HR_K \longrightarrow \tilde{E}\mathcal{F}.$$

Proof: The case of $HP_{\mathbb{T}}$ is straightforward: we kill homotopy groups in positive degrees to obtain a map $S^0 \longrightarrow HP_{\mathbb{T}}$, and observe its fibre is $\mathcal{F}$-contractible. The same idea works for HP_K: kill homotopy groups in positive degrees to obtain a map $E\langle K\rangle \longrightarrow HP_K$ and observe its fibre is $\mathcal{F}$-contractible.

The first cofibre sequence for HR_K follows from the fact that σ_K^0 is a two stage Postnikov tower. The second follows from the defining extension $0 \longrightarrow P_K \longrightarrow R_K \longrightarrow L[\mathbb{Q}] \longrightarrow 0$ for R_K, as in A.10. □

One may obtain a cell picture for an arbitrary Eilenberg-MacLane spectrum HM by taking a minimal projective resolution $0 \longrightarrow P_1 \longrightarrow P_0 \longrightarrow M \longrightarrow 0$ and using the associated cofibre sequence $HP_1 \longrightarrow HP_0 \longrightarrow HM$.

To complete the analysis of HM in our framework we need to analyse $HM \wedge E\mathcal{F}_+$ and $HM \wedge \tilde{E}\mathcal{F}$. We let $M(\mathcal{F}) = \lim_{\to H} M(H) = \oplus_H M^e(H)$.

LEMMA 12.2.2. *For any Mackey functor M there is an equivalence*

$$HM \wedge E\mathcal{F}_+ \simeq \bigvee_H E\langle H\rangle \wedge S^0[M^e(H)],$$

and hence $HM \wedge E\mathcal{F}_+$ corresponds to the $\mathcal{O}_{\mathcal{F}}$-module $M(\mathcal{F}) \otimes_{(\mathcal{O}_{\mathcal{F}})_0} \Sigma\mathbb{I} = \bigoplus_H M^e(H) \otimes \Sigma\mathbb{I}(H)$.

Proof: By construction, $(HM \wedge E\mathcal{F}_+)(K) = HM \wedge E\langle K\rangle$. Recall that $E\langle K\rangle = \sigma_K^0 \cup \sigma_K^0 \wedge e^2 \cup \sigma_K^0 \wedge e^4 \cup \ldots$; now consider the spectral sequence obtained by applying $[S^0, HM \wedge \bullet]_*^{\mathbb{T}}$ to the skeletal filtration. Because of the suspension in the Wirthmüller isomorphism we find $[S^0, HM \wedge \sigma_K^0]_*^{\mathbb{T}} = \Sigma M^e(K)$, and thus the spectral sequence collapses to show that the homotopy groups are $M^e(K)$ in each positive odd dimension.

It remains to show that c_K gives an isomorphism $\pi_{2n+1}^{\mathbb{T}}(HM \wedge E\langle K\rangle) \longrightarrow \pi_{2n-1}^{\mathbb{T}}(HM \wedge E\langle K\rangle)$ for $n \geq 1$. For this we use the long exact sequence of the cofibering

$$HM \wedge E\langle K\rangle \wedge \sigma_K^0 \longrightarrow HM \wedge E\langle K\rangle \wedge S^0 \xrightarrow{e} HM \wedge E\langle K\rangle \wedge S^{V(K)}$$

in which e induces multiplication by c_K. The result follows from the fact that $HM \wedge E\langle K\rangle$ is K-equivariantly an Eilenberg-MacLane spectrum, so that $\pi_*^{\mathbb{T}}(HM \wedge E\langle K\rangle \wedge \sigma_K^0)$ is concentrated in degree 1. □

The homotopy type of the rational spectrum $\Phi^{\mathbb{T}}X$ is determined by its homotopy groups $\pi_*(\Phi^{\mathbb{T}}X) \cong \pi_*^{\mathbb{T}}(\tilde{E}\mathcal{F} \wedge X)$. With due caution about duality we may calculate $\Phi^{\mathbb{T}}HM$ for any Mackey functor M.

LEMMA 12.2.3.

$$\pi_k(\Phi^{\mathbb{T}}HM) = \begin{cases} M(\mathbb{T}) & \text{if } k = 0 \\ M(\mathcal{F}) & \text{if } k \text{ is even and positive} \\ 0 & \text{otherwise} \end{cases}$$

Proof: This follows immediately from the cofibre sequence $E\mathcal{F}_+ \longrightarrow S^0 \longrightarrow \tilde{E}\mathcal{F}$ and 12.2.2. □

Note in particular that $\Phi^{\mathbb{T}}HM$ is an Eilenberg-MacLane spectrum only if M is an $\tilde{\mathcal{F}}$-injective. By contrast, the idempotent description of the geometric fixed point functor given in Theorem 8.4.1 shows that $\Phi^K HM$ is an Eilenberg-MacLane spectrum for all finite K.

COROLLARY 12.2.4. *The map $HM \wedge \tilde{E}\mathcal{F} \longrightarrow HM \wedge \Sigma E\mathcal{F}_+$ is classified by the $\mathcal{O}_{\mathcal{F}}$-morphism*

$$q : \pi_*(\Phi^{\mathbb{T}}HM) \otimes t_*^{\mathcal{F}} \longrightarrow M(\mathcal{F}) \otimes_{(\mathcal{O}_{\mathcal{F}})_0} \Sigma^2\mathbb{I}$$

described as follows, in which $\epsilon : t_*^{\mathcal{F}} \longrightarrow \Sigma^2\mathbb{I}$ denotes the quotient. If $n \geq 1$ and $x \in \pi_{2n}(\Phi^{\mathbb{T}}HM) = M(\mathcal{F})$ then $q(x \otimes y) = x \otimes \epsilon(y)$; if $x \in \pi_0(\Phi^{\mathbb{T}}HM) = M(\mathbb{T})$, then $q(x \otimes y) = res_{\mathcal{F}}^{\mathbb{T}}(x) \otimes \epsilon(y)$.

Note that $res_{\mathcal{F}}^{\mathbb{T}}$ in the statement is a map $M(\mathbb{T}) \longrightarrow \prod_H M^e(H)$, whose codomain is the product, but that $\epsilon(y)$ is only nonzero in finitely many coordinates.

Proof: Suppose first that x is of positive degree. From the proof of 12.2.3, we know that $q(x \otimes 1) = x$ and the result follows since non-zero elements of $\mathbb{I}(H)$ are uniquely divisible.
For x of degree zero it is enough to show that $q(x \otimes c_H^{-1}) = res_H^{\mathbb{T}}(x)$. For this we examine the map

$$HM \wedge \tilde{E}\mathcal{F} \wedge \sigma^{-V(H)} \longrightarrow HM \wedge \Sigma E\langle H\rangle \wedge \sigma^{-V(H)}$$

in $[S^0, \cdot]^{\mathbb{T}}$. Transposing $\sigma^{-V(H)}$ across to its dual in the domain, and replacing the spaces in the codomain by suitable skeleta we must examine

$$[\sigma^{V(H)}, HM \wedge \tilde{E}\langle H\rangle^{(2)}]^{\mathbb{T}} \longrightarrow [\sigma^{V(H)}, HM \wedge \sigma_H^1]^{\mathbb{T}}.$$

Now recall that $\tilde{E}\langle H\rangle^{(2)} = \sigma^{V(H)}$, and note that if $x \in M(\mathbb{T})$, it is represented in the first group by $x \wedge 1 : S^0 \wedge \sigma^{V(H)} \longrightarrow HM \wedge \sigma^{V(H)}$. Its image in the second is obtained by composing with the map $HM \wedge \sigma^{V(H)} \longrightarrow HM \wedge \sigma_H^1$. This corresponds to $res_H^{\mathbb{T}}(x) \in M(\sigma_H^0) = e_H M(H)$ as required. □

12.3. coMackey functors and spectra representing ordinary homology.

A coMackey functor is the algebraic object providing coefficients for ordinary homology. It is therefore a covariant additive functor $N : h\mathcal{SO} \longrightarrow \mathbf{Ab}$. By exactly the same argument as for Mackey functors, we see that the category of coMackey functors is equivalent to the category of $h\mathcal{SB}^{op}$-modules. An $h\mathcal{SB}^{op}$-module W is thus described by vector spaces $W(\mathbb{T})$ and $W(H)$ together with induction maps $W(H) \longrightarrow W(\mathbb{T})$ for each finite subgroup H. The $h\mathcal{SB}^{op}$-module corresponding to the coMackey functor N will be written N^e, and this is defined by $N^e(\mathbb{T}) = N(\mathbb{T})$ and $N^e(H) = e_H N(H)$. As for Mackey functors, it is much simpler to conduct proofs with the associated $h\mathcal{SB}^{op}$-modules.

The treatment of Mackey functors in Appendix A is the algebraic template for the theory of co-Mackey functors, so the reader should refer to it as necessary. In view of the similarity to the structure of Mackey functors, we omit the proofs of the following facts.

One may make explicit the condition that a coMackey functor is injective or projective, and give canonical resolutions.

LEMMA 12.3.1. *(i) A coMackey functor N is projective if and only if the map*

$$\bigoplus_H N^e(H) \longrightarrow N^e(\mathbb{T})$$

is a monomorphism.
(ii)A coMackey functor N is injective if and only if the maps $N^e(H) \longrightarrow N^e(\mathbb{T})$ are all epimorphisms. □

Note in particular that if $N(H) = 0$ for all H then N is projective, and if $N(\mathbb{T}) = 0$ then N is injective. The principal projectives are (a) $Q_{\mathbb{T}}$ defined by $Q_{\mathbb{T}}(H) = 0$ for all H and $Q_{\mathbb{T}}(\mathbb{T}) = \mathbb{Q}$ and (b) Q_H defined by $Q^e_H(\mathbb{T}) = \mathbb{Q}$, $Q^e_H(H) = \mathbb{Q}$ (with induction being the identity), and $Q^e_H(K) = 0$ if $H \neq K$. It is not hard to see how to construct canonical projective and injective resolutions of length 1.

The existence and uniqueness of $\mathbb{T}$-spectra JN representing homology with coefficients in N is guaranteed by Brown representability. For any coMackey functor N we let $N|_{\mathcal{F}}$ denote the largest quotient which is zero at $\mathbb{T}$; this has zero structure maps involving $\mathbb{T}$ and can therefore be regarded as a Mackey functor. In fact JN is a two stage Postnikov tower with

$$\underline{\pi}^{\mathbb{T}}_k(JN) = \begin{cases} N|_{\mathcal{F}} & \text{if } k = -1 \\ L[N(\mathbb{T})] & \text{if } k = 0 \\ 0 & \text{otherwise} \end{cases},$$

where $L[N(\mathbb{T})]$ is the Mackey functor concentrated at $\mathbb{T}$ where it takes the value $N(\mathbb{T})$. There is thus a cofibre sequence

$$\tilde{E}\mathcal{F} \wedge S^0[N(\mathbb{T})] \longrightarrow JN \longrightarrow \Sigma^{-1} H(N|_{\mathcal{F}}).$$

Dually, the cofibre sequence

$$\tilde{E}\mathcal{F} \wedge S^0[M(\mathbb{T})] \longleftarrow HM \longleftarrow H(M|_{\mathcal{F}})$$

can be regarded as a decomposition of HM in terms of spectra JN.

Finally we place JN in the torsion model.

LEMMA 12.3.2. *For any coMackey functor N there is an equivalence*

$$JN \wedge E\mathcal{F}_+ \simeq \bigvee_H \Sigma^{-1} E\langle H\rangle \wedge S^0[N^e(H)],$$

and hence $JN \wedge E\mathcal{F}_+$ corresponds to the injective $\mathcal{O}_{\mathcal{F}}$-module $N(\mathcal{F}) \otimes_{(\mathcal{O}_{\mathcal{F}})_0} \mathbb{I} = \bigoplus_H N^e(H) \otimes \mathbb{I}(H)$.

Proof: The proof follows that of 12.2.2, but is slightly simpler. □

The behaviour of the homological Eilenberg-MacLane spectra JN is rather different from their more familiar cohomological counterparts, and this is manifested in the fixed points. Suppose N is a coMackey functor with corresponding $h\mathcal{SB}^{op}$-module N^e, and let $N(\mathcal{F}) = \lim_{\to H} N(H) = \bigoplus_H N^e(H)$. Let us define the vector spaces $D(N)$ and $E(N)$ by the exact sequence

$$0 \longrightarrow D(N) \longrightarrow N(\mathcal{F}) \longrightarrow N(\mathbb{T}) \longrightarrow E(N) \longrightarrow 0.$$

Thus $E(N)$ is measures the failure of an $\mathcal{F}$-induction theorem for N, and $D(N)$ measures the precision of such a theorem.

LEMMA 12.3.3.

$$\pi_k(\Phi^{\mathbb{T}} JN) = \begin{cases} E(N) & \text{if } k = 0 \\ D(N) & \text{if } k = 1 \\ N(\mathcal{F}) & \text{if } k \text{ is at least 3 and odd} \\ 0 & \text{otherwise} \end{cases}$$

Proof: This follows immediately from the cofibre sequence $E\mathcal{F}_+ \longrightarrow S^0 \longrightarrow \tilde{E}\mathcal{F}$ and 12.3.2. □

Note in particular that $\Phi^{\mathbb{T}} JN$ is an Eilenberg-MacLane spectrum only if N is a $\tilde{\mathcal{F}}$-projective.

Because $N(\mathcal{F}) \otimes_{(\mathcal{O}_{\mathcal{F}})_0} \mathbb{I}$ is uniquely divisible on all non-zero elements the assembly map is easily identified in this case.

COROLLARY 12.3.4. *The map $JN \wedge \tilde{E}\mathcal{F} \longrightarrow JN \wedge \Sigma E\mathcal{F}_+$ is classified by the $\mathcal{O}_{\mathcal{F}}$-morphism*

$$q : \pi_*(\Phi^{\mathbb{T}} JN) \otimes t_*^{\mathcal{F}} \longrightarrow N(\mathcal{F}) \otimes_{(\mathcal{O}_{\mathcal{F}})_0} \Sigma\mathbb{I}$$

defined by $q(x \otimes y) = q(x) \otimes \epsilon(y)$, where $\epsilon : t_*^{\mathcal{F}} \longrightarrow \Sigma^2\mathbb{I}$ is the quotient map. □

12.4. Brown-Comenetz spectra.

Another interesting class of spectra arising from a Mackey functor may be constructed as follows. If I is any injective Mackey functor we may define the functor $hI^*_{\mathbb{T}}(\bullet)$ by

$$hI^n_{\mathbb{T}}(X) = \mathrm{Hom}(\underline{\pi}^{\mathbb{T}}_n(X), I).$$

Since I is injective, this is an exact functor of X and hence it is represented by a spectrum hI, which we call the Brown-Comenetz spectrum with coefficients in I. The reader may find the homotopy groups of hI more complicated than expected, and in particular hI is very rarely bounded below. For further information on Mackey functors, and associated notation the reader is referred to Appendix A. In particular, A.8 shows that any injective is a sum of those of two types, and it is enough to deal with $\mathcal{F}$-injectives and $\tilde{\mathcal{F}}$-injectives.

LEMMA 12.4.1. *(i) If I is an $\mathcal{F}$-injective Mackey functor*

$$\underline{\pi}^{\mathbb{T}}_k(hI) = \begin{cases} I & \text{if } k = 0 \\ I|_{\mathcal{F}} & \text{if } k = -1 \\ 0 & \text{otherwise,} \end{cases}$$

where $I|_{\mathcal{F}}$ is the largest subfunctor of I which is zero at $\mathbb{T}$.
(ii) If $I = L[U]$ is an $\tilde{\mathcal{F}}$-injective Mackey functor

$$\underline{\pi}^{\mathbb{T}}_*(hI) = \begin{cases} L[U] & \text{if } k = 0 \\ C\mathcal{F}[U] & \text{if } k = -1 \\ L[U^{\mathcal{F}}] & \text{if } k \text{ is odd and } \leq -3 \\ 0 & \text{otherwise.} \end{cases}$$

Here $U^{\mathcal{F}}$ is the vector space of U-valued functions on $\mathcal{F}$, and $C\mathcal{F}[U]$ is the $\mathcal{F}$-injective defined by $C\mathcal{F}[U](\mathbb{T}) = U^{\mathcal{F}}$ and $C\mathcal{F}[U](H) = U$ with restriction maps being the relevant projections.

Proof: First recall the homotopy functors of cells from A.9. Since $\underline{\pi}^{\mathbb{T}}_0(S^0) = P_{\mathbb{T}}$, and $\underline{\pi}^{\mathbb{T}}_0(\sigma^0_H) = P_H$, it follows from the Yoneda lemma that $\underline{hI}^0_{\mathbb{T}} = I$ whichever sort of injective I is.

On the other hand, in positive degrees $\underline{\pi}^{\mathbb{T}}_*(S^0)$ is $L[\mathbb{Q}\mathcal{F}]$ in each odd degree and 0 in each even degree. It is easy to verify that $\mathrm{Hom}(L[U], M) = \mathrm{Hom}(U, C(M))$ where the core is defined by $C(M) = ker(M(\mathbb{T}) \longrightarrow \prod_H M(H))$. Observe that $C(L[U]) = U$ and $C(I) = 0$ if I is a $\mathcal{F}$-injective.

For each finite subgroup we have $\underline{\pi}^{\mathbb{T}}_1(\sigma^0_H) = R_H$ (see A.9), and an extension

$$0 \longrightarrow P_H \longrightarrow R_H \longrightarrow L[\mathbb{Q}] \longrightarrow 0.$$

Since $\mathrm{Hom}(P_H, L[U]) = 0$ and $\mathrm{Hom}(L[\mathbb{Q}], I) = 0$ if I is $\mathcal{F}$-injective, this is enough to calculate the homotopy groups. In the case that I is $\mathcal{F}$-injective this also determines the functors. Finally we need to understand $I = L[U]$ and in particular $I^* := \underline{\pi}^{\mathbb{T}}_{-1}(hI)$, or more precisely the restriction maps $I^*(\mathbb{T}) = \mathrm{Hom}(\mathbb{Q}\mathcal{F}, U) \longrightarrow U = I^*(H)$.

Unravelling definitions we see that the restriction map is induced by applying $\mathrm{Hom}(\ , L[U])$ to the map $\pi_* : R_H = \underline{\pi}^{\mathbb{T}}_1(\sigma^0_H) \longrightarrow \underline{\pi}^{\mathbb{T}}_1(S^0) = L[\mathbb{Q}\mathcal{F}]$ induced by the projection $\sigma^0_H \longrightarrow S^0$. Since $\mathrm{Hom}(M, L[U]) = \mathrm{Hom}(M(\mathbb{T}), U)$, it is only the part of π_* at $\mathbb{T}$ that is relevant, and this is dealt with in the following lemma. □

LEMMA 12.4.2. *The map* $\pi_* : \mathbb{Q} = \pi_1^{\mathbb{T}}(\sigma_H^0) \longrightarrow \pi_1^{\mathbb{T}}(S^0) = \mathbb{Q}\mathcal{F}$ *is the inclusion of the Hth factor.*

Proof: The tom Dieck splitting isomorphism states that a certain natural map from a direct sum to $\pi_1^{\mathbb{T}}(X)$ is an isomorphism, for any space X. The Hth factor is $\pi_1(E(\mathbb{T}/H_+) \wedge_{\mathbb{T}/H} \Sigma X^H) \cong \pi_0(X^H)$, and the map $\Phi^H(\sigma_H^0) \longrightarrow (G/H_+)^H \longrightarrow (S^0)^H$ induces a bijection of π_0. □

Note in particular that $\underline{\pi}_0^{\mathbb{T}}(hI) = I$, so that if M is an arbitrary Mackey functor one may define a spectrum hM by taking a resolution $0 \longrightarrow M \longrightarrow I_0 \longrightarrow I_1 \longrightarrow 0$ of Mackey functors and realizing it as a cofibre sequence $\mathrm{hM} \longrightarrow hI_0 \longrightarrow hI_1$. It is easy to see there is a short exact sequence

$$0 \longrightarrow \mathrm{Ext}(\underline{\pi}_n^{\mathbb{T}}(\Sigma X), M) \longrightarrow hM_{\mathbb{T}}^n(X) \longrightarrow \mathrm{Hom}(\underline{\pi}_n^{\mathbb{T}}(X), M) \longrightarrow 0.$$

We remark that it is extremely easy to construct an Adams resolution by spectra hJ, and its convergence is immediate from the ordinary Whitehead theorem. Its E_2 term is also reasonably computable. However the spectral sequence does not seem useful except perhaps for maps from bounded below spectra to finite Postnikov towers.

LEMMA 12.4.3. *(i) For any $\mathcal{F}$-injective Mackey functor I with associated $h\mathcal{SB}$-module I^e there is an equivalence*

$$hI \wedge E\mathcal{F}_+ \simeq \bigvee_H \sigma_H^{-1} \wedge S^0[I^e(H)].$$

We might reasonably write $\sigma_{\mathcal{F}}^{-1}[I]$ for this spectrum.
(ii) For any $\tilde{\mathcal{F}}$-injective Mackey functor $I = L[U]$ there is an equivalence

$$hI \wedge E\mathcal{F}_+ \simeq \Sigma^{-1} E\mathcal{F}_+[U].$$

Proof: Note that

$$[S^0, hI \wedge \sigma_H^0]_k^{\mathbb{T}} = [\sigma_H^{-1}, hI]_k^{\mathbb{T}} = \underline{\pi}_{k-1}^{\mathbb{T}}(hI)(\sigma_H^0),$$

and refer to 12.4.1 for the values this takes.

For Part (ii) we argue as in the proof of 12.2.2. If I is an $\tilde{\mathcal{F}}$-injective the displayed group is only nonzero for $k = 0$. The homotopy groups follow as before, and the action of c_H follows from the fact that $hI \wedge E\langle H \rangle$ is H-equivariantly an Eilenberg-MacLane spectrum.

For Part (i) the shape is slightly different. As above we see that $\pi_*^{\mathbb{T}}(hI \wedge \sigma_H^0) = I^e(H) \oplus \Sigma I^e(H)$. Now the attaching maps $\sigma_H^{2n+1} \longrightarrow \sigma_H^{2n}$ in $E\langle H \rangle$ are non-trivial in $\pi_*^{\mathbb{T}}$, and hence so are their duals. It follows from the definition of $hI_{\mathbb{T}}^*(\bullet)$ that the map

$$\begin{array}{ccc}
[S^{2n+1}, hI \wedge \sigma_H^{2n+1}]^{\mathbb{T}} & \longrightarrow & [S^{2n+1}, hI \wedge \sigma_H^{2n}]^{\mathbb{T}} \\
\| & & \| \\
\mathrm{Hom}(\underline{\pi}_1^{\mathbb{T}}(\sigma_H^0), I) & \longrightarrow & \mathrm{Hom}(\underline{\pi}_0^{\mathbb{T}}(\sigma_H^0), I) \\
\| & & \| \\
I^e(H) & \longrightarrow & I^e(H)
\end{array}$$

is an isomorphism. Hence, by induction on n, the limit $\pi^{\mathbb{T}}_*(hI \wedge E\langle H\rangle)$ is concentrated in degree 0, where it is $I^e(H)$. □

LEMMA 12.4.4. *(i) For any $\mathcal{F}$-injective Mackey functor I with associated $h\mathcal{SB}$-module I^e we have*

$$\pi_k(\Phi^{\mathbb{T}} hI) = \begin{cases} \prod_H I^e(H)/\bigoplus_H I^e(H) & \text{if } k = 0 \\ 0 & \text{otherwise} \end{cases}$$

(ii) For any $\tilde{\mathcal{F}}$-injective Mackey functor $I = L[U]$ we have

$$\pi_k(\Phi^{\mathbb{T}} hI) = \begin{cases} U\mathcal{F} & \text{if } k \text{ is odd and } \geq 3 \\ \overline{U\mathcal{F}} & \text{if } k = 1 \\ U^{\mathcal{F}} & \text{if } k \text{ is odd and } \leq -1 \\ 0 & \text{otherwise} \end{cases}$$

where $\overline{U\mathcal{F}}$ is the kernel of the natural map $U\mathcal{F} \longrightarrow U$.

Proof: In both cases we consider the cofibre sequence $E\mathcal{F}_+ \longrightarrow S^0 \longrightarrow \tilde{E}\mathcal{F}$, and use 12.4.3 and 12.4.1 to give the groups in the long exact sequence in $\pi^{\mathbb{T}}_*$. The connecting maps are identified using the definition of hI, since it applies to identify the map $hI^0_{\mathbb{T}}(\sigma_H^{-1}) \longrightarrow hI^0_{\mathbb{T}}(S^0)$ as in the proof of 12.4.3 (i). □

COROLLARY 12.4.5. *If I is an $\mathcal{F}$-injective, then in the notation of 12.4.3*

$$hI \simeq \sigma_{\mathcal{F}}^{-1}[I] \vee \tilde{E}\mathcal{F} \wedge S^0[\prod_H I^e(H)/\bigoplus_H I^e(H)].$$

In particular, if $I^e(H)$ is only nonzero for finitely many H, then $hI \simeq \sigma_{\mathcal{F}}^{-1}[I]$ is an $\mathcal{F}$-spectrum.

Proof: Since $t^{\mathcal{F}}_*$ and $\pi^{\mathbb{T}}_*(\sigma_{\mathcal{F}}^{-1}[I])$ are in even degrees, the assembly map is detected as an element $f \in \mathrm{Hom}(t^{\mathcal{F}}_* \oplus \prod_H I^e(H)/\bigoplus_H I^e(H), \bigoplus_H I^e(H))$. But if $f \neq 0$ with $f(x)$ nonzero in the Hth component then $f(e_H x) \neq 0$; but $e_H x = c_H \tilde{x}$ for some $\tilde{x}$ with degree 2, and by dimension $f(\tilde{x}) = 0$. □

COROLLARY 12.4.6. *If $I = L[U]$ the map $hI \wedge \tilde{E}\mathcal{F} \longrightarrow hI \wedge \Sigma E\mathcal{F}_+$ is classified by the $\mathcal{O}_{\mathcal{F}}$-morphism*

$$q : \pi_*(\Phi^{\mathbb{T}} hI) \otimes t^{\mathcal{F}}_* \longrightarrow U \otimes \Sigma\mathbb{I}$$

defined by the fact that $q(x \otimes c_H^k)$ is $x(H)$ if it lies in a nonzero group; this applies whether x is in $U\mathcal{F}, \overline{U\mathcal{F}}$ or $U^{\mathcal{F}}$.

Proof: We apply the strategy of 6.1.2.

The statement of the corollary follows from the immediately preceeding discussion unless k is negative. Furthermore, since $U \otimes \Sigma\mathbb{I}$ is uniquely divisible in positive degrees, the result follows unless x is of negative degree.

Let us therefore suppose $x \in U^{\mathcal{F}}$ comes from $\pi_{-2k+1}(\Phi^{\mathbb{T}}hI)$, with $k \geq 1$. It suffices to show that $q(x \otimes c_H^{-k})$ is $x(H)$. For this we must examine the map

$$hI \wedge \tilde{E}\mathcal{F} \wedge \sigma^{-kV(H)} \longrightarrow hI \wedge \Sigma^1 E\langle H\rangle \wedge \sigma^{-kV(H)}$$

in $\pi^{\mathbb{T}}_{-2k+1} = [S^{-2k+1}, \cdot]^{\mathbb{T}}$. Transposing $\sigma^{-kV(H)}$ across into its dual in the domain, and using the fact that $\tilde{E}\langle H\rangle \longrightarrow E\langle H\rangle$ is the direct limit of maps $\sigma^{lV(H)} \longrightarrow \Sigma E\langle H\rangle^{(2l-2)}$ with dual $\sigma^{-lV(H)} \longleftarrow \Sigma^{-2l}E\langle H\rangle^{(2l-2)}$, we must examine the direct system

$$[\sigma^{kV(H)} \wedge \sigma^{-lV(H)}, \Sigma^{2k-1}hI]^{\mathbb{T}} \longrightarrow [\sigma^{kV(H)} \wedge \Sigma^{-2l}E\langle H\rangle^{(2l-2)}, \Sigma^{2k-1}hI]^{\mathbb{T}}.$$

Of course $[X, \Sigma^{2k-1}hI] = \mathrm{Hom}(\pi^{\mathbb{T}}_{2k-1}(X), U)$ because $I = L[U]$, so we only need to understand

$$\sigma^{(k-l)V(H)} \simeq \sigma^{kV(H)} \wedge \sigma^{-lV(H)} \xleftarrow{f_l} \sigma^{kV(H)} \wedge \Sigma^{-2l}E\langle H\rangle^{(2l-2)} \simeq \Sigma^{2k-2l}E\langle H\rangle^{(2l-2)}$$

in $\pi^{\mathbb{T}}_{2k-1}$. It is more convenient to consider the previous term in the cofibre sequence $\sigma^{kV(H)} \wedge S^{-1}$, and observe this has zero homotopy in positive odd degrees. The map $\pi^{\mathbb{T}}_{2k-1}(f_l)$ is thus surjective for each l and hence also in the limit. □

We may describe hI for an $\tilde{\mathcal{F}}$-injective I in more familiar terms.

LEMMA 12.4.7. *If $I = L[U]$ there is a cofibre sequence*

$$\tilde{E}\mathcal{F} \wedge S^0[U] \longrightarrow hI \longrightarrow \Sigma^{-1}F(E\mathcal{F}_+, S^0[U]).$$

Proof: By 12.4.3 we have a map

$$hI \wedge E\mathcal{F}_+ = \Sigma^{-1}E\mathcal{F}_+ \wedge S^0[U] \longrightarrow \Sigma^{-1}S^0[U],$$

and hence, by adjunction a map $hI \longrightarrow \Sigma^{-1}F(E\mathcal{F}_+, S^0[U])$. By construction this is an $\mathcal{F}$-equivalence, so that the fibre is an $\mathcal{F}$-contractible spectrum. The result follows provided we check that the map $hI \longrightarrow \Sigma^{-1}F(E\mathcal{F}_+, S^0[U])$ is an isomorphism in negative dimensional homotopy. This is easily verified by adjunction, since diagrams $S^k \longrightarrow hI \longrightarrow \Sigma^{-1}F(E\mathcal{F}_+, S^0[U])$ correspond to diagrams $S^k \wedge E\mathcal{F}_+ \longrightarrow hI \wedge E\mathcal{F}_+ \longrightarrow \Sigma^{-1}S^0[U]$. □

CHAPTER 13

Classical miscellany.

The sections in this chapter are independent of each other; each answers a natural question about rational $\mathbb{T}$-spectra.

In Section 13.1 we give a complete analysis of the behaviour of the Atiyah-Hirzebruch spectral sequence for $\mathcal{F}$-spectra, generalizing the study in Section 1.4. Section 13.2 sets up a calculable spectral sequence for calculating maps between $\mathbb{T}$-spectra from a cellular decomposition, based on the graded orbit category. Section 13.3 shows how the existence of tom Dieck splitting makes the models of suspension spectra very special. Section 13.4 finally returns to complex $\mathbb{T}$-equivariant K-theory, and identifies its place in the torsion model, showing that it is formal. Finally, Section 13.5 identifies the functional dual $DE\mathbb{T}_+$, giving the rational analogue of the geometric equivariant Segal conjecture.

13.1. The collapse of the Atiyah-Hirzebruch spectral sequence.

The purpose of this section is to analyse the Atiyah-Hirzebruch spectral sequence for $\mathcal{F}$-spectra.

We suppose given an arbitrary $\mathbb{T}$-equivariant cohomology theory $K^*_{\mathbb{T}}(\cdot)$, and consider the Atiyah-Hirzebruch spectral sequence

$$E_2^{s,t} = H^s_{\mathbb{T}}(X; \underline{K}^t_{\mathbb{T}}) \Longrightarrow K^*_{\mathbb{T}}(X).$$

This may be constructed either by using the skeletal filtration of X or, in complete generality, by using the Postnikov filtration of K [**14**, Appendix B]. It is conditionally convergent if X is bounded below.

First, let us suppose that $K = E\mathbb{T}_+^{(2n)}$; we observe that, if X is free, the only relevant part of the Mackey functor $\underline{K}^*_{\mathbb{T}}$ is the nonequivariant homotopy of K. Thus the spectral sequence is concentrated on the lines $q = 1$ and $q = 2n+2$. There are therefore no differentials except

$$d_{2n+2} : H^p(X/\mathbb{T}) = H^p_{\mathbb{T}}(X; \underline{K}^{2n+2}_{\mathbb{T}}) \longrightarrow H^{p+2n+2}_{\mathbb{T}}(X; \underline{K}^1_{\mathbb{T}}) = H^{p+2n+2}(X/\mathbb{T}).$$

If we take the special case $X = E\mathbb{T}_+^{(2m)}$, we see that (up to multiplication by a non-zero scalar) the differential must be given by multiplication by c^{n+1} in order to give the correct answer, as calculated by the Adams short exact sequence. In particular, the differential is non-zero if and only if $m > n$.

THEOREM 13.1.1. *If X is an $\mathcal{F}$-spectrum then the Atiyah-Hirzebruch spectral collapses at E_2 if $\pi^{\mathbb{T}}_*(K \wedge E\mathcal{F}_+)$ is injective over $\mathcal{O}_{\mathcal{F}}$. Conversely, if the Atiyah-Hirzebruch Spectral sequence collapses at E_2 for all $\mathcal{F}$-spectra X, then $\pi^{\mathbb{T}}_*(K \wedge E\mathcal{F}_+)$ is injective.*

More precisely, any differential d_{2i+1} is zero, and the nonzero differentials d_{2n+2} are all explained by the above example, in a sense to be made precise in the proof.

Proof: The first observation is that if X is an $\mathcal{F}$-spectrum then $K^*_{\mathbb{T}}(X) = [X,K]^*_{\mathbb{T}} = [X, K \wedge E\mathcal{F}_+]^*_{\mathbb{T}}$. Since the spectral sequence is natural, we may suppose that both X and $K \wedge E\mathcal{F}_+$ have homotopy in even degrees. We argue that if $\pi^{\mathbb{T}}_*(K \wedge E\mathcal{F}_+)$ is injective then the E_2 term is entirely in even total degrees, and hence the spectral sequence collapses.

First note that by 2.2.3 and 3.1.1

$$K^*_{\mathbb{T}}(T) = \mathrm{Hom}(\pi^{\mathbb{T}}_*(T), \pi^{\mathbb{T}}_*(K \wedge E\mathcal{F}_+)),$$

for any $\mathcal{F}$-spectrum T. Thus, taking $T = \sigma^0_H$, we see that, for any finite subgroup H, the H-equivariant basic homotopy groups of K are purely in odd degrees, since $\pi^{\mathbb{T}}_*(\sigma^0_H) = \Sigma\mathbb{Q}$ is odd. Thus the part of the graded Mackey functor $\underline{K}^*_{\mathbb{T}}$ over $\mathcal{F}$ is entirely in odd degrees. On the other hand, for any Mackey functor M, since X is an $\mathcal{F}$-spectrum, $[X, HM]^*_{\mathbb{T}} = [X, HM \wedge E\mathcal{F}_+]^*_{\mathbb{T}}$, and $HM \wedge E\mathcal{F}_+$ is a wedge of copies of $E\langle H \rangle$, with one factor for each basis element of $M^e(H) = e_H M(H)$ by 12.2.2. Thus, if we let $\mathbb{I} \otimes M = \bigoplus_H \mathbb{I}(H) \otimes M^e(H)$ we have

$$H^*_{\mathbb{T}}(X; M) = \mathrm{Hom}(\pi^{\mathbb{T}}_*(X), \Sigma\mathbb{I} \otimes M),$$

which is entirely in odd degrees. Since $\pi^H_*(K)$ is in odd degrees, we conclude $E_2^{*,*} = H^*_{\mathbb{T}}(X; \underline{K}^*_{\mathbb{T}})$ is in even total degrees as claimed.

Now suppose that the Atiyah-Hirzebruch spectral sequence does not collapse, and that $x \in E_r^{p,q}$ supports a non-zero differential $d_r(x) = y \neq 0$. We shall prove that $r = 2n+2$ for some n, and that the differential is explained by naturality and the differentials described before the statement of the theorem.

We may pick a representative $x' \in E_1^{p,q} = [X^{(p)}/X^{(p-1)}, K]^{p+q}_{\mathbb{T}}$ for x. This shows that x' is supported on a map $\sigma^p_H \longrightarrow X/X^{(p-1)} \longrightarrow \Sigma^{p+q}K$. Replacing X by $X/X^{(p-1)}$ and suspending, we may assume that X is (-1)-connected and $p = 0$. We may therefore replace K by its connective cover K_0^∞ without changing the fate of x in the spectral sequence. Now, letting K_m^n denote the Postnikov section of K with non-zero homotopy groups in degree i

with $m \leq i \leq n$, we consider the Postnikov tower of K:

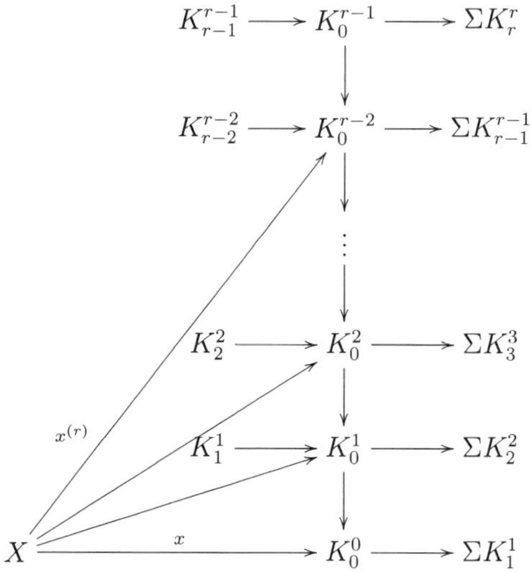

By hypothesis, $x : X \longrightarrow K^0_0$ lifts to $x^{(r)} : X \longrightarrow K^{r-2}_0$ so that the composite $x^{(r)} : X \longrightarrow K^{r-2}_0 \longrightarrow \Sigma K^{r-1}_{r-1}$ is essential and represents y. Since X is an $\mathcal{F}$-spectrum, the behaviour is unaltered if the diagram is smashed with $E\mathcal{F}_+$. Since $HM \wedge E\mathcal{F}_+$ is injective (12.2.2), all maps at the E_2-term are detected by their d-invariant, and it is thus appropriate to examine the effect of taking homotopy of the above diagram smashed with $E\mathcal{F}_+$. The basic fact is 12.2.2, stating that

$$\pi^{\mathbb{T}}_*(HM \wedge E\mathcal{F}_+) = \Sigma \mathbb{I} \otimes M,$$

which is in odd degrees. Furthermore, x is detected by $\pi^{\mathbb{T}}_1$, and we need only look at the odd graded part. This immediately shows that all odd differentials are zero, so that $r = 2n + 2$ for some n.

Next, we note that the maps

$$\pi^{\mathbb{T}}_*(K^{2s+1}_0 \wedge E\mathcal{F}_+) \longrightarrow \pi^{\mathbb{T}}_*(K^{2s}_0 \wedge E\mathcal{F}_+)$$

are injective in odd degrees, whilst the maps

$$\pi^{\mathbb{T}}_*(K^{2s+2}_0 \wedge E\mathcal{F}_+) \longrightarrow \pi^{\mathbb{T}}_*(K^{2s+1}_0 \wedge E\mathcal{F}_+)$$

are surjective in odd degrees. Furthermore, the image of the composite consists of elements divisible by c. We thus find the diagram

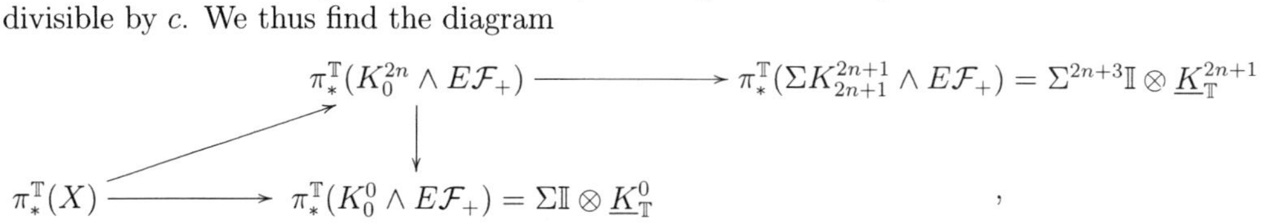

and we know that some element $\tilde{z}$ of $\pi^{\mathbb{T}}_{2n+3}(X)$ maps to $z \in \pi^{\mathbb{T}}_{2n+3}(K^{2n}_0 \wedge E\mathcal{F}_+)$, and $c^{n+1}z$ detects x, whilst the image of z in $\pi^{\mathbb{T}}_{2n+3}(\Sigma K^{2n+1}_{2n+1} \wedge E\mathcal{F}_+)$ detects y. We therefore find a map

$$\pi_*^{\mathbb{T}}(E\langle H\rangle^{(2n+2)}) = \Sigma\mathbb{I}_0^{2n+2} \longrightarrow \pi_*^{\mathbb{T}}(X)$$

with $\tilde{z}$ as the image of the top class, and $\tilde{x}$ as the image of the bottom class. By the Adams short exact sequence 3.1.1, this is realized by a map $E\langle H\rangle^{(2n+2)} \longrightarrow X$. □

13.2. Orbit category resolutions.

Integrally one expects the cellular decomposition to be unhelpful in global calculations because one does not know the stable homotopy groups of spheres. Rationally, everything is much simpler. For finite groups, cells are Eilenberg-MacLane spectra, and hence the cellular decomposition is simply another way of viewing the complete splitting [**14**]. In the present $\mathbb{T}$-equivariant context, cells are not all Eilenberg-MacLane spectra, but one may understand the entire graded category of natural or basic cells. We shall concentrate on the graded category $h\mathcal{SB}_*$ of basic cells (i.e. the full subcategory of the graded stable category with the basic cells as objects).

One thus views the entire homotopy functor $\underline{\pi}_*^{\mathbb{T}}(X)$ of X as a module over the graded category $h\mathcal{SB}_*$. By the Yoneda lemma, the case when X is a cell plays the role of a free object, and a resolution is a form of cellular approximation. We understand maps of degree 0 from the discussion of Mackey functors presented in Appendix A. Referring to 2.1.6, we see that composition in $h\mathcal{SB}_*$ is usually zero for dimensional reasons. In fact, there are no maps of degree 2 between any pair of objects, and the only case with maps of degree more than 1 is $[\sigma_{\mathbb{T}}^0, \sigma_{\mathbb{T}}^0]_*^{\mathbb{T}} = \mathbb{Q} \oplus \Sigma(\mathbb{Q}\mathcal{F}[c^{-1}])$. For maps of degree 1 we have $[\sigma_{\mathbb{T}}^0, \sigma_{\mathbb{T}}^0]_1^{\mathbb{T}} = \mathbb{Q}\mathcal{F}$, $[\sigma_{\mathbb{T}}^0, \sigma_H^0]_1^{\mathbb{T}} = \mathbb{Q}$, $[\sigma_H^0, \sigma_{\mathbb{T}}^0]_1^{\mathbb{T}} = 0$ and $[\sigma_H^0, \sigma_H^0]_1^{\mathbb{T}} = \mathbb{Q}$; for maps of degree 0 we have $[\sigma_{\mathbb{T}}^0, \sigma_{\mathbb{T}}^0]_0^{\mathbb{T}} = \mathbb{Q}$, $[\sigma_{\mathbb{T}}^0, \sigma_H^0]_0^{\mathbb{T}} = 0$, $[\sigma_H^0, \sigma_{\mathbb{T}}^0]_0^{\mathbb{T}} = \mathbb{Q}$ and $[\sigma_H^0, \sigma_H^0]_0^{\mathbb{T}} = \mathbb{Q}$. It therefore remains to deal with the composite of a degree 1 morphism (which must be of form $x_{\mathbb{T}}^{\mathbb{T}} : \sigma_{\mathbb{T}}^1 \longrightarrow \sigma_{\mathbb{T}}^0$, $x_H^H : \sigma_H^1 \longrightarrow \sigma_H^0$ or a multiple of the transfer $tr_H^{\mathbb{T}} : \sigma_{\mathbb{T}}^1 \longrightarrow \sigma_H^0$) and a degree 0 morphism (which is either a multiple of the relevant identity or a multiple of the projection $pr_{\mathbb{T}}^H : \sigma_H^0 \longrightarrow \sigma_{\mathbb{T}}^0$). We now deal with these remaining cases.

LEMMA 13.2.1. *The composites are as follows.*
(i) $x_{\mathbb{T}}^{\mathbb{T}} pr_{\mathbb{T}}^H = 0$ *and* $pr_{\mathbb{T}}^H x_H^H = 0$,
(ii) $tr_H^{\mathbb{T}} pr_{\mathbb{T}}^H \neq 0$ *and* $pr_{\mathbb{T}}^H tr_H^{\mathbb{T}}$ *corresponds to the inclusion of the* H*th factor in* $\mathbb{Q}\mathcal{F} = [\sigma_{\mathbb{T}}^0, \sigma_{\mathbb{T}}^0]_1^{\mathbb{T}}$.

Proof: Part (i) is clear since the composites lie in the zero group. The first fact in Part (ii) follows from the explicit geometric construction of the transfer. The second fact in Part (ii) follows by construction of the isomorphism in tom Dieck splitting [**19**, V.11]. □

It is thus natural to take $\tau_H = tr_H^{\mathbb{T}} pr_{\mathbb{T}}^H$ as the basic generator of $[\sigma_H^0, \sigma_H^0]_1^{\mathbb{T}}$ and $\delta_H = pr_{\mathbb{T}}^H tr_H^{\mathbb{T}}$ as a standard basis element of $[\sigma_{\mathbb{T}}^0, \sigma_{\mathbb{T}}^0]_1^{\mathbb{T}}$.

Now, suppose given any $\mathbb{T}$-spectrum X. We consider $[\cdot, X]_*^{\mathbb{T}}$ as a contravariant functor on the graded orbit category. As such, we may form a projective resolution, and we may realize it. In fact we may construct a map $P_0 \longrightarrow X$, which is surjective in graded equivariant homotopy for all subgroups of $\mathbb{T}$, and in which P_0 is a wedge of cells. Now let X_1 be the cofibre of this, and iterate to form the diagram

$$\begin{array}{ccccccc} X = X_0 & \longrightarrow & X_1 & \longrightarrow & X_2 & \longrightarrow & \cdots \\ \uparrow & & \uparrow & & \uparrow & & \\ P_0 & & P_1 & & P_2 & & \end{array}$$

By construction, all the maps $X_s \longrightarrow X_{s+1}$ are zero in H-equivariant homotopy for all subgroups H, and so $\underrightarrow{\mathrm{holim}}_s\, X_s$ is contractible by the Whitehead theorem.

It is convenient to form the dual diagram with $X^{p-1} = \mathit{fibre}(X \longrightarrow X_p)$, so that $X^{-1} = *$ and $X \simeq \underrightarrow{\mathrm{holim}}\, X^p$:

$$\begin{array}{ccccccc} * = X^{-1} & \longrightarrow & X^0 & \longrightarrow & X^1 & \longrightarrow & \cdots \\ \uparrow & & \uparrow & & \uparrow & & \\ \Sigma^{-1}P_0 & & \Sigma^{-1}P_1 & & \Sigma^{-1}P_2 & & \end{array}$$

Replacing the maps by inclusions, we view this as a filtration of X with subquotients $X^p/X^{p-1} = P_p$. We may now construct a spectral sequence by applying $[\cdot, Y]^*_{\mathbb{T}}$ to the diagram. It has $E_1^{p,q} = [P_p, Y]^{p+q}_{\mathbb{T}}$, and $D_1^{p,q} = [X^p, Y]^{p+q}_{\mathbb{T}}$. The spectral sequence lies in the right half-plane, and the differentials are cohomological, so that $d_r : E_r^{p,q} \longrightarrow E_r^{p+r,q-r+1}$, and it is evidently conditionally convergent. Finally, we see by construction that the sequence

$$\cdots \longrightarrow \underline{\pi}^{\mathbb{T}}_*(\Sigma^{-2}P_2) \longrightarrow \underline{\pi}^{\mathbb{T}}_*(\Sigma^{-1}P_1) \longrightarrow \underline{\pi}^{\mathbb{T}}_*(P_0) \longrightarrow \underline{\pi}^{\mathbb{T}}_*(X) \longrightarrow 0$$

is exact. Hence we may identify the E_2 term as an Ext group, and the spectral sequence takes the form

$$E_2^{p,q} = \mathrm{Ext}^{p,q}_{h\mathcal{SB}_*}(\underline{\pi}^{\mathbb{T}}_*(X), \underline{\pi}^{\mathbb{T}}_*(Y)) \Longrightarrow [X, Y]^{p+q}_{\mathbb{T}}.$$

The fact that the category of Mackey functors is one dimensional gives a vanishing line if X is bounded below. For definiteness, suppose X is (-1)-connected. Since cells are also (-1)-connected, we may ensure P_0 is (-1)-connected, and hence that X_1 is 0-connected. This formality proves that if the resolution is dimensionally minimal, X_p is $(p-1)$-connected. However, if we use *basic* cells and the fact that the category of Mackey functors is of projective dimension 1, we may ensure that X_2 is 2-connected. By iteration we see that X_{2s} and P_{2s} are $(3s-1)$-connected; similarly X_{2s+1} and P_{2s+1} are $3s$-connected. Thus the map $X^{2s-1} \longrightarrow X$ is $(3s-2)$-connected, and the map $X^{2s} \longrightarrow X$ is $(3s-1)$-connected. Unfortunately this only seems to be useful if the homotopy of the spectrum Y is bounded *above*. Thus if $Y^n_H = 0$ for $n \leq -1$ and all H, then the nonzero entries are in the first quadrant and lie on or above the line $q = (p-1)/2$.

13.3. Suspension spectra.

In this section we suppose given a based $\mathbb{T}$-space Z, and attempt to identify the place of its suspension spectrum in our classification. We follow our usual convention of omitting notation for the suspension spectrum functor, and using the notation $\Psi^{\mathbb{T}}$ for the Lewis-May fixed point functor.

The basic tool is tom Dieck splitting [**19**, V.11.1], which states that the Lewis-May fixed point spectrum of the suspension spectrum of Z is

$$\Psi^{\mathbb{T}} Z = Z \vee \bigvee_K E\mathbb{T}/K_+ \wedge_{\mathbb{T}/K} \Sigma Z^K;$$

furthermore this is natural, and applies to stable retracts of spaces. The crucial simplification for spaces is that there is a map $Z^{\mathbb{T}} \longrightarrow Z$ of $\mathbb{T}$-spectra, and hence a diagram

$$\begin{array}{ccc} \tilde{E}\mathcal{F} \wedge Z^{\mathbb{T}} & \longrightarrow & \Sigma E\mathcal{F}_+ \wedge Z^{\mathbb{T}} \\ \simeq\downarrow & & \downarrow \\ \tilde{E}\mathcal{F} \wedge Z & \longrightarrow & \Sigma E\mathcal{F}_+ \wedge Z. \end{array}$$

The structure map $\tilde{E}\mathcal{F} \wedge Z \longrightarrow \Sigma E\mathcal{F}_+ \wedge Z$ thus factors through the corresponding map for $Z^{\mathbb{T}}$, which we understand completely, since $Z^{\mathbb{T}}$ is rationally a wedge of spheres.

On the other hand, by naturality of tom Dieck splitting, we find

$$\pi_*^{\mathbb{T}}(Z \wedge E\langle H\rangle) = \pi_*(E\mathbb{T}/H_+ \wedge_{\mathbb{T}/H} \Sigma Z^H) = H_*(E\mathbb{T}/H_+ \wedge_{\mathbb{T}/H} \Sigma Z^H),$$

which we regard as computable.

SUMMARY 13.3.1. The algebraic model of the suspension spectrum of a space Z in the torsion model has homology

$$t_*^{\mathcal{F}} \otimes H_*(Z^{\mathbb{T}}) \longrightarrow$$
$$\Sigma^2 \mathbb{I} \otimes H_*(Z^{\mathbb{T}}) = \Sigma^2 \bigoplus_H H_*(E\mathbb{T}/H_+ \wedge_{\mathbb{T}/H} Z^{\mathbb{T}}) \longrightarrow \Sigma^2 \bigoplus_H H_*(E\mathbb{T}/H_+ \wedge_{\mathbb{T}/H} Z^H).$$

Note that this need not be formal, so this does not always determine the model exactly. The first map in the displayed composite necessarily has zero e-invariant, and is simply induced by the quotient $t_*^{\mathcal{F}} \longrightarrow \Sigma^2\mathbb{I}$. However the second map is induced by the inclusion $X^{\mathbb{T}} \longrightarrow X$, and may have non-zero d and e invariant. We may be satisfied that the d-invariant is given by a homology calculation. For the e invariant, since the above discussion applies to retracts of spaces, we may assume that $X^{\mathbb{T}}$ and $E\mathcal{F}_+ \wedge X$ are of pure parity, and then identify the e invariant with the Borel homology of the cofibres $X^H/X^{\mathbb{T}}$ as in [**1**].

There remains the question of whether this characterizes suspension spectra. More precisely, we cannot distinguish $\mathbb{T}$-fixed suspensions so that we are asking if every model of the above sort is the model of a spectrum $\Sigma^n\Sigma^\infty X$ for some $\mathbb{T}$-space X and some integer n. We do not have available the option of identifying the suspension spectrum functor, since no algebraic model of rational $\mathbb{T}$-spaces is available. One might hope to realize finitely generated examples by explicit construction, but one would expect a certain amount of suspension to be necessary to achieve stability in each case. To obtain a global realization one would need a bound on these suspensions; since the model contains no data relevant to the achievement of stability, there is no ready way to do this.

13.4. K-theory revisited.

Let us consider the structure of the $\mathbb{T}$-spectrum K representing rational equivariant K-theory. We know that $K_*^H = R(H)[\beta, \beta^{-1}]$ for all subgroups H, where $R(H)$ is the rationalization of the complex representation ring and $\beta \in K_{-2}$ is the Bott element. More explicitly, $R = R(\mathbb{T}) = \mathbb{Q}[z, z^{-1}]$, and $R(H) = \mathbb{Q}[z]/(z^n - 1)$ when H is of order n. The restriction maps are implicit in the notation here, and the induction maps to $\mathbb{T}$ are zero (holomorphic induction maps are not part of the structure). Now, by Bott periodicity we have K-theory Euler classes $c(H)$ of degree -2 for each finite subgroup H, obtained by applying the Thom isomorphism to the image of $e(V(H))$ in K-theory. We may apply Bott periodicity to obtain the usual K-theory Euler class $\lambda(H) \in R(\mathbb{T})$. In other words if H is of order n we have

$$\lambda(H) = 1 - z^n \text{ and } \chi(H) = \beta\lambda(H).$$

Furthermore

$$\lambda(H) = \prod_{d|n} \Phi_d,$$

where Φ_d is the dth cyclotomic polynomial. Let S be the multiplicative set generated by the Euler classes $1 - z^n$, and T be generated by the cyclotomic polynomials Φ_d; in practice the geometry localizes so as to invert S, which is algebraically the same as inverting T, and the latter is easier to understand. Let $F = S^{-1}R = T^{-1}R$, so that

$$\pi_*^{\mathbb{T}}(K \wedge \tilde{E}\mathcal{F}) \cong F[\beta, \beta^{-1}] \text{ and } \pi_*^{\mathbb{T}}(K \wedge \Sigma E\mathcal{F}_+) \cong (F/R)[\beta, \beta^{-1}].$$

Since the cyclotomic polynomials are coprime, an element of F/R can be written uniquely as a sum of terms $f_d(z)/\Phi_d(z)^n$ for $n \geq 1$ where $f_d(z) \in R$ is not divisible by $\Phi_d(z)$. Hence

$$F/R = \bigoplus_n R[1/\Phi_n]/R.$$

We should relate this to our geometric decompositions.

LEMMA 13.4.1. *If $K(H)$ denotes the part of K in $\mathbb{T}$-Spec/H as usual, then*

$$\pi_*^{\mathbb{T}}(K(H)) = (R[1/\Phi_n]/R)[\beta, \beta^{-1}]$$

and c_H acts as multiplication by $\beta\Phi_n$ times an automorphism.

REMARK 13.4.2. Our map c_H is defined up to a non-zero rational number, whilst the K-theory Euler class is defined absolutely. The automorphism is therefore well defined up to a non-zero scalar, but its exact description is not relevant to us at present. Crabb studies the relationship in greater detail [**5**].

Proof: The only part requiring proof is that the action of c_H is as claimed. We shall show that c_H acts as $c(V(H))$ times an automorphism on $K \wedge E\langle H\rangle$. To do this, we let $V = V(H)$, and unravel definitions.

We have a K-theory Thom isomorphism $t : K \wedge S^V \xrightarrow{\cong} K \wedge S^{|V|}$, and the $\mathcal{F}$-spectrum Thom isomorphism $\tau : S^V \wedge E\langle H\rangle \xrightarrow{\cong} S^{|V|} \wedge E\langle H\rangle$; we need to know they are compatible in the sense that the composites

$$K \wedge S^0 \wedge E\langle H\rangle \xrightarrow{e(V)} K \wedge S^V \wedge E\langle H\rangle \underset{1\wedge\tau}{\overset{t\wedge 1}{\rightrightarrows}} K \wedge S^{|V|} \wedge E\langle H\rangle$$

are unit multiples of each other. Now, we observe that both are maps of K-module spectra, and hence it is sufficient to show both induce the same map $S^0 \wedge E\langle H\rangle \longrightarrow K \wedge S^{|V|} \wedge E\langle H\rangle$.

Maps of this form are classified by $K_{\mathbb{T}}^{|V|}(E\langle H\rangle) \cong \lim_{\leftarrow n} K^{|V|}(E\langle H\rangle^{(2n)})$, and hence by what they induce in homotopy:

$$[E\langle H\rangle \wedge K, S^{|V|} \wedge E\langle H\rangle \wedge K]_0^{K,\mathbb{T}} \xrightarrow{\cong} \mathrm{Hom}_R(R/\Phi_n^\infty, R/\Phi_n^\infty).$$

We know the K-theory Thom isomorphism gives multiplication by Φ_n. We may express the action of c_H in its Φ_n-adic expansion as multiplication by $x_0 + x_1\Phi_n + x_2\Phi_n^2 + \cdots$, where $x_i \in R$. It suffices to prove that the $\mathcal{F}$-spectrum Euler class is multiplication by $\lambda\Phi_n$ mod Φ_n^2 for a non-zero $\lambda \in \mathbb{Q}$, i.e. that $x_0 = 0$ and $x_1 = \lambda \neq 0$.

First, we know that the map $\sigma_H^0 \longrightarrow \sigma_H^0 \wedge K$ induces the permutation module map $\mathbb{Q} = e_H A(H) \longrightarrow e_H R(H) = R/\Phi_n$. Next, we need to understand something of the map $E\langle H\rangle^{(2n)} \longrightarrow K \wedge E\langle H\rangle^{(2n)}$, and we can infer enough by considering the diagram

$$\begin{array}{ccccc}
0 & & \mathbb{Q} & & \mathbb{Q} \\
\| & & \| & & \| \\
\pi_{2k+1}^{\mathbb{T}}(E\langle H\rangle^{(2k-2)}) & \longrightarrow & \pi_{2k+1}^{\mathbb{T}}(E\langle H\rangle^{(2k)}) & \xrightarrow{\cong} & \pi_{2k+1}^{\mathbb{T}}(\sigma_H^{2k}) \\
\downarrow & & \downarrow & & \downarrow \\
\pi_{2k+1}^{\mathbb{T}}(E\langle H\rangle^{(2k-2)} \wedge K) & \longrightarrow & \pi_{2k+1}^{\mathbb{T}}(E\langle H\rangle^{(2k)} \wedge K) & \longrightarrow & \pi_{2k+1}^{\mathbb{T}}(\sigma_H^{2k} \wedge K) \\
\| & & \| & & \| \\
R/\Phi_n^k & & R/\Phi_n^{k+1} & & R/\Phi_n.
\end{array}$$

This shows that the image g_k of the generator of $\pi_{2k+1}^{\mathbb{T}}(E\langle H\rangle^{(2k)})$ in $\pi_{2k+1}^{\mathbb{T}}(E\langle H\rangle^{(2k)} \wedge K) = R/\Phi_n^{k+1}$ is λ_k modulo Φ_n^k for a non-zero rational number λ_k.

Now consider the diagram

$$\begin{array}{ccc}
\mathbb{Q} & & \mathbb{Q} \\
\| & & \| \\
\pi_{2k+1}^{\mathbb{T}}(E\langle H\rangle^{(2k)}) & \xrightarrow{\cong} & \pi_{2k+1}^{\mathbb{T}}(\Sigma^{kV} E\langle H\rangle^{(0)}) \\
\cong\downarrow & & \downarrow\cong \\
\pi_{2k+1}^{\mathbb{T}}(E\langle H\rangle) & \xrightarrow{c_H^k} & \pi_{2k+1}^{\mathbb{T}}(\Sigma^{kV} E\langle H\rangle) \\
\downarrow & & \downarrow \\
\pi_{2k+1}^{\mathbb{T}}(E\langle H\rangle \wedge K) & \longrightarrow & \pi_{2k+1}^{\mathbb{T}}(\Sigma^{kV} E\langle H\rangle \wedge K) \\
\| & & \| \\
R/\Phi_n^\infty & & R/\Phi_n^\infty.
\end{array}$$

We see that c_H^k takes the image of g_k in $R/\Phi_n^{k+1} \subseteq R/\Phi_n^\infty$ to the image of $1 \in R/\Phi_n$. Now g_k is mapped to λ_k/Φ_n^{k+1} modulo elements annihilated by Φ_n^k. We conclude from the case $k = 1$ that $x_1 = \lambda_1$ as required. The general case shows that $\lambda_k = \lambda_1^k$. □

COROLLARY 13.4.3. *The $\mathbb{Q}[c_H]$-module $\pi_*^{\mathbb{T}}(K(H))$ is injective. Indeed, if L_n^i is the space of Laurent polynomials in z with poles of order at most i at a primitive nth root of unity, then multiplication by Φ_n gives an isomorphism $L_n^{i+1}/L_n^i \xrightarrow{\cong} L_n^i/L_n^{i-1}$. Thus $\pi_*^{\mathbb{T}}(K(H)) \cong \mathbb{I}(H) \otimes (L_n^1/L_n^0)[\beta,\beta^{-1}]$, and hence*

$$K(H) \simeq E\langle H\rangle \wedge S^0[\,(L_n^1/L_n^0)[\beta,\beta^{-1}]\,]. \quad \square$$

Accordingly K is characterized by $t_*^{\mathcal{F}} \otimes \pi_*^{\mathbb{T}}(K \wedge \tilde{E}\mathcal{F})$, $\pi_*^{\mathbb{T}}(K \wedge \Sigma E\mathcal{F}_+)$ and the homomorphism between them. To make sense of the following statement, note that $(F/R)[\beta, \beta^{-1}]$ is a module over $\mathcal{O}_{\mathcal{F}}$, since it is $\mathcal{F}$-finite: more explicitly, if $x \in R[1/\Phi_n]/R$ and H is of order n, then $c_H x$ is a unit multiple of $\Phi_n x \beta$ as mentioned above. Multiplication by Φ_n^{-1} is not defined on F/R, but it makes sense for F. In the following statement, $\overline{x}$ denotes the image of x modulo R.

THEOREM 13.4.4. *The $\mathbb{T}$-spectrum K is the unique $\mathbb{T}$-spectrum for which*

$$t_*^{\mathcal{F}} \otimes \pi_*^{\mathbb{T}}(K \wedge \tilde{E}\mathcal{F}) \longrightarrow \pi_*^{\mathbb{T}}(K \wedge \Sigma E\mathcal{F}_+)$$

is the map

$$\hat{q}_K : t_*^{\mathcal{F}} \otimes F[\beta, \beta^{-1}] \longrightarrow (F/R)[\beta, \beta^{-1}]$$

described as follows. For $x \in \mathcal{O}_{\mathcal{F}}$ of degree $-2k$,

$$\hat{q}_K(x \otimes f\beta^l) = x\overline{f}\beta^{l+k} \text{ and } \hat{q}_K(c_H^{-k} \otimes f\beta^l) = \overline{\Phi_n^{-k} f}\beta^{l-k}.$$

Proof: The value of $\hat{q}_K(x \otimes f\beta^l)$ is immediate from our method of calculating $\pi_*^{\mathbb{T}}(K \wedge \Sigma E\mathcal{F}_+)$.

For $\hat{q}_K(c_H^{-k} \otimes f\beta^l)$ we apply 6.1.2, using the compatibility statement in 13.4.1 to relate it to our present naming of elements. Consider the diagram

$$\begin{array}{ccccccc}
K & \xleftarrow{1\wedge e} & K \wedge S^{-kV(H)} & \xrightarrow{\simeq} & K \wedge S^{-2k} & \xrightarrow{\simeq} & K \\
\downarrow & & \downarrow & & \downarrow & & \downarrow \\
K \wedge \tilde{E}\mathcal{F} & \xleftarrow{\simeq} & K \wedge \tilde{E}\mathcal{F} \wedge S^{-kV(H)} & \xrightarrow{\simeq} & K \wedge \tilde{E}\mathcal{F} \wedge S^{-2k} & \xrightarrow{\simeq} & K \wedge \tilde{E}\mathcal{F} \\
\downarrow & & r\downarrow & & \downarrow & & \downarrow \\
K \wedge \Sigma E\mathcal{F}_+ & \longleftarrow & K \wedge \Sigma E\mathcal{F}_+ \wedge S^{-kV(H)} & \xrightarrow{\simeq} & K \wedge \Sigma E\mathcal{F}_+ \wedge S^{-2k} & \xrightarrow{\simeq} & K \wedge \Sigma E\mathcal{F}_+.
\end{array}$$

The first horizontal in each row is induced by $e : S^{-kV(H)} \longrightarrow S^0$, the second is the K-theory Thom class, and the third is multiplication by the integer degree Bott class. By 6.1.2 the relevant map is induced by r.

Applying $\pi_*^{\mathbb{T}}$ we see by definition that the composite from right to left in the first row is multiplication by $\lambda(kV(H))$, and hence this is also true in the second row. Since the bottom right hand vertical induces projection, we identify the second vertical r in the lower ladder as stated. □

It is tempting to rewrite the description of $\hat{q}_K$ above as $\hat{q}_K(x \otimes f\beta^l) = \overline{xf}\beta^{l+k}$ for all $x \in t_*^{\mathcal{F}}$, but this makes no sense, since F is not an $\mathcal{O}_{\mathcal{F}}$-module. This suggests we should perform some algebraic completion to F. This can be achieved geometrically by replacing K with $F(E\mathcal{F}_+, K)$. This is a reasonable thing to do since the fibre of the completion map is the $\mathcal{F}$-contractible spectrum $F(\tilde{E}\mathcal{F}, K)$, which is determined by its homotopy groups. On the other hand this method does not appear to add to our knowledge.

The fact that K theory has injective dimension 1 in the torsion model, shows that $K_{\mathbb{T}}^*(X)$ only depends on the homomorphism $t_*^{\mathcal{F}} \otimes \pi_*(\Phi^{\mathbb{T}} X) \longrightarrow \pi_*^{\mathbb{T}}(\Sigma E\mathcal{F}_+ \wedge X)$. Combining this with the results of 13.3, we obtain a partial substitute for the Chern isomorphism.

COROLLARY 13.4.5. *If X is a $\mathbb{T}$-space, then $K^*_{\mathbb{T}}(X)$ only depends on the map*

$$\mathbb{I} \otimes H_*(X^{\mathbb{T}}) = H_*(E\mathcal{F}_+ \wedge_{\mathbb{T}} X^{\mathbb{T}}) \longrightarrow H_*(E\mathcal{F}_+ \wedge_{\mathbb{T}} X)$$

induced by the inclusion $X^{\mathbb{T}} \longrightarrow X$.

It is not hard to do the relevant homological algebra, to make the dependence explicit.

13.5. The geometric equivariant rational Segal conjecture for $\mathbb{T}$.

In this section we aim to analyse the functional dual $DE\mathbb{T}_+ = F(E\mathbb{T}_+, S^0)$; the title is something of a misnomer since neither Segal nor anyone else has made a conjecture about $DE\mathbb{T}_+$. It completes the description given in [**8, 11**] of the integral functional dual. We shall show in Part IV (specifically, in 24.2.4) that it is possible to give an entirely algebraic treatment using models of function spectra but here we give a direct treatment.

Since $E\mathbb{T}_+ = E\langle 1\rangle$, we should really discuss the more general question of identifying $DE\langle H\rangle$. Indeed, we should also consider $DE\mathcal{F}_+ \simeq D(\bigvee_H E\langle H\rangle) \simeq \prod_H DE\langle H\rangle$. Since the initial stages of the analysis are easier to understand for $E\mathcal{F}_+$, we begin with that.

We already know $E\mathcal{F}_+ \wedge DE\mathcal{F}_+ \simeq E\mathcal{F}_+$, and that $\pi^{\mathbb{T}}_*(DE\mathcal{F}_+) = t^{\mathcal{F}}_*$; accordingly we have the cofibre sequence

$$DE\mathcal{F}_+ \longrightarrow \tilde{E}\mathcal{F} \wedge S^0[t^{\mathcal{F}}_*] \stackrel{a}{\longrightarrow} \Sigma E\mathcal{F}_+,$$

where the induced map $a_* : t^{\mathcal{F}}_* \longrightarrow \Sigma^2\mathbb{I}$ is projection onto the positive dimensional part. Since $\Sigma E\mathcal{F}_+$ is injective, this is determined by the element $\hat{a} \in \mathrm{Hom}_{\mathcal{O}_{\mathcal{F}}}(t^{\mathcal{F}}_* \otimes t^{\mathcal{F}}_*, \Sigma^2\mathbb{I})$ given by smashing a with $DE\mathcal{F}_+$ and looking in $\pi^{\mathbb{T}}_*$. Equivalently, $DE\mathcal{F}_+$ is formal and is described in the torsion model by the object

$$t^{\mathcal{F}}_* \otimes t^{\mathcal{F}}_* \stackrel{\hat{a}}{\longrightarrow} \Sigma^2\mathbb{I}.$$

For definiteness we emphasize that the second copy of $t^{\mathcal{F}}_*$ is the vertex.

LEMMA 13.5.1. *The map $\hat{a} : t^{\mathcal{F}}_* \otimes t^{\mathcal{F}}_* \longrightarrow \Sigma^2\mathbb{I}$ is given by $\hat{a}(x \otimes y) = a_*(xy)$.*

Note that $\hat{a}$ realizes Tate duality 5.2.2 between negative and positive parts of $t^{\mathcal{F}}_*$.

Proof: We apply the method summarized in 6.1.2.

Given an $\mathcal{O}_{\mathcal{F}}$-map $\theta : t^{\mathcal{F}}_* \longrightarrow M$, so that $\theta(1) = m$ the value of θ on $\mathcal{O}_{\mathcal{F}} \subseteq t^{\mathcal{F}}_*$ follows, and if the components of m are uniquely divisible by the relevant c_H then θ is determined. The result follows provided x is of positive degree. The real content of the lemma is that the formula is valid also when $a_*(x) = 0$.

To begin with we note how a nontrivial map $\tilde{E}\mathcal{F} \longrightarrow \Sigma^{2k+1}E\mathcal{F}_+$ is detected. As motivation, we observe that the obvious example is the quotient of $\tilde{E}\mathcal{F} = S^{\infty V(\mathcal{F})}$ by $S^{kV(\mathcal{F})}$. If it were possible to simply desuspend by smashing with a putative spectrum $S^{(-k-1)V(\mathcal{F})}$, then the map would be detected in homotopy. Since the spectrum $S^{V(\mathcal{F})}$ is not invertible we must be slightly less direct by concentrating on a single subgroup H at a time, and using $\sigma^{kV(H)}$ instead of $S^{kV(\mathcal{F})}$. This part of the analysis is given in the proof of 13.5.2 below. □

The following identifies $DE\langle H\rangle$, and the special case $H = 1$ gives $DE\mathbb{T}_+$.

PROPOSITION 13.5.2. *There is a cofibre sequence*

$$DE\langle H\rangle \longrightarrow \tilde{E}\mathcal{F} \wedge S^0[\,\mathbb{Q}[c_H, c_H^{-1}]\,] \xrightarrow{b} \Sigma E\langle H\rangle,$$

where the induced map $b_* : \mathbb{Q}[c_H, c_H^{-1}] \longrightarrow \Sigma^2\mathbb{I}(H)$ *is projection onto the positive dimensional part. The cofibre sequence is determined by the fact that*

$$\hat{b} \in \mathrm{Hom}_{\mathcal{O}_\mathcal{F}}(t_*^\mathcal{F} \otimes \mathbb{Q}[c_H, c_H^{-1}], \Sigma^2\mathbb{I}(H)) = \mathrm{Hom}_{\mathbb{Q}[c_H]}(\mathbb{Q}[c_H, c_H^{-1}] \otimes \mathbb{Q}[c_H, c_H^{-1}], \Sigma^2\mathbb{I}(H))$$

is given by $\hat{b}(x \otimes y) = b_*(xy)$, *which represents a perfect duality of* $\mathbb{Q}[c_H, c_H^{-1}]$. *Equivalently,* $DE\langle H\rangle$ *is formal and represented in the torsion model by the object*

$$t_*^\mathcal{F} \otimes \mathbb{Q}[c_H, c_H^{-1}] \xrightarrow{\hat{b}} \Sigma^2\mathbb{I}(H).$$

Proof: We smash the standard cofibre sequence $E\mathcal{F}_+ \longrightarrow S^0 \longrightarrow \tilde{E}\mathcal{F}$ with $DE\langle H\rangle$; the terms are identified with those in the statement by the following lemma.

LEMMA 13.5.3. *(i) There is an equivalence*

$$E\mathcal{F}_+ \wedge DE\langle H\rangle \simeq E\langle H\rangle.$$

(ii) There is an isomorphism

$$\pi_*^\mathbb{T}(\Phi^\mathbb{T} DE\langle H\rangle) \cong \mathbb{Q}[c_H, c_H^{-1}].$$

Remark: Part (i) of 13.5.3 with $H = 1$ corrects statements in the rational analysis of [**8**]. More precisely, the space $E\mathcal{F}_+$ should be replaced by EG_+ in 1.6, Theorem B, 4.8, and on the right hand side of 4.5 and page 359 line -3. The correction is discussed in more detail in [**11**].

Proof of 13.5.3: (i) The equivalence $E\mathcal{F}_+ \wedge DE\langle H\rangle \simeq E\langle H\rangle$ follows since $E\mathcal{F}_+ \wedge DE\langle H\rangle$ is a retract of $E\mathcal{F}_+ \wedge DE\mathcal{F}_+ \simeq E\mathcal{F}_+$. Indeed $E\mathcal{F}_+ \simeq \bigvee_H E\langle H\rangle$, and the idempotent for all subgroups $K \neq H$ annihilates $DE\langle H\rangle$. Hence $E\mathcal{F}_+ \wedge DE\langle H\rangle$ is a retract of $E\langle H\rangle$, and by homotopy groups it is an equivalence.
(ii) The identification of $\Phi^\mathbb{T} DE\langle H\rangle$ is immediate from 2.4.1. □

It remains to show that $\hat{b}(x \otimes y) = b_*(xy)$; this follows when x is of positive degree exactly as in 13.5.1. Now suppose $x = c_H^{-k}$ for $k \geq 0$; the verification that $\hat{b}(c_H^{-k} \otimes y) = b_*(c_H^{-k} y)$ in this case will also complete the proof of 13.5.1.

We take the cofibre sequence in the statement and smash it with $\sigma^{nV(H)}$. Since $\sigma^{nV(H)}$ is formed from S^0 and various basic cells with isotropy H, we have $\sigma^{nV(H)} \wedge \tilde{E}\mathcal{F} \simeq \tilde{E}\mathcal{F}$; by the Thom isomorphism 2.3.7(a) we have $E\langle H\rangle \wedge \sigma^{nV(H)} \simeq E\langle H\rangle \wedge S^{2n}$, and also

$$\sigma^{nV(H)} \wedge DE\langle H\rangle \simeq D(E\langle H\rangle \wedge \sigma^{-nV(H)}) \simeq D(E\langle H\rangle \wedge S^{-2n}).$$

Thus the cofibre sequence becomes

$$S^{2n} \wedge DE\langle H\rangle \longrightarrow \tilde{E}\mathcal{F} \wedge S^0[\,\mathbb{Q}[c_H, c_H^{-1}]\,] \xrightarrow{1 \wedge b} \Sigma^{2n+1} E\langle H\rangle.$$

Because the final term is $2n$-connected and the first has $\pi_*^\mathbb{T}$ only in degrees $\leq 2n$ it follows that $\pi_*^\mathbb{T}(1 \wedge b)$ is surjective for all n. Taking $n \leq -k-1$ establishes the required formula for $\hat{b}$. □

CHAPTER 14

Cyclic and Tate cohomology.

This is a short chapter, but fits naturally between its neighbours. The first section identifies rational cyclic cohomology, the second gives an algebraic description of the Tate construction on rational spectra, and the third is exceptional in this book in that it gives an *integral* description of the Tate spectrum of *integral* complex equivariant K-theory.

14.1. Cyclic cohomology.

In this section we consider periodic cyclic cohomology. We begin by observing that the rationalization of the integral cyclic cohomology is the cyclic cohomology of the rationals, so that the two possible interpretations coincide.

It was proved in [**14**] that the representing spectrum for cyclic cohomology with coefficients in an abelian group A is the Tate spectrum $t(HA) = F(E\mathbb{T}_+, HA) \wedge \tilde{E}\mathbb{T}$.

LEMMA 14.1.1. *For any Mackey functor A the rationalization of $t(HA)$ is $t(H(A \otimes \mathbb{Q}))$.*

Proof: The essential point is that $F(E\mathbb{T}_+, HA) = \operatorname{holim}_{\leftarrow n} F(E\mathbb{T}_+^{(n)}, HA)$, and that the maps induced in $[X, \cdot]^{\mathbb{T}}$ by those of the inverse system are ultimately isomorphisms for each finite X. This inverse limit therefore commutes with direct limit under degree zero selfmaps of HA.

Thus

$$\begin{aligned} t(HA) \wedge S^0\mathbb{Q} &= \operatorname{holim}_{\rightarrow m} (\ F(E\mathbb{T}_+, HA) \wedge \tilde{E}\mathbb{T}, m!\ ) \\ &\simeq F(E\mathbb{T}_+, \operatorname{holim}_{\rightarrow m} (HA, m!)) \wedge \tilde{E}\mathbb{T} \\ &\simeq F(E\mathbb{T}_+, HA \otimes \mathbb{Q}) \wedge \tilde{E}\mathbb{T} = t(H(A \otimes \mathbb{Q})). \quad \square \end{aligned}$$

Henceforth we suppose A is rational.

LEMMA 14.1.2. *Provided A is rational, $t(HA)$ is $\mathcal{F}$-contractible. We therefore have an equivalence*

$$t(HA) \simeq \tilde{E}\mathcal{F} \wedge S^0[\pi_*^{\mathbb{T}}(t(HA))].$$

Proof: This is a special case of 14.2.1 below, but admits a more elementary proof. The Tate construction commutes with restriction: $t(HA)|_H = t(HA|_H)$. However the ordinary Tate cohomology of any finite group with rational coefficients is 0. $\square$

For any abelian group A we have $\pi_*^{\mathbb{T}}(t(HA)) = \hat{H}C_* \otimes A = A[c, c^{-1}]$.

14.2. Rational Tate spectra.

In this section we discuss the Tate construction of [**14**], which generalizes the periodic cyclic cohomology discussed in the previous section. Recall that the Tate construction on a $\mathbb{T}$-spectrum X is defined by $t(X) = F(E\mathbb{T}_+, X) \wedge \tilde{E}\mathbb{T}$. This simplifies considerably in the rational case, and it seems worth giving a complete description of the Tate construction in the category of rational $\mathbb{T}$-spectra.

We begin with the warning that if X is integral, the map $t(X) \longrightarrow t(X \wedge S^0\mathbb{Q})$ need not be a rational equivalence, so that Lemma 14.1.1 above is special to suitably bounded theories like HA. An example is given by complex K-theory, since $t(K\mathbb{Z})|_H$ is non-trivial and rational for all non-trivial finite subgroups H [**9, 14, 15**]; the following lemma shows this is false for $t(K\mathbb{Q})$.

We revert to our global assumption that all spectra are rational.

LEMMA 14.2.1. *The natural map*

$$t(X) = F(E\mathbb{T}_+, X) \wedge \tilde{E}\mathbb{T} \longrightarrow F(E\mathbb{T}_+, X) \wedge \tilde{E}\mathcal{F}$$

is an equivariant equivalence. Thus $t(X)$ is an $\mathcal{F}$-contractible spectrum determined by its homotopy groups:

$$t(X) \simeq \tilde{E}\mathcal{F} \wedge S^0[t(X)^{\mathbb{T}}_*].$$

Proof: We give two proofs. For the first proof, note that the Tate construction commutes with restriction: $t(X)|_H = t(X|_H)$. The lemma follows from the fact that the Tate construction is trivial on rational spectra for finite groups. One way of seeing this is to use the fact that if H is finite and $e \in A(H)$ is the idempotent with support 1 then $EH_+ = eS^0$ and $\tilde{E}H = (1-e)S^0$.

For the second proof, we compare the cofibre sequence $E\mathbb{T}_+ \longrightarrow S^0 \longrightarrow \tilde{E}\mathbb{T}$ with $E\mathcal{F}_+ \longrightarrow S^0 \longrightarrow \tilde{E}\mathcal{F}$. We see that the lemma is equivalent to showing that the natural map $f : F(E\mathbb{T}_+, X) \wedge E\mathbb{T}_+ \longrightarrow F(E\mathbb{T}_+, X) \wedge E\mathcal{F}_+$ is an equivalence. However, the cofibre of f is a wedge of terms $F(E\mathbb{T}_+, X) \wedge E\langle H \rangle$ with $H \neq 1$; this is contractible, as one sees from the fact that $F(E\mathbb{T}_+, X) \wedge \sigma^0_H \simeq *$ by using cofibre sequences and passing to direct limits. □

PROPOSITION 14.2.2. *If X is a rational $\mathbb{T}$-spectrum with associated module $M = \pi^{\mathbb{T}}_*(X \wedge E\mathbb{T}_+)$ over $\mathbb{Q}[c_1]$, then $t(X)$ is the $\mathcal{F}$-contractible spectrum with homotopy groups*

$$\hat{H}^0_{(c_1)}(M) \oplus \Sigma \hat{H}^{-1}_{(c_1)}(M)$$

where $\hat{H}^*_{(c_1)}$ denotes local Tate cohomology in the sense of [**9**].

REMARK 14.2.3. There are two methods for calculating the local Tate cohomology; since (c_1) is principal both are very simple. The second description simplifies further because M is torsion. In fact, the local Tate cohomology is only non-zero in codegrees 0 and -1, and for these cases we have

$$\hat{H}^{-i}_{(c_1)}(M) = (L^{(c_1)}_i M)[1/c_1] = \lim_{\leftarrow}{}^i(M, c_1),$$

where $L_*^{(c_1)}$ denotes the left derived functors of completion at (c_1). Note that this is only likely to be equal to (c_1)-adic completion when M is finitely generated. However this case is trivial since M is torsion and therefore already complete, so the Tate cohomology vanishes. This fact is familiar geometrically.

Proof: Observe $F(E\mathbb{T}_+, X) \simeq F(E\mathbb{T}_+, X \wedge E\mathbb{T}_+)$, so that if $M = \pi_*^{\mathbb{T}}(X \wedge E\mathbb{T}_+)$, there is an exact sequence

$$0 \longrightarrow \mathrm{Ext}(\Sigma^2\mathbb{I}, M) \longrightarrow [E\mathbb{T}_+, X]_*^{\mathbb{T}} \longrightarrow \mathrm{Hom}(\Sigma\mathbb{I}, M) \longrightarrow 0.$$

This is precisely parallel to the algebraic situation. We may split X into even and odd parts, and thus the exact sequence splits. Therefore $F(E\mathbb{T}_+, X)$ is modelled by the complex $\mathrm{Hom}(PK(c_1), M)$ where $PK(c_1)$ is a complex of projectives approximating the stable Koszul complex $\mathbb{Q}[c_1] \longrightarrow \mathbb{Q}[c_1, c_1^{-1}]$; the homology of this complex gives the left derived functors of c_1-completion [**13**]. In particular, when X is even, $[E\mathbb{T}_+, X]_*^{\mathbb{T}}$ is $L_0^{(c_1)}M$ in even degrees and $\Sigma L_1^{(c_1)}M$ in odd degrees.

Now conclude that there is a split exact sequence

$$0 \longrightarrow \mathrm{Ext}(\Sigma^2\mathbb{I}, M)[1/c] \longrightarrow t(X)_*^{\mathbb{T}} \longrightarrow \mathrm{Hom}(\Sigma\mathbb{I}, M)[1/c] \longrightarrow 0.$$

Therefore, if $TT_{(c_1)}(M)$ is the complex of the second avatar in the notation of [**9**], $t(X)$ is modelled by the corresponding torsion free model, $e(TT_{(c_1)}(M))$. Thus, if X is even, $t(X)_*^{\mathbb{T}}$ is $\hat{H}^0_{(c_1)}(M)$ in even degrees and $\Sigma\hat{H}^{-1}_{(c_1)}(M)$ in odd degrees. □

14.3. The integral $\mathbb{T}$-equivariant Tate spectrum for complex K-theory.

In this section we apply the general theory to identify the Tate spectrum of complex equivariant K-theory *integrally*. For emphasis we write $K\mathbb{Z}$ for the representing spectrum usually written K or KU. However, we note that $t(K\mathbb{Z})$ is not rational, and its rationalization is not $t(K\mathbb{Q})$, so this is not an application of the previous section.

Before we state the theorem, recall that the representation ring $R(\mathbb{T}) = \mathbb{Z}[z, z^{-1}]$, and that the Euler class of the representation z^n is $1-z^n$. In particular, we let $\chi = 1-z$ and find $R(\mathbb{T})^\wedge_{(\chi)} = \mathbb{Z}[[\chi]]$; indeed, $z = 1-\chi$ is invertible in $\mathbb{Z}[\chi]/(\chi^n)$, so that $\mathbb{Z}[\chi] \longrightarrow \mathbb{Z}[\chi, z^{-1}] = \mathbb{Z}[z, z^{-1}]$ induces an isomorphism of (χ)-completions. We write $\mathbb{Z}((\chi))$ for the localization $\mathbb{Z}[[\chi]][\chi^{-1}]$, and S for the mulitiplicative set generated by the Euler classes. Note that if $n \geq 2$ the Euler class $1-z^n$ is a multiple of χ. However, although the multiplier is a unit in $\mathbb{Q}((\chi))$, it is not a unit in $\mathbb{Z}((\chi)) \otimes \mathbb{Q}$.

In fact, the Tate spectrum $t(K\mathbb{Z})$ is $\mathcal{F}$-equivalent to a rational spectrum, and is thus determined by the homotopy type of $\Phi^{\mathbb{T}}K\mathbb{Z}$ and its rational type.

THEOREM 14.3.1. *The Tate spectrum $t(K\mathbb{Z})$ is determined by a cofibre sequence*

$$t(K\mathbb{Z}) \longrightarrow \tilde{E}\mathcal{F} \wedge KS^{-1}\mathbb{Z}((\chi)) \xrightarrow{q} \bigvee_{H \neq 1} E\langle H\rangle[W],$$

where W is a graded vector space of the cardinality of the continuum in each even degree. More precisely,

(i) There is an equivalence of $K\mathbb{Z}$-module spectra

$$\Phi^{\mathbb{T}} K\mathbb{Z} \simeq KS^{-1}\mathbb{Z}((\chi)),$$

where the right hand side represents K-theory with coefficients in the abelian group $S^{-1}\mathbb{Z}((\chi))$.
(ii) The spectrum $t(K\mathbb{Z}) \wedge E\mathcal{F}_+$ is rational and injective; it has trivial free part but $t(K\mathbb{Z}) \wedge E\langle H\rangle$ is non-trivial if $H \neq 1$.
(iii) The rational spectrum $t(K\mathbb{Z}) \wedge S^0\mathbb{Q}$ is formal, and classified in the torsion model by the rationalization of the map

$$t_*^{\mathcal{F}} \otimes S^{-1}\mathbb{Z}((\chi))[\beta, \beta^{-1}] \longrightarrow S^{-1}\mathbb{Z}((\chi))/\mathbb{Z}((\chi))[\beta, \beta^{-1}].$$

Here the codomain is the rationalization of

$$S^{-1}\mathbb{Z}((\chi))/\mathbb{Z}((\chi))[\beta, \beta^{-1}] = \bigoplus_{n\neq 1} \left(\mathbb{Z}((\chi))[1/\Phi_n]/\mathbb{Z}((\chi))\right)[\beta, \beta^{-1}],$$

where Φ_n is the nth cyclotomic polynomial in $z = 1 - \chi$, and the structure map is described as in 13.4.4.

Proof: First note that $t(K\mathbb{Z}) = F(E\mathbb{T}_+, K\mathbb{Z}) \wedge \tilde{E}\mathbb{T}$, so that we may calculate its coefficient ring from the Atiyah-Segal completion theorem. First, equivariant K-theory has coefficients $R(\mathbb{T})[\beta, \beta^{-1}]$, with $R(\mathbb{T}) = \mathbb{Z}[z, z^{-1}]$, and the K-theory Euler class of z^n is $1 - z^n$. By the Atiyah-Segal completion theorem, $\pi_*^{\mathbb{T}}(F(E\mathbb{T}_+, K\mathbb{Z})) = R(\mathbb{T})^\wedge_{(\chi)}[\beta, \beta^{-1}]$.

Consider the cofibre sequence

$$t(K\mathbb{Z}) \longrightarrow t(K\mathbb{Z}) \wedge \tilde{E}\mathcal{F} \longrightarrow t(K\mathbb{Z}) \wedge \Sigma E\mathcal{F}_+.$$

We shall identify $t(K\mathbb{Z}) \wedge \tilde{E}\mathcal{F}$, $t(K\mathbb{Z}) \wedge \Sigma E\mathcal{F}_+$, and the map between them in turn.

Firstly, since χ is an Euler class,

$$\pi_*^{\mathbb{T}}(t(K\mathbb{Z}) \wedge \tilde{E}\mathcal{F}) = S^{-1}\mathbb{Z}((\chi))[\beta, \beta^{-1}],$$

where S is the multiplicative set generated by $1 - z^n$ for $n \geq 1$. Now $S^{-1}\mathbb{Z}((\chi))$ is flat over $\mathbb{Z}$, and hence the coefficients of $\Phi^{\mathbb{T}} t(K\mathbb{Z})$ are the same as those of K-theory with coefficients in $S^{-1}\mathbb{Z}((\chi))$.

We may now prove Part (i). First note that $\Phi^{\mathbb{T}} t(K\mathbb{Z})$ is a module over $K\mathbb{Z}$. Now let $MS^{-1}\mathbb{Z}((\chi))$ be a non-equivariant Moore spectrum, and construct a map $f : MS^{-1}\mathbb{Z}((\chi)) \longrightarrow t(K\mathbb{Z})$ inducing an isomorphism in $\pi_0^{\mathbb{T}}$. Now form the composite

$$KS^{-1}\mathbb{Z}((\chi)) = K \wedge MS^{-1}\mathbb{Z}((\chi)) \longrightarrow K \wedge \Phi^{\mathbb{T}} t(K\mathbb{Z}) \longrightarrow \Phi^{\mathbb{T}} t(K\mathbb{Z})$$

in which the first map is obtained from f by applying $K \wedge \Phi^{\mathbb{T}}(\cdot)$, and the second uses the module structure. By construction this induces an isomorphism in homotopy, and is therefore an equivalence.

Part (ii) follows from the fact that, $t(K\mathbb{Z})|_H$ is rational for all finite subgroups H **[9, 14, 15]**. Indeed, we know that $t(K\mathbb{Z}) \wedge \mathbb{T}/H_+$ is induced from $t(K\mathbb{Z})|_H$. Rationally $E\mathcal{F}_+ \simeq \bigvee_H E\langle H\rangle$, so $t(K\mathbb{Z}) \wedge E\mathcal{F}_+$ has a corresponding splitting.

Finally we turn to Part (iii). Consider $t(K\mathbb{Z}) \wedge \Sigma E\langle H\rangle$. From the identification of c_H, we find $\pi_*^{\mathbb{T}}(t(K\mathbb{Z}) \wedge E\langle H\rangle)$ is $\mathbb{Z}((\chi))/\Phi_n^\infty$ in each even degree, when H is of order n. In particular $\Phi_1 = 1 - z = \chi$, so this is zero if $H = 1$. In general c_H a unit multiple of $\Phi_n\beta$ by 13.4.1, and in particular the $\mathbb{Q}[c_H]$-module is injective.

Finally, the map $t(K\mathbb{Z}) \wedge \tilde{E}\mathcal{F} \longrightarrow t(K\mathbb{Z}) \wedge \Sigma E\mathcal{F}_+$ factors through the rationalization $t(K\mathbb{Z}) \wedge \tilde{E}\mathcal{F} \longrightarrow t(K\mathbb{Z}) \wedge \tilde{E}\mathcal{F} \wedge S^0\mathbb{Q}$, and, by 5.3.2, the resulting map $t(K\mathbb{Z}) \wedge \tilde{E}\mathcal{F} \wedge S^0\mathbb{Q} \longrightarrow t(K\mathbb{Z}) \wedge \Sigma E\mathcal{F}_+$ is classified by its d-invariant since the codomain is injective. The map from $t_*^{\mathcal{F}} \otimes S^{-1}\mathbb{Z}((\chi))[\beta, \beta^{-1}]$ is the analogue of that in Theorem 13.4.4. □

CHAPTER 15

Cyclotomic spectra and topological cyclic cohomology.

In this chapter, we study various $\mathbb{T}$-spectra arising from algebraic K-theory. Various constructions are used to define suitable targets for trace maps from algebraic K-theory, and the most sophisticated takes Bökstedt's Topological Hochschild homology of a ring, and forms the associated topological cyclic spectrum in the sense of Bökstedt-Hsiang-Madsen [**2**]. Madsen has recently given a very helpful general survey [**21**].

The topological cyclic construction can be applied to any $\mathbb{T}$-spectrum with appropriate extra structure, and we begin in Section 15.1 by identifying the extra structure involved in specifying such a 'cyclotomic' spectrum. In the following section, we illustrate this by considering the basic examples: free loop spaces on a $\mathbb{T}$-fixed space, and topological Hochschild homology of a functor with smash products. In Section 15.3 we define the topological cyclic homology of a cyclotomic spectrum. Finally, in Section 15.4 we generalize a theorem of Goodwillie's [**7**] by giving a complete analysis of the topological cyclic construction on rational cyclotomic spectra.

15.1. Cyclotomic spectra.

We must begin by recalling the definition of a cyclotomic spectrum from [**18**]. The basic idea is that it is a spectrum X with the property analogous to that of the free loop space ΛZ, on a $\mathbb{T}$-fixed based space Z, namely that for any finite subgroup K the fixed point set $(\Lambda Z)^K$ is equivalent to the original space ΛZ. The analogue should be that any fixed point spectrum $\Phi^K X$ is equivalent to X again. Of course $\Phi^K X$ is really a $\mathbb{T}/K$-spectrum, so we must begin by explaining exactly how we interpret it as a $\mathbb{T}$-spectrum indexed on the original universe. Since we want to avoid redundant structure, we require that the resulting equivalences are transitive.

Consider the group $\mathbb{T}$ and all its quotients $\overline{\mathbb{T}} = \mathbb{T}/K$ by finite subgroups K. We want transitive systems of structure, so we first let $\rho = \rho_K : \mathbb{T} \xrightarrow{\cong} \overline{\mathbb{T}}$ be the isomorphism given by taking the $|K|$th root. If we index our $\mathbb{T}$-spectra on a complete universe $\mathcal{U}$, we index our $\overline{\mathbb{T}}$-spectra on the complete universe $\mathcal{U}^K$. However we want these universes to be comparable, so we say that a complete $\mathbb{T}$-universe $\mathcal{U}$ is *cyclotomic* if it is provided with isomorphisms $\mathcal{U} \xrightarrow{\cong} \rho_K^* \mathcal{U}^K$. Identifying $\mathbb{T}$ and $\overline{\mathbb{T}}$ via ρ_K, this also specifies isomorphisms $\mathcal{U}^L \longrightarrow \rho_{K/L}^*(\mathcal{U}^L)^{K/L}$. We require that these are transitive in the sense that if $L \subseteq K$ then the composite

$$\mathcal{U} \longrightarrow \rho_L^* \mathcal{U}^L \longrightarrow \rho_L^*(\rho_{K/L}^* \mathcal{U}^L)^{K/L} = \rho_K^* \mathcal{U}^K$$

is the isomorphism for K. One such cyclotomic universe is the direct sum $\mathcal{U} = \oplus_{n\in\mathbb{Z}} \mathcal{U}_n$ where $\mathcal{U}_n = \bigoplus_{i\in\mathbb{N}} z^n$, and z is the natural representation. The cyclotomic isomorphisms are those suggested by the indexing.

Suppose then that X is a $\mathbb{T}$-spectrum indexed on the cyclotomic universe $\mathcal{U}$. Thus, for any finite subgroup K, $\Phi^K X$ is a $\overline{\mathbb{T}}$-spectrum indexed on $\mathcal{U}^K$; by pullback along the isomorphism ρ_K we obtain a $\mathbb{T}$-spectrum $\rho_K^* \Phi^K X$ indexed on $\rho_K^* \mathcal{U}^K$, which may be viewed as a $\mathbb{T}$-spectrum $\rho_K^! \Phi^K X$ indexed on $\mathcal{U}$ by using the cyclotomic structure of the universe. A *cyclotomic* structure on X consists of a transitive system of $\mathbb{T}$-equivalences

$$r_K : \rho_K^! \Phi^K X \xrightarrow{\simeq} X.$$

By transitivity, the essential pieces of the structure come from the cases that K is of prime order.

Although this structure is really designed to capture profinite information, there is enough residue rationally to make it worthwhile identifying the cyclotomic objects in the algebraic model of rational $\mathbb{T}$-spectra. The essential idea is that in a cyclotomic spectrum all finite subgroups behave in an analogous way, differing only in the multiplicity with which information occurs. There is no significant constraint on total fixed points $\Phi^{\mathbb{T}} X$. The first step of our general analysis was to split $\mathcal{F}$-spectra into the parts over different subgroups, so it is easy to describe the cyclotomic structure in these terms. It turns out that a spectrum X is cyclotomic if we have specified equivalences $X(C_1) \simeq X(C_2) \simeq X(C_3) \simeq \ldots$. This uniformity itself imposes constraints on the assembly map of a $\mathbb{T}$-spectrum.

Let us now describe the algebraic model for cyclotomic spectra more precisely. It is useful to bear in mind the torsion model rather than the standard model.

DEFINITION 15.1.1. The ring of cyclotomic operations is the polynomial ring $\mathbb{Q}[c_0]$ on a single generator c_0 of degree -2. The standard injective $\mathbb{I}_0$ is defined by the exact sequence $0 \longrightarrow \mathbb{Q}[c_0] \longrightarrow \mathbb{Q}[c_0, c_0^{-1}] \longrightarrow \Sigma^2 \mathbb{I}_0 \longrightarrow 0$. The *cyclotomic torsion model* is the category $\mathcal{C}_t$ with objects $(\Sigma^2 \mathbb{I}_0 \otimes V \longrightarrow T_0)$ where V is a graded vector space and T_0 is a torsion $\mathbb{Q}[c_0]$-module. The morphisms are given by commutative squares as usual. □

REMARK 15.1.2. For comparison with the torsion model $\mathcal{A}_t$ (see 6.2.1), note that the category $\mathcal{C}_t$ is equivalent to the category of objects $(\mathbb{Q}[c_0, c_0^{-1}] \otimes V \longrightarrow T_0)$ with T_0 a torsion $\mathbb{Q}[c_0]$-module and $1 \otimes V$ mapping to zero. □

LEMMA 15.1.3. *The category $\mathcal{C}_t$ is abelian and 2 dimensional. Hence we may form the derived category $D\mathcal{C}_t$.*

Proof: The proof is precisely analogous to 6.3.2 for the torsion model category $\mathcal{A}_t$. The relevant injectives are $e_t^{\mathcal{C}}(V) = (\Sigma^2 \mathbb{I}_0 \otimes V \longrightarrow 0)$ and $f_t^{\mathcal{C}}(I) = (\Sigma^2 \mathbb{I}_0 \otimes \mathrm{Hom}(\Sigma^2 \mathbb{I}_0, I) \longrightarrow I)$. □

Although we make no use of it here, it is again convenient to have a 1-dimensional model. The analogue of the standard model is considerably simplified in the present context.

DEFINITION 15.1.4. The *standard cyclotomic model* is the category $\mathcal{C}$ with objects, $(S_0 \longrightarrow \Sigma^2 \mathbb{I}_0 \otimes V)$ where V is a graded vector space and S_0 is a torsion $\mathbb{Q}[c_0]$-module. The morphisms are given by commutative squares as usual.

REMARK 15.1.5. For comparison with the standard model $\mathcal{A}$ (see 5.4.1), note that the category $\mathcal{C}$ is equivalent to the category $\mathcal{C}'$ with objects $\mathbb{Q}[c_0]$-maps $(N_0 \longrightarrow \mathbb{Q}[c_0, c_0^{-1}] \otimes V)$ which become isomorphisms when c_0 is inverted, and so that the pullback $i^* N_0 \longrightarrow \mathbb{Q}[c_0] \otimes V$ along $i : \mathbb{Q}[c_0] \otimes V \longrightarrow \mathbb{Q}[c_0, c_0^{-1}] \otimes V$ is also an isomorphism. The diagram

$$\begin{array}{ccccccccc} 0 & \longrightarrow & i^*N_0 & \longrightarrow & N_0 & \longrightarrow & S_0 & \longrightarrow & 0 \\ & & \downarrow \cong & & \downarrow & & \downarrow & & \\ 0 & \longrightarrow & \mathbb{Q}[c_0] \otimes V & \xrightarrow{i} & \mathbb{Q}[c_0, c_0^{-1}] \otimes V & \longrightarrow & \Sigma^2 \mathbb{I}_0 \otimes V & \longrightarrow & 0. \end{array}$$

suggests the functors giving the equivalence. □

LEMMA 15.1.6. *The cyclotomic category $\mathcal{C}$ is abelian and 1-dimensional. Hence we may form the derived category $D\mathcal{C}$.*

Proof: The proof is precisely analogous to that for the standard model category $\mathcal{A}$, 5.5.2. The relevant injectives are $e^{\mathcal{C}}(V) = (\Sigma^2 \mathbb{I}_0 \otimes V \longrightarrow \Sigma^2 \mathbb{I}_0 \otimes V)$ and $f^{\mathcal{C}}(I) = (I \longrightarrow 0)$. □

LEMMA 15.1.7. *There is an equivalence of derived categories,*

$$D\mathcal{C} \simeq D\mathcal{C}_t,$$

given by passage to fibre $dg\mathcal{C}_t \longrightarrow dg\mathcal{C}$ and passage to cofibre $dg\mathcal{C} \longrightarrow dg\mathcal{C}_t$.

Proof: The proof is similar to the case 6.5.1 of the standard model, but with the great simplification that the cofibre functor arrives in the correct category before passing to homology. □

Now define a functor

$$\Delta : \mathcal{C}_t \longrightarrow \mathcal{A}_t$$

as follows. For an object we define $\Delta(\Sigma^2 \mathbb{I}_0 \otimes V \xrightarrow{s_0} T_0)$ to be the composite

$$(t_*^{\mathcal{F}} \otimes V \longrightarrow \Sigma^2 \mathbb{I} \otimes V = \bigoplus_H \Sigma^2 \mathbb{I}_0 \otimes V_0 \xrightarrow{\oplus_H s_0} \bigoplus_H T_0).$$

Here the first map is induced by the quotient $t_*^{\mathcal{F}} \longrightarrow t_*^{\mathcal{F}} / \mathcal{O}_{\mathcal{F}} = \Sigma^2 \mathbb{I}$, and the second is the direct sum of countably many copies of s_0 made into a $\mathcal{O}_{\mathcal{F}}$-module in the obvious way.

The functor is obviously exact and hence induces a functor

$$\Delta : D\mathcal{C}_t \longrightarrow D\mathcal{A}_t.$$

We may now state a precise theorem.

THEOREM 15.1.8. *A $\mathbb{T}$-spectrum admits the structure of a cyclotomic spectrum if and only if it corresponds to an object of $D\mathcal{A}_t$ equivalent to one in the image of Δ.*

Note that the condition in the theorem gives a rather satisfactory characterization of cyclotomic spectra. It essentially says that a cyclotomic spectrum is one that has two properties. Firstly, the structure map "factors through that for its geometric fixed point spectrum" (as happens for suspension spectra) and secondly, that all finite subgroups behave alike.

If a spectrum admits a cyclotomic structure then a structure is imposed by choosing particular equivalences between the idempotent pieces of the torsion part of the model. Note that for a spectrum X with torsion model $s : t_*^{\mathcal{F}} \otimes V \longrightarrow T$ admitting a cyclotomic structure, the corresponding cyclotomic spectrum is simply $\Sigma^2 \mathbb{I}_0 \otimes V \longrightarrow T_0$ where $T_0 = e_1 T = \pi_*^{\mathbb{T}}(E\mathbb{T}_+ \wedge X)$ and the map is obtained by factoring s through the projection and applying the idempotent e_1.

Proof: We have explained how to put a cyclotomic structure on an object in the image of Δ. Any imprecision will be eliminated in the course of the proof in the reverse direction.

Suppose then that X is a cyclotomic spectrum with cyclotomic structure maps $r_K : \rho_K^! \Phi^K X \xrightarrow{\simeq} X$ as required. We already know from Section 8.4 the effect of passage to geometric fixed points. Indeed, by Theorem 8.4.1, if M is the model of X then eM is the model for $\Phi^K X$ where e is the idempotent supported on the subgroups containing K. Here $\overline{\mathcal{O}}_{\overline{\mathcal{F}}}$ is identified with $e\mathcal{O}_{\mathcal{F}}$ by letting a subgroup $\overline{H}$ of $\overline{\mathbb{T}}$ correspond to its inverse image in $\mathbb{T}$. The effect of $\rho_K^!$ is to identify subgroups of $\mathbb{T}$ with those of the same order in $\overline{\mathbb{T}}$.

Define $n = n_K : \overline{\mathcal{F}} \longrightarrow \mathcal{F}$ by letting $n(\overline{H})$ be the subgroup of $\mathbb{T}$ with the same order as $\overline{H}$, and consider the induced ring isomorphism $n_K^* : \mathcal{O}_{\mathcal{F}} \xrightarrow{\cong} \overline{\mathcal{O}}_{\overline{\mathcal{F}}}$.

LEMMA 15.1.9. *The functor* $\rho_K^! : \overline{\mathbb{T}}$*-Spec* $\longrightarrow \mathbb{T}$*-Spec corresponds to pullback along* n_K^* *in the usual sense that the diagram*

$$\begin{array}{ccc} \overline{\mathbb{T}}\text{-}Spec & \xrightarrow{\rho_K^!} & \mathbb{T}\text{-}Spec \\ \simeq\downarrow & & \downarrow\simeq \\ D\overline{\mathcal{A}} & \xrightarrow{n_\#^K} & D\mathcal{A} \end{array}$$

commutes up to natural isomorphism, and similarly for torsion models. □

It is then clear (for example by using the cyclotomic structure for K itself) that the part $e_K M$ of the model over any subgroup K will be the same as the piece $e_1 M$ over the trivial subgroup.

The equivalence along $n_\#^K$ also forces the map to factor as specified. Indeed, if M is an object of the torsion model with zero differential, we see that the structure map must be zero on any element of the form $1 \otimes v$; otherwise it would have nonzero image in $e_H T$ for some H, and hence for all finite subgroups H. This contradicts $\mathcal{F}$-finiteness of T. Since this argument passes to an injective resolution, it applies to all differential graded objects. □

15.2. Free loop spaces and THH.

This section is devoted to considering the two best known examples of cyclotomic spectra from our point of view.

EXAMPLE 15.2.1. For a based space Z, we attempt to identify the place of the free loop space ΛZ in the present scheme. One benefit is that we may then study the group $K^*_{\mathbb{T}}(\Lambda Z)$, which is a conjectural approximation to $Ell^*(Z)$.

We restrict attention to the case that $Z = \Sigma Y$ is a suspension. Here, Carlsson and Cohen [**4**] prove the splitting

$$E\mathbb{T}_+ \wedge_{\mathbb{T}} \Lambda\Sigma Y = \bigvee_n (EC_n)_+ \wedge_{C_n} Y^{\wedge n}.$$

Hence, rationally we have

$$\pi^{\mathbb{T}}_*(E\mathbb{T}_+ \wedge \Lambda\Sigma Y) = \Sigma \bigoplus_n \{H_*(Y)^{\otimes n}\}^{C_n}$$

with trivial $H^*(B\mathbb{T})$ action. This leaves us to describe the structure map

$$\Sigma^2\mathbb{I}_0 \otimes H_*(\Sigma Y) \longrightarrow \Sigma^2 \bigoplus_n \{H_*(Y)^{\otimes n}\}^{C_n}$$

for the cyclotomic torsion model.

LEMMA 15.2.2. *The structure map is induced by* $E\mathbb{T}_+ \wedge_{\mathbb{T}} \Sigma Y \longrightarrow E\mathbb{T}_+ \wedge_{\mathbb{T}} \Lambda\Sigma Y$.

Proof: As for suspension spectra (Section 13.3), the lemma follows from naturality and the $\mathbb{T}$-map $\Sigma Y \longrightarrow \Lambda\Sigma Y$. □

It follows that the structure map has zero d-invariant. By duality it is sufficient to consider cohomology, and the domain has torsion free cohomology whilst the codomain has torsion cohomology.

Now, exactly as in the case of suspension spectra, the e invariant is the element of

$$e_{\Lambda\Sigma Y} \in \mathrm{Ext}(\Sigma^4\mathbb{I}_0 \otimes H_*(Y), \bigoplus_n \Sigma^2\{H_*(Y)^{\otimes n}\}^{C_n})$$

corresponding to the extension obtained by taking rational homology of the cofibre sequence

$$E\mathbb{T}_+ \wedge_{\mathbb{T}} \Lambda\Sigma Y \longrightarrow E\mathbb{T}_+ \wedge_{\mathbb{T}} (\Lambda\Sigma Y)/\Sigma Y \longrightarrow E\mathbb{T}_+ \wedge_{\mathbb{T}} \Sigma^2 Y. \quad \square$$

There is another important example of cyclotomic spectra.

EXAMPLE 15.2.3. **Topological Hochschild Homology:**

Suppose that F is a functor with smash products in the sense of Bökstedt. One may define a cyclotomic spectrum $THH(F)$, which comes with a spectral sequence

$$HH_*(F(S^0)_*) \Longrightarrow THH(F)_*$$

for calculating its homotopy groups. One may then hope to calculate $\pi^{\mathbb{T}}_*(E\mathbb{T}_+ \wedge_{\mathbb{T}} THH(F))$ using the skeletal filtration of $E\mathbb{T}_+$.

It is always the case that $\Phi^{\mathbb{T}}THH(F) \simeq S^0$, and so the structure map of the cyclotomic spectrum $THH(F)$ takes the form

$$\Sigma^2\mathbb{I}_0 \longrightarrow \Sigma^2 THH(F)^*_{h\mathbb{T}},$$

where the subscript $h\mathbb{T}$ denotes the homotopy quotient. By definition, we always have a map from the identity functor to F and hence a cyclotomic map $S^0 = THH(I) \longrightarrow THH(F)$. This gives the diagram

$$\begin{array}{ccc} \tilde{E}\mathcal{F} \wedge DE\mathcal{F}_+ & \longrightarrow & \Sigma E\mathcal{F}_+ \\ \simeq\downarrow & & \downarrow\simeq \\ \tilde{E}\mathcal{F} \wedge DE\mathcal{F}_+ \wedge THH(I) & \longrightarrow & \Sigma E\mathcal{F}_+ \wedge THH(I) \\ \simeq\downarrow & & \downarrow \\ \tilde{E}\mathcal{F} \wedge DE\mathcal{F}_+ \wedge THH(F) & \longrightarrow & \Sigma E\mathcal{F}_+ \wedge THH(F) \end{array}$$

The map for S^0 has zero d-invariant, so we deduce that the structure map for an arbitrary functor F has zero d-invariant. It would be interesting to understand its e-invariant more precisely.

One case of particular interest is when the FSP arises from a ring R. In this case $\pi^{\mathbb{T}}_*(E\mathbb{T}_+ \wedge THH(R)) = \Sigma HC_*(R)$, which can be calculated by the algebraic Loday-Quillen double complex. It remains to identify the torsion model structure map, but we can obtain information by naturality from the unit $\mathbb{Q} \longrightarrow R$, which gives the above diagram with I replaced by $\mathbb{Q}$ and F by R. Thus we only need to understand the algebraic map $\Sigma^2\mathbb{I}_0 = \Sigma^2 HC_*(\mathbb{Q}) \longrightarrow \Sigma^2 HC_*(R)$. This is in fact either zero or injective: this follows from the Tate spectral sequence by naturality. Indeed $t(THH(R))$ is a module over $t(THH(\mathbb{Q}))$, and hence over the ring spectrum $t(H\mathbb{Q})$, whose coefficients are $\mathbb{Q}[c_0, c_0^{-1}]$: thus the behaviour is completely determined by the image of the unit. If R is augmented, then of course the map is injective.

This special case is not too far from the general case, because any rational FSP arises from a simplicial ring, and $\pi^{\mathbb{T}}_*(E\mathbb{T}_+ \wedge THH(R_\bullet))$ can be calculated algebraically, since there is a $\mathbb{T}$-map $THH(R_\bullet) \longrightarrow HH(R_\bullet)$ which is a non-equivariant rational equivalence. □

15.3. The definition of topological cyclic homology.

In the first instance, the topological cyclic homology of a ring is designed to be the target of a refined trace map from the algebraic K-theory. Hesselholt and Madsen have shown that the cyclotomic trace is very close to being an isomorphism in many cases **[18]**. The construction of the topological cyclic homology in this case begins with the topological Hoschschild homology, and the definition of a cyclotomic spectrum abstracts precisely what is required to make the construction.

Goodwillie has identified the topological cyclic homology of the topological Hoschschild homology of a rational functor with smash products **[7]**, and we generalize this to an arbitrary cyclotomic spectrum. This is not a deep result, but it demonstrates the character of the topological cyclic homology and illustrates the effectiveness of the present theory. The author is grateful to L. Hesselholt for many helpful discussions.

We must begin by describing the construction. For any $\mathbb{T}$-spectrum X, if $L \subseteq K$ we have an inclusion of the Lewis-May fixed points $F_L^K : \Psi^K X \longrightarrow \Psi^L X$; the letter F is chosen because it corresponds to the Frobenius map in algebraic K-theory. To avoid confusion, the reader should ignore for the duration of the present section the fact that F_L^K induces the restriction map from K-equivariant to L-equivariant homotopy groups. The cyclotomic structure supplies a second set of maps $R_L^K : \Psi^K X \longrightarrow \Psi^L X$ defined as follows. First we let L^* be the unique subgroup of K with order $|K/L|$. Now consider the inclusion $X \longrightarrow X \wedge \tilde{E}[\not\supseteq L^*]$; applying Lewis-May L^*-fixed points we obtain a map $\Psi^{L^*} X \longrightarrow \Phi^{L^*} X$. Applying $\rho^!_{L^*}$ and the cyclotomic structure we obtain

$$\rho^!_{L^*} \Psi^{L^*} X \longrightarrow \rho^!_{L^*} \Phi^{L^*} X \xrightarrow{r_{L^*}} X;$$

finally we apply L-fixed points and obtain the required map

$$\rho^!_{K/L} \Psi^K X = \rho^!_{K/L} \Psi^{K/L^*} \Psi^{L^*} X = \Psi^L(\rho^!_{L^*} \Psi^{L^*} X) \longrightarrow \Psi^L X.$$

Again, the letter R is chosen because the induced map is the restriction map in algebraic K-theory.

To simplify notation, we index F and R simply by the order of the quotient K/L. Thus we find $F_1 = R_1 = 1$, $F_r F_s = F_{rs}$ and $R_r R_s = R_{rs}$. It turns out that the Frobenius and restriction maps also commute.

The most familiar version of the topological cyclic homology construction is simply to take the the homotopy inverse limit of the system of non-equivariant fixed point spectra under the restriction and Frobenius maps:

$$TC'(X) = \underset{\leftarrow}{\text{holim}}(\Psi^K X, R)^{hF} = \underset{\leftarrow}{\text{holim}}(\Psi^K X, F)^{hR}.$$

It may help later motivation to view this as the homotopy fixed point object of an 'action of a category'. It turns out that the intermediate object

$$TR'(X) = \underset{\leftarrow \; K}{\text{holim}}(\Psi^K X, R),$$

has significance of its own, so we prefer the first description $TC'(X) = TR'(X)^{hF}$. Furthermore, we note that the above construction shows that the map $R_L^K : \rho^!_K \Psi^K X \longrightarrow \rho^!_L \Psi^L X$ is a map of $\mathbb{T}$-spectra so

$$TR(X) = \underset{\leftarrow \; K}{\text{holim}}(\rho^!_K \Psi^K X, R)$$

is a $\mathbb{T}$-spectrum with underlying spectrum $TR'(X)$. However, we warn that the identification of all terms with $TR(X)$ means that the Frobenius maps are not maps of $\mathbb{T}$-spectra. We shall identify the relevant equivariance below.

For non-profinite work, Goodwillie points out that the diagram given by the restriction and Frobenius maps should be augmented by adding in the circle action; we may now think of an action by a topological category. Since R commutes with the Frobenius, passing to limits under R, we obtain a diagram with a copy of $TR(X)$ for each finite subgroup, and Frobenius maps relating them; the quotient category acts on $TR(X)$. For the present we view all objects as the same and hence we think of having an action of the monoid M

occurring in a split exact sequence $1 \longrightarrow \mathbb{T} \longrightarrow M \longrightarrow \mathbb{Z}_{>0} \longrightarrow 1$; in fact if $w, z \in \mathbb{T}$ then $(wF_r)(zF_s) = wz^r F_{rs}$. This leads to the definition

$$TC(X) = TR(X)^{hM} \simeq (TR(X)^{h\mathbb{T}})^{hF}.$$

15.4. Topological cyclic homology of rational spectra.

The following result may simply be regarded as evidence that the definition is a reasonable one: rationally, the topological cyclic construction is a complicated way of doing something familiar.

THEOREM 15.4.1. (Goodwillie) *For any rational cyclotomic spectrum X we have an equivalence of rational spectra*

$$TC(X) \simeq X^{h\mathbb{T}},$$

so that the topological cyclic homology agrees with the Borel cohomology.

Goodwillie proves this in the case that $X = THH(F)$ for a rational functor F with smash products [**7**, 14.2].

Proof: The first step is to note that homotopy fixed points commute with homotopy inverse limits, and that the homotopy fixed point spectrum of a non-equivariantly contractible spectrum is contractible. Thus

$$\begin{aligned} TR(X)^{h\mathbb{T}} &= (\underset{\underset{K}{\leftarrow}}{\text{holim}}\ \rho_K^! \Psi^K X)^{h\mathbb{T}} \\ &= \underset{\underset{K}{\leftarrow}}{\text{holim}}\ ((\rho_K^! \Psi^K X)^{h\mathbb{T}}) \\ &= \underset{\underset{K}{\leftarrow}}{\text{holim}}\ ((E\mathbb{T}_+ \wedge \rho_K^! \Psi^K X)^{h\mathbb{T}}) \end{aligned}$$

This shows that it is really only necessary to understand $X(1) = E\mathbb{T}_+ \wedge X$. Since the end result is simply a non-equivariant rational spectrum, it is only necessary to calculate homotopy groups.

We showed in Part I (for example it follows from 3.1.1) that the $\mathbb{T}$-free spectrum $E\mathbb{T}_+ \wedge \rho^! \Psi^K X$ is determined by its homotopy groups as modules over $\mathbb{Q}[c_1]$.

LEMMA 15.4.2. *If X is a cyclotomic spectrum with $\pi_*^{\mathbb{T}}(E\mathbb{T}_+ \wedge X) = T_0$ then*

$$\pi_*^{\mathbb{T}}(E\mathbb{T}_+ \wedge \rho_K^! \Psi^K X) = \bigoplus_{L \subseteq K} T_0$$

and hence

$$E\mathbb{T}_+ \wedge \rho_K^! \Psi^K X = \bigvee_{L \subseteq K} X(1).$$

Proof: This is immediate from 15.1.9 together with our exact identification of Lewis-May fixed points in Theorem 10.3.2, (or more directly from 10.4.2). □

The relevant inverse system thus has Kth term given as a wedge of copies of the spectrum $X(1)$ indexed by the subgroups of K. It will perhaps be clearest if we think of this as the set of functions from the finite set $[\subseteq K]$ of subgroups of K to $X(1)$. The advantage is that it permits a helpful notation for maps: any function $f : A \longrightarrow B$ of finite sets induces $f^* : X(1)^B \longrightarrow X(1)^A$. If f is an inclusion, the map f^* is simply projection.

LEMMA 15.4.3. *When $L \subseteq K$, the restriction map R^K_L induces the projection corresponding to the function $\lambda^K_L : [\subseteq L] \longrightarrow [\subseteq K]$ defined by requiring $\lambda^K_L(H)$ to have order $|H| \cdot |K/L|$, in the sense that the diagram*

$$\begin{array}{ccc} E\mathbb{T}_+ \wedge \rho^!_K \Psi^K X & \stackrel{1\wedge R^K_L}{\longrightarrow} & E\mathbb{T}_+ \wedge \rho^!_L \Psi^L X \\ \simeq\downarrow & & \downarrow\simeq \\ \bigvee_{H\subseteq K} X(1) & \stackrel{(\lambda^K_L)^*}{\longrightarrow} & \bigvee_{H\subseteq L} X(1) \end{array}$$

commutes.

Proof: Recall that L^* denotes the subgroup of K with order $|K/L|$. The effect of the map $\Psi^{L^*}X \longrightarrow \Psi^{L^*}(X \wedge \tilde{E}[\not\subseteq L^*]) = \Phi^{L^*}X$ follows from our account of the Lewis-May fixed points. Now we just need to rename subgroups using $\rho^!_{L^*}$ 15.1.9, and apply L-fixed points as described in 10.4.2. □

COROLLARY 15.4.4. *We have an equivalence*

$$TR(X)^{h\mathbb{T}} \simeq \prod_H X(1)^{h\mathbb{T}}. \quad \square$$

Notice that the above argument could also be used to identify the $\mathbb{T}$-spectrum $TR(X)$ exactly in the algebraic model. Indeed the maps $\rho^!_K \Psi^K X \longrightarrow \rho^!_L \Psi^L X$ are all identified exactly, and, using the identification of products in Chapter 21 of Part IV, we can identify the homotopy inverse limit in the standard model. However, since inverse limits do not preserve $\mathcal{F}$-free objects, the answer is not very attractive. Our present purpose requires much less; indeed, since $TR(X)^{h\mathbb{T}}$ is just a rational spectrum, it remains only to understand the action of F on homotopy groups.

LEMMA 15.4.5. *When $L \subseteq K$ the Frobenius map F^K_L induces the projection corresponding to the inclusion $i^K_L : [\subseteq L] \longrightarrow [\subseteq K]$ in the sense that the diagram*

$$\begin{array}{ccc} E\mathbb{T}_+ \wedge \Psi^K X & \stackrel{1\wedge F^K_L}{\longrightarrow} & E\mathbb{T}_+ \wedge \Psi^L X \\ \simeq\downarrow & & \downarrow\simeq \\ \bigvee_{H\subseteq K} X(1) & \stackrel{(i^K_L)^*}{\longrightarrow} & \bigvee_{H\subseteq L} X(1) \end{array}$$

commutes.

Proof: The first necessity is to understand the statement. We begin with a map $\Psi^K X \longrightarrow \Psi^L X$, which we may view as a map of $\mathbb{T}$ spectra indexed on $\mathcal{U}^K$. Once we have smashed with $E\mathbb{T}_+$ the universe may be replaced by a complete one, and we obtain a map in the category for which we have a model.

To understand the map we factor $\Psi^K X \longrightarrow \Psi^L X$ as $\Psi^K X \longrightarrow \mathrm{inf}\Psi^K X \longrightarrow \Psi^L X$. If we view these as maps of $\mathbb{T}/L$ spectra, the second map is the counit of the K/L fixed point adjunction, completely understood by 10.3.2 and the contents of Section 9.2. The first map has the property that it is a nonequivariant equivalence. The result now follows from our description of the adjunction between inflation and Lewis-May fixed points in 10. □

It remains only to index the terms so that the relevant structure is visible, and to verify that the circle action does not get in the way.

We begin by replacing subgroups by their orders, and defining a category as follows. The object set $\mathbb{Z}_{>0} \times \mathbb{Z}_{>0}$ consists of integer points in the strictly positive quadrant. There are morphisms $(\phi_s, \rho_t) : (m,n) \longrightarrow (ms, nt)$ for $s, t \in \mathbb{Z}_{>0}$. Next, consider the diagram D of divisors defined on objects by $D(m,n) = \{d \mid d \text{ divides } mn\}$ and on morphisms by letting $\phi_s : D(m,n) \longrightarrow D(ms,n)$ be inclusion, and $\rho_t : D(m,n) \longrightarrow D(m,nt)$ be multiplication by t. Finally, for an object Y we consider the contravariant functor Y^D defined by taking functions from D into Y; thus on objects, $Y^D(m,n) := Y^{D(m,n)}$. The connection with the restriction and Frobenius diagram is immediate from 15.4.5 and 15.4.3. The maps will be clearest if we replace $D(m,n)$ by the set of rational numbers i/j where i divides m and j divides n. It is easy to check that these fractions are in bijective correspondence to divisors of mn: if d divides mn the relevant fraction is d/n. With this indexing, both R and F simply drop irrelevant coordinates. Now, passing to limits under restriction maps we obtain $Y^D(m) := \lim\limits_{\leftarrow n} Y^D(m,n)$, which simply consists of sequences $(y_{i/j})$ with i dividing m. The map $F_s : Y^D(ms) \longrightarrow Y^D(m)$ again simply drops coordinates with numerator dividing ms but not m. In other words, if we now identify $Y^D(m)$ with $Y^D(1)$ by dividing the coordinate indexes by m we find F_s is the shift map specified by multiplying indices by s and ignoring fractions with an integer numerator bigger than 1. The system consists of surjections, so $\lim\limits_{\leftarrow m}{}^1 (Y^D(1), F) = 0$, and evidently the only compatible families are those with all coordinates equal: $\lim\limits_{\leftarrow m} (Y^D(1), F) = Y$. This description suggests that we should have a means for discussing subgroups of $\mathbb{T}$ with fractional orders, which suggests we should be considering the solenoid $SS := \lim\limits_{\leftarrow}(\mathbb{T}, \psi_s)$ which is the inverse limit of copies of the circle under the power maps ψ_s .

We must now check that the fact we have taken homotopy $\mathbb{T}$-fixed points between the R and F stages does not invalidate the above procedure. The time has come to be precise about the equivariance of the Frobenius maps. First, note that although we have the behaviour $F_s z = z^s F_s$ in the monoid M, so that F_s is identified with the map $\psi_s : E\mathbb{T}_+ \longrightarrow \psi_s^* E\mathbb{T}_+$, we expect the reverse type of behaviour for the objects acted upon.

LEMMA 15.4.6. *The Frobenius map induces a map of $\mathbb{T}$-spectra along ψ_s, in the sense that $F_s : \psi_s^* TR(X) \longrightarrow TR(X)$ is a map of naive $\mathbb{T}$-spectra.*

Proof: We must remember that the map F_s arose from the inclusions $\Psi^K X \longrightarrow \Psi^L X$, with K/L of order s. This inclusion is a map of $\mathbb{T}$-spectra. However, when we apply $\rho^!$ in the appropriate way, we must insert the power map ψ_s to retrieve the equivariance. □

The relevant map $TR(X)^{h\mathbb{T}} \longrightarrow TR(X)^{h\mathbb{T}}$ is then obtained by passage to fixed points from

$$F(\psi_s, F_s) : \psi_s^* F(E\mathbb{T}_+, TR(X)) = F(\psi_s^* E\mathbb{T}_+, \psi_s^* TR(X)) \longrightarrow F(E\mathbb{T}_+, TR(X)).$$

The relevant untwisting result is as follows.

LEMMA 15.4.7. *The sth power map $\psi_s : E\mathbb{T}_+ \longrightarrow \psi_s^* E\mathbb{T}_+$ is a stable rational equivalence.* $\square$

Let $Y = \prod_n X(1)$, and consider the map $F_s : \psi_s^* Y \longrightarrow Y$ of $\mathbb{T}$-spectra. The commutative diagram

$$\begin{array}{ccc} F(E\mathbb{T}_+, \psi_s^* Y) & \stackrel{F(1,F_s)}{\longrightarrow} & F(E\mathbb{T}_+, Y) \\ F(\psi_s, 1) \uparrow\simeq & & \uparrow= \\ F(\psi_s^* E\mathbb{T}_+, \psi_s^* Y) & \stackrel{F(\psi_s,F_s)}{\longrightarrow} & F(E\mathbb{T}_+, Y) \end{array}$$

has an equivalence in its left hand vertical. Hence we can untwist the action on $\prod_n X(1)^{h\mathbb{T}}$.

COROLLARY 15.4.8. *Rationally, we may identify the system of copies of $TR(X)^{h\mathbb{T}}$ under the Frobenius map with $\prod_{n>0} X(1)^{h\mathbb{T}}$ and with the Frobenius F_s acting via multiplication by s shifts.* $\square$

Theorem 15.4.1 now follows. $\square$

Part IV

Tensor and Hom in algebra and topology.

CHAPTER 16

Introduction.

16.1. General outline.

The main topological purpose of Part IV is to identify the models of the smash product and the function spectrum, but we also give a satisfying account of the Lewis-May fixed point functor as a byproduct. As one would expect, these models are given by the appropriate derived category versions of tensor product and Hom functors. Accordingly we must begin by identifying these algebraic functors, and this turns out to require all of the first seven chapters of Part IV, and topology only enters the discussion in the final chapter, in which it is shown that the algebraic and topological constructions do indeed correspond on objects. Because of this necessary delay, we shall outline the arguments in the introduction, hoping that this will help sustain the reader's interest in the algebra.

The ideal proof that the standard model is correct would proceed by a string of Quillen equivalences. It would then follow immediatedly that, tensor and Hom are the correct models for smash and function spectra. If such a proof is found, some of Chapter 24 will be bypassed, but the rest of the material will remain essential for calculation.

Since most of our work is algebraic, and since we are a long way from Part I, we begin by summarizing the algebraic categories that are relevant. We work almost exclusively over the graded ring $\mathcal{O}_{\mathcal{F}}$, which we may view as obtained from the ring $(\mathcal{O}_{\mathcal{F}})_0 = C(\mathcal{F}, \mathbb{Q}) \cong \prod_H \mathbb{Q}$ by adjoining an indeterminate c of degree -2. We let e_H denote the idempotent of $(\mathcal{O}_{\mathcal{F}})_0$ corresponding to the projection onto the Hth factor, and we let $c_H = e_H c$ in $e_H \mathcal{O}_{\mathcal{F}} = \mathbb{Q}[c_H]$. We also need Euler classes 4.6.1 generalizing the elements $c_1^k \in H^*(B\mathbb{T}_+)$ for $k \geq 0$. For any function $v : \mathcal{F} \longrightarrow \mathbb{Z}_{\geq 0}$ with finite support we have an associated Euler class c^v whose Hth coordinate is $c^{v(H)}$. These Euler classes are not homogeneous elements of $\mathcal{O}_{\mathcal{F}}$, but they acts on a $\mathcal{O}_{\mathcal{F}}$-module M as follows. Indeed if $\phi = \{H_1, H_2, \ldots, H_n\}$ is the support of v and $e_\phi \in \mathcal{O}_{\mathcal{F}}$ is the idempotent with this support, we have

$$M = M(H_1) \oplus M(H_2) \oplus \cdots \oplus M(H_n) \oplus (1 - e_\phi)M,$$

where $M(H) = e_H M$, and c^v acts as multiplication by $c^{v(H_i)}$ on the i th factor, and as the identity on $(1 - e_\phi)M$. Thus the result of inverting c^v on M is again a graded module. It is therefore reasonable to consider the multiplicative set

$$\mathcal{E} = \{c^v | v : \mathcal{F} \longrightarrow \mathbb{Z}_{\geq 0} \text{ of finite support}\}$$

of Euler classes, and allow ourselves to invert its elements.

Next, an $\mathcal{O}_{\mathcal{F}}$-module M is said to be $\mathcal{F}$-finite if the natural inclusion is an isomorphism $\bigoplus_H e_H M \cong M$, and it is torsion if $M[c^{-1}] = 0$. It is not hard to see (4.6.6) that M is an $\mathcal{F}$-finite torsion module if and only if $\mathcal{E}^{-1}M = 0$.

The final ingredient is the twisting module $t_*^{\mathcal{F}} = \mathcal{E}^{-1}\mathcal{O}_{\mathcal{F}}$, for which we have the short exact sequence 5.2.1

$$0 \longrightarrow \mathcal{O}_{\mathcal{F}} \longrightarrow t_*^{\mathcal{F}} \longrightarrow \Sigma^2\mathbb{I} \longrightarrow 0$$

with $\mathbb{I} = \bigoplus_H \mathbb{I}_H$, and $\Sigma^2\mathbb{I}_H = \mathbb{Q}[c_H, c_H^{-1}]/\mathbb{Q}[c_H]$.

There are five algebraic categories that we use in this part; they are all abelian and of finite injective dimension.

The category of torsion $k[c]$-modules: The algebra in this case is largely familiar and useful for motivation. When $k = \mathbb{Q}$ the derived category gives a model for the category of $\mathbb{T}$-spectra over H by Theorem 3.1.1.

The category of $\mathcal{F}$-finite torsion $\mathcal{O}_{\mathcal{F}}$-modules: The derived category gives a model of $\mathcal{F}$-spectra by Theorem 4.5.3.

The torsion model $\mathcal{A}_t$: The objects are maps $t_*^{\mathcal{F}} \otimes V \longrightarrow T$ of $\mathcal{O}_{\mathcal{F}}$-modules with T an $\mathcal{F}$-finite torsion module. Morphisms are required to be the identity on the tensor factor $t_*^{\mathcal{F}}$. The derived category gives a model of all $\mathbb{T}$-spectra by Theorems 6.5.1 and 5.6.1.

The semifree standard model: The objects are maps $N \longrightarrow t \otimes V$ of $k[c]$-modules, which become an isomorphism when c is inverted, where $t = k[c, c^{-1}]$. Morphisms are required to be the identity on the tensor factor $t_*^{\mathcal{F}}$. When $k = \mathbb{Q}$ the derived category gives a model of semifree $\mathbb{T}$-spectra.

The full standard model, $\mathcal{A}$: The objects are maps $N \longrightarrow t_*^{\mathcal{F}} \otimes V$ of $\mathcal{O}_{\mathcal{F}}$-modules, which become an isomorphism when $\mathcal{E}$ is inverted. Morphisms are required to be the identity on the tensor factor $t_*^{\mathcal{F}}$. The derived category gives a model of all $\mathbb{T}$-spectra, by Theorem 5.6.1.

The torsion model is somewhat exceptional, partly because it has injective dimension 2 whilst the other examples have injective dimension 1, but also because we are not able to construct a good tensor product before passing to derived categories.

We begin the introduction with a summary of the topological content of Part IV, and then turn to the algebraic highlights.

16.2. Modelling of the smash product and the function spectrum.

In the course of Chapters 17 to 24 we identify the algebraic model of the smash product functor and its right adjoint, the function spectrum functor. In essence the smash product is modelled by the left derived tensor product, and the function spectrum by its right adjoint. However there are several important caveats. Firstly, the left derived tensor product does not exist in all the categories modelling spectra: one may instead interpret it as the right derived torsion product. The nomenclature is justified since the construction is the left derived tensor product in a larger category. Similarly, in some cases the right adjoint of the left derived tensor product is *not* the right derived internal Hom functor. This fact is a consequence of the caveat about left derived tensor products. Both complications are essential in order to get models of the geometric constructions. Finally, we are only able to describe the models on objects, and not as functors; neither the smash product nor its adjoint preserve pure parity objects in general, so one cannot get a firm grip on morphisms. This means that once we have shown the smash product is modelled by the left derived tensor product it does not follow formally that the function spectrum is modelled by its adjoint. Nonetheless it is true, which seems strong evidence that our models are in fact functorial. A more satisfactory resolution would be to prove the equivalence of homotopy categories from a string of equivalences at the level of model categories, following the example of Quillen [**22, 23**]. However this approach appears inaccessible at present.

For the smash product, the first observation is that it preserves $\mathcal{F}$-spectra and $\mathcal{F}$-contractible spectra. More precisely $X\wedge Y\wedge E\mathcal{F}_+ \simeq (X\wedge E\mathcal{F}_+)\wedge(Y\wedge E\mathcal{F}_+)$ and $\Phi^{\mathbb{T}}(X\wedge Y) \simeq \Phi^{\mathbb{T}}X\wedge\Phi^{\mathbb{T}}Y$. Since $\Phi^{\mathbb{T}}(X\wedge Y) = \Phi^{\mathbb{T}}(X)\wedge\Phi^{\mathbb{T}}(Y)$, the Künneth theorem for rational homology shows that

$$\pi_*(\Phi^{\mathbb{T}}(X\wedge Y)) = \pi_*(\Phi^{\mathbb{T}}X)\otimes\pi_*(\Phi^{\mathbb{T}}Y),$$

and we shall show that $\pi^{\mathbb{T}}_*(E\mathcal{F}_+\wedge X\wedge Y)$ is the 'left derived tensor product' of $\pi^{\mathbb{T}}_*(E\mathcal{F}_+\wedge X)$ and $\pi^{\mathbb{T}}_*(E\mathcal{F}_+\wedge Y)$. This might suggest the use of the torsion model, but note that $\Sigma E\mathcal{F}_+\wedge\Sigma E\mathcal{F}_+ \simeq \Sigma^2 E\mathcal{F}_+$; the extra suspension puts the gluing map out of reach. We therefore use the standard model. However, since $t^{\mathcal{F}}_* = \mathcal{E}^{-1}\mathcal{O}_{\mathcal{F}}$ it is clear that the definition

$$(M\longrightarrow t^{\mathcal{F}}_*\otimes U)\otimes(N\longrightarrow t^{\mathcal{F}}_*\otimes V) = \big(M\otimes_{\mathcal{O}_{\mathcal{F}}}N\longrightarrow(t^{\mathcal{F}}_*\otimes U)\otimes_{\mathcal{O}_{\mathcal{F}}}(t^{\mathcal{F}}_*\otimes V) = t^{\mathcal{F}}_*\otimes U\otimes V\big)$$

preserves objects of $\mathcal{A}$. There are enough flat objects in $\mathcal{A}$, and the left derived torsion product $\circledast$ can be calculated using these.

When Y is $\mathcal{F}$-contractible, the function spectrum it is easy to see that

$$F(X,Y)\simeq F(\Phi^{\mathbb{T}}X,\Phi^{\mathbb{T}}Y)\wedge\tilde{E}\mathcal{F}.$$

Similarly if X is $\mathcal{F}$-contractible, so is the function spectrum, and hence $F(X,Y)$ is described by its homotopy groups. However in the general case, even if X and Y are both $\mathcal{F}$-spectra, $F(X,Y)$ is not: in fact the right adjoint of the smash product with an $\mathcal{F}$-spectrum X (regarded as a functor from $\mathcal{F}$-spectra to $\mathcal{F}$-spectra) is $Y\longmapsto F(X,Y)\wedge E\mathcal{F}_+$. This expression should prepare for the algebraic surprise in modelling function spectra.

The function spectrum is recognized as the right adjoint of the smash product, and hence the natural expectation (which proves to be correct on objects) is that the model should be given by the right adjoint to the total left derived functor $\circledast$. This is correct even when this right adjoint is not the total right derived functor of the internal Hom functor.

To illustrate the complications in the simplest context, we continue the discussion by concentrating on free spectra, modelled by torsion $\mathbb{Q}[c]$-modules. To begin with, we must identify the internal Hom functor on the category of torsion modules. This functor is characterized by the adjunction

$$\mathrm{Hom}(R\otimes S,T) = \mathrm{Hom}(R,\mathrm{IntHom}(S,T)),$$

and hence the internal Hom functor between torsion modules S and T is given by

$$\mathrm{IntHom}(S,T) = \Gamma_c\,\mathrm{Hom}(S,T),$$

where Γ_c is the c-power torsion functor. In fact the right adjoint of the left derived tensor product $\circledast$ is $R'\mathrm{Hom} := R\Gamma_c\,R\mathrm{Hom}$; since there are not enough flat torsion $\mathbb{Q}[c]$-modules, this is different from $R\mathrm{IntHom}$. In retrospect this is a natural counterpart of the familiar topological fact mentioned above, but it bewildered the author for some time. Just as the function object is more straightforward in the category of all $\mathbb{T}$-spectra, so too there are enough flat objects in the standard model, and therefore $R\hat{\Gamma} R\hat{\mathbf{Hom}} = R\mathbf{IntHom}$ in that case.

Here is an example where $R\mathrm{IntHom}(S, T)$ does not model the function spectrum whilst $R\Gamma_c\, R\mathrm{Hom}(S, T)$ does. This shows in particular that the two algebraic functors are different, and concentrates our attention on the correct one.

EXAMPLE 16.2.1. Let $X = Y = E\mathbb{T}_+$, so that we have models $S = T = \Sigma\mathbb{I}$. We observe that $R\mathrm{IntHom}(S, T) = \Gamma_c \mathbb{Q}[c] = 0$. On the other hand $\mathrm{Hom}(S, T) = \mathbb{Q}[c]$ is not injective, so that to calculate $R\Gamma_c\, R\mathrm{Hom}(S, T) = R\Gamma_c\, \mathrm{Hom}(S, T)$ we must take an injective approximation $fibre(\mathbb{Q}[c, c^{-1}] \longrightarrow \Sigma^2\mathbb{I})$ to $\mathbb{Q}[c]$ before applying Γ_c. We thus obtain $R\Gamma_c\, R\mathrm{Hom}(S, T) = \Sigma\mathbb{I}$, in agreement with $F(E\mathbb{T}_+, E\mathbb{T}_+) \wedge E\mathbb{T}_+ \simeq E\mathbb{T}_+$. □

The explanation of this is that the internal Hom functor is only recognized as such by virtue of the intermediate category of all modules. In fact we consider the pairs

$$\text{tors } \mathbb{Q}[c]\text{-mod} \underset{\Gamma_c}{\overset{i}{\rightleftarrows}} \mathbb{Q}[c]\text{-mod} \underset{\mathrm{Hom}(S,\cdot)}{\overset{\otimes S}{\rightleftarrows}} \mathbb{Q}[c]\text{-mod}$$

of adjoint functors (with left adjoints displayed at the top). The composite of the two lower functors is $\mathrm{IntHom}(S, \cdot) = \Gamma_c\, \mathrm{Hom}(S, \cdot)$ with its domain extended to all $\mathbb{Q}[c]$-modules. In fact the right adjoint of $\circledast S : D(\mathbb{Q}[c]) \longrightarrow D(\mathbb{Q}[c])$ is $R'\mathrm{Hom}(S, \cdot) := R\Gamma_c\, R\mathrm{Hom}(S, \cdot)$. And if S and T model the free spectra X and Y, then $R'\mathrm{Hom}(S, T)$ models $F(X, Y) \wedge E\mathbb{T}_+$.

16.3. Torsion Functors.

Chapters 17 to 20 are occupied with the discussion of torsion functors, and the reader may wonder what they are and why they occupy such a prominent place. The easiest example is the construction of the product in the category $\mathbb{A}$ of torsion $k[c]$-modules. If B_α is a torsion $k[c]$-module then $\Pi_\alpha B_\alpha$ need not be a torsion module, and the product in the category of torsion modules is $\Gamma_c \Pi_\alpha B_\alpha$, where $\Gamma_c M$ consists of the c-power torsion elements of M. In other words we have constructed the product in the category $\hat{\mathbb{A}}$ of all $k[c]$-modules, and then applied the right adjoint $\hat{\Gamma}$ to the inclusion $i : \mathbb{A} \longrightarrow \hat{\mathbb{A}}$. We call these right adjoints torsion functors by analogy with Γ_c. Note too that Γ_c is not an exact functor, so nor is the product functor. We therefore need to study the right derived functors of $\hat{\Gamma}$. Evidently the functors $\hat{\Gamma}$ play a universal role in the construction of right adjoints. This explains the need to spend so much space on them, and we can only reassure the impatient reader that use of torsion functors enormously clarifies the arguments.

Thus when $\mathbb{A}$ is any one of our five standard examples we must identify the appropriate larger category $\hat{\mathbb{A}}$, and then construct the torsion functor. This is fairly straightforward except in the most important cases: the semifree and full standard models. Even if these were our only interest, it would be important to consider the other examples, because features which are easily visible in the simpler cases recur in a less transparent form in the standard model.

The interesting cases are the semifree and full standard models. In the semifree case we give three constructions of the torsion functor. Firstly a very direct one, designed for calculation. Secondly, one analogous to the description $\Gamma_I M = \mathrm{Hom}(A/I^k, M)$ for the I-power torsion functor on an A-module M in commutative algebra, and thirdly its dual

$$\hat{\Gamma} C = \big(\mathrm{Hom}(S^0, \Delta_n \otimes C) \longrightarrow \mathrm{Hom}(S^0, \Delta_{tv} \otimes C)\big)$$

in which Δ_n and Δ_{tv} are suitable objects of the semifree model. The better behaviour of the internal Hom functor (ie that its right derived functor is the right adjoint to the left derived tensor product) results from a flabbiness result: although $\hat{\mathbf{Hom}}(B, C)$ need not be injective, it is flabby in the sense that $(R^1\hat{\Gamma})\hat{\mathbf{Hom}}(B, C) = 0$ and hence

$$R\hat{\Gamma} R\hat{\mathbf{Hom}}(B, C) = R\mathbf{IntHom}(B, C).$$

This shows that it is not logically necessary to introduce the functor $R\hat{\Gamma}$.

The full standard model is closely analogous to the semifree case except that the most obvious enveloping category $\hat{\mathbb{A}}$ is not of finite homological dimension. This means that we have not constructed its derived category. However, the flabbiness of Hom shows that it is only necessary to consider the right derived functors of the internal Hom functor, and it is no longer necessary to use the derived category of $\hat{\mathbb{A}}$ as a stepping stone. This does not prevent us from discussing the right derived functors of $\hat{\Gamma}$ as a tool for calculation.

16.4. Modelling of the product spectrum.

Once the torsion functors have been defined it is rather straightforward to identify the product. In some suitable category it has the character of a set theoretic product, and it is then only necessary to adapt it slightly and then finally apply the torsion functor.

Thus the algebra is rather simple. The reason we spend time on it, is that the product is right adjoint to the diagonal functor. It is therefore automatic that the right derived product models the product of spectra. This identification of the model of a product of spectra in the derived standard model is an essential tool in the analysis of the function spectrum. The point is that we can identify the model of the product of spectra as a functor, because the product is right adjoint to the diagonal functor. This gives us control over $F(\bigvee_i Y_i, Z) \simeq \prod_i F(Y_i, Z)$ when $F(Y_i, Z)$ is known for all i.

16.5. Modelling the Lewis-May fixed point functor.

With the machinery of internal Hom we can give give a very satisfactory description of the inflation functor and the Lewis-May K-fixed point functor on the standard model. This will not surprise readers used to thinking of inflation as an external smash product with a suitable sphere. We use bars to denote objects associated to the quotient $\overline{\mathbb{T}} = \mathbb{T}/K$, so that $\mathcal{A}$ is the standard model for $\mathbb{T}$-spectra and $\overline{\mathcal{A}}$ for $\overline{\mathbb{T}}$-spectra. The result is then that

$$\mathrm{inf}_{\overline{\mathcal{A}}}^{\mathcal{A}}(B) = S^0_{\mathcal{A}} \otimes_{\overline{\mathcal{O}}_{\mathcal{F}}} B \text{ and } \Psi^K(C) = R\mathbf{IntHom}_{\overline{\mathcal{O}}_{\mathcal{F}}}(S^0_{\mathcal{A}}, C).$$

For inflation this is only a notational change from the definition of Part II, but for the Lewis-May fixed point functor this is the first definition we have given on the standard model which arises from a functor at the level of abelian categories.

16.6. Genera of small objects.

In the course of developing the algebra we need to understand the tensor and Hom functors in the derived category, a number of interesting phenomena have emerged. One of some interest is a classification of finite semifree spectra X with $\Phi^{\mathbb{T}}X$ a wedge of even spheres and $E\mathbb{T}_+ \wedge X$ a wedge of even suspensions of $E\mathbb{T}_+$. The interesting fact is how many of them there are.

If $\Phi^{\mathbb{T}}X$ is a wedge of d spheres, then $E\mathbb{T}_+ \wedge X$ is also a wedge of d suspended copies of $E\mathbb{T}_+$. If we specify the the dimensions of the spheres, and the dimensions in which the terms $E\mathbb{T}_+$ occur then if $d = 1$ or 2 there are only finitely many objects X. However, when $d \geq 3$ there are infinitely many inequivalent indecomposable spectra X. For example there are infinitely many indecomposable finite semifree spectra X with $\Phi^{\mathbb{T}}X \simeq S^0 \vee S^2 \vee S^4$ and $E\mathbb{T}_+ \wedge X \simeq E\mathbb{T}_+ \wedge (S^0 \vee S^2 \vee S^4)$. The only significance of $0, 2$ and 4 is that they are distinct.

CHAPTER 17

Torsion functors.

This chapter is the first step towards understanding products and internal Hom objects in our various algebraic categories of interest. It is often easy to construct right adjoints in two steps: first one makes a crude construction without worrying about finiteness, and then one applies a torsion functor to obtain the final version. This motivation is formalized in Section 17.1; the remaining sections of this chapter deal in turn with each of the more straightforward algebraic categories. The semifree and full standard models are more complicated, and treated in subsequent chapters.

17.1. Context.

When constructing products (or other right adjoints) in the category of torsion $k[c]$-modules it is often convenient to first construct them in the category of all $k[c]$-modules, and then take the c-power torsion subobject. Thus the calculation

$$\text{tors-}k[c]\text{-mod}(A, \Gamma_c \Pi_\alpha B_\alpha) = k[c]\text{-mod}(A, \Pi_\alpha B_\alpha) = \Pi_\alpha k[c]\text{-mod}(A, \Pi_\alpha B_\alpha)$$

shows $\Gamma_c \Pi_\alpha B_\alpha$ is the product *in the category of torsion* $k[c]$*-modules.* It follows that the product of torsion $k[c]$-modules is not an exact functor, so we also need control when we pass to derived categories.

This situation occurs repeatedly, often in a less transparent way, so we establish the context. We suppose given a category $\mathbb{A}$ (such as tors-$k[c]$-mod), and we aim to identify a category $\hat{\mathbb{A}}$ (such as $k[c]$-mod) of which $\mathbb{A}$ is a full subcategory and so that the inclusion $i : \mathbb{A} \longrightarrow \hat{\mathbb{A}}$ has a right adjoint $\hat{\Gamma}$, which we refer to as a torsion functor. Furthermore, we shall want to assume that the categories are homologically well behaved.

CONTEXT 17.1.1. An adjoint pair of functors

$$\mathbb{A} \underset{\hat{\Gamma}}{\overset{i}{\rightleftarrows}} \hat{\mathbb{A}}$$

with the left adjoint on top so that

1. i is the inclusion of a full subcategory.
2. i is exact
3. $\mathbb{A}$ is of finite injective dimension.
4. $\hat{\mathbb{A}}$ is of finite injective dimension.
5. i preserves injectives.

□

Note first that because i is full and faithful, $\hat{\Gamma} i = 1$. It is worth recording one other general result.

LEMMA 17.1.2. *In Context 17.1.1, we have* $R^0\hat{\Gamma} = \hat{\Gamma}$ *in general, and*

$$R^n\hat{\Gamma}(iB) = \begin{cases} B & \textit{if } n = 0 \\ 0 & \textit{if } n \neq 0 \end{cases}$$

Proof: The first statement follows since $\hat{\Gamma}$ is a right adjoint, and hence left exact. Since i preserves injectives and exactness, it preserves an injective resolution of B in $\mathbb{A}$. Since $\hat{\Gamma} i = 1$, the result follows. □

In the three categories involving infinitely many finite subgroups, some of our constructions takes us outside the category $\hat{\mathbb{A}}$ into a larger category in which we have no control over the homological algebra. We therefore also need a functor to bring us back: in practice this extra functor has only a minor effect, and its formal existence is the significant fact.

CONTEXT 17.1.3. Two adjoint pairs of functors

$$\mathbb{A} \underset{\hat{\Gamma}}{\overset{i}{\rightleftarrows}} \hat{\mathbb{A}} \underset{\phi}{\overset{j}{\rightleftarrows}} \hat{\hat{\mathbb{A}}}$$

with the left adjoints on top so that

1. i and $\hat{\Gamma}$ give Context 17.1.1
2. j is the inclusion of a full subcategory.
3. j is exact

□

The following sections will each treat one of the following examples, taken in ordering of increasing complexity.

EXAMPLE 17.1.4. $\mathbb{A}$ is the category of torsion $k[c]$-modules, for a field k and $\hat{\mathbb{A}}$ is the category of all $k[c]$-modules. Both categories have injective dimension 1. This is treated in Section 17.2, and we recommend it as an introduction to all other cases.

EXAMPLE 17.1.5. $\mathbb{A}$ is the category of $\mathcal{F}$-finite torsion $\mathcal{O}_{\mathcal{F}}$-modules, $\hat{\mathbb{A}}$ is the category of all $\mathcal{F}$-finite $\mathcal{O}_{\mathcal{F}}$-modules, and $\hat{\hat{\mathbb{A}}}$ is the category of all $\mathcal{O}_{\mathcal{F}}$-modules. The categories $\mathbb{A}$ and $\hat{\mathbb{A}}$ have injective dimension 1. This is treated in Section 17.3, and we recommend it as an introduction to the torsion model and the full standard model.

EXAMPLE 17.1.6. $\mathbb{A}$ is the torsion model category $\mathcal{A}_t$ of objects $(t_*^{\mathcal{F}} \otimes V \longrightarrow T)$ with T an $\mathcal{F}$-finite torsion $\mathcal{O}_{\mathcal{F}}$-module, $\hat{\mathbb{A}}$ is the category in which T is only required to be $\mathcal{F}$-finite and $\hat{\hat{\mathbb{A}}}$ is the category in which T is arbitrary. The categories $\mathbb{A}$ and $\hat{\mathbb{A}}$ have injective dimension 2. There is a semifree analogue of this example that we need not make explicit. This is treated in Section 17.4, but may be omitted by those only interested in the standard model.

EXAMPLE 17.1.7. $\mathbb{A}$ is the semifree standard model category $\mathcal{A}_1$ of $k[c]$-morphisms $(N \stackrel{s}{\longrightarrow} t \otimes V)$ which become an isomorphism when c is inverted, where k is a field and $t = k[c, c^{-1}]$. The larger category $\hat{\mathbb{A}}$ consists of all maps $(N \longrightarrow t \otimes V)$. Both categories have injective dimension 1. This example is substantially more complicated than the previous ones and therefore has Chapters 18 and 19 devoted to it. The first of these chapters is an essential introduction to the full standard model.

EXAMPLE 17.1.8. $\mathbb{A}$ is the standard model category $\mathcal{A}$ of objects $(N \stackrel{s}{\longrightarrow} t^{\mathcal{F}}_{*} \otimes V)$ which become an isomorphism when $\mathcal{E}$ is inverted. It is natural to take $\hat{\mathcal{A}}$ to be the category of all such maps. However this fails to be an example of Context 17.1.1 because $\hat{\mathcal{A}}$ is not of finite injective dimension. This means that we have not constructed the derived category of $\hat{\mathcal{A}}$. We must therefore avoid mention of $R\hat{\Gamma}$, and this is possible because the standard model has finite flat dimension. The treatment of the torsion functor in Chapter 20 is thus very similar to the semifree case explained in Chapter 18.

17.2. Torsion $k[c]$-modules.

Here we show that $\mathbb{A}$ = tors-$k[c]$-mod and $\hat{\mathbb{A}}$ = $k[c]$-mod give an example of Context 17.1.1.

It is well known that tors-$k[c]$-mod and $k[c]$-mod are of injective dimension 1, and since injectives in each are characterized as the c-divisible modules it follows that inclusion preserves injectives. It is obvious that i is exact. In this case the torsion functor requires little introduction, but we record the result.

LEMMA 17.2.1. *The right adjoint of the inclusion i : tors-$k[c]$-mod $\longrightarrow$ $k[c]$-mod is the c-power torsion functor Γ_c defined by*

$$\Gamma_c(M) = \{x \in M \mid c^n x = 0 \textit{ for } n >> 0\}. \quad \square$$

We must also record the relevant homological behaviour.

LEMMA 17.2.2. *(i) For all $i \geq 0$, and any $k[c]$-module B we have*

$$R^i\Gamma_c(B[1/c]) = 0.$$

(ii) The right derived functors of Γ_c are calculated by the exact sequence

$$0 \longrightarrow \Gamma_c B \longrightarrow B \longrightarrow B[1/c] \longrightarrow R^1\Gamma_c(B) \longrightarrow 0.$$

Proof: First note by 17.1.2 that $R^*\Gamma_c(iB) = B$. Part (ii) is clear since $B[1/c]$ is a torsion free injective.

Now, for an arbitrary $k[c]$-module B, consider the natural map $B \longrightarrow B[1/c]$. Note that $B[1/c]$ and the cokernel $B[1/c]/B$ are both c-divisible and hence injective. Now we have the short exact sequences $0 \longrightarrow \Gamma_c B \longrightarrow B \longrightarrow B' \longrightarrow 0$ and $0 \longrightarrow B' \longrightarrow B[1/c] \longrightarrow B[1/c]/B \longrightarrow 0$. Since both $\Gamma_c B$ and $B[1/c]/B$ are torsion modules the result follows. $\square$

17.3. $\mathcal{F}$-finite torsion $\mathcal{O}_{\mathcal{F}}$-modules.

First note that $\mathbb{A} = \text{tors-}\mathcal{O}_{\mathcal{F}}^f\text{-mod}$, $\hat{\mathbb{A}} = \mathcal{O}_{\mathcal{F}}^f\text{-mod}$ and $\hat{\hat{\mathbb{A}}} = \mathcal{O}_{\mathcal{F}}\text{-mod}$ give an example of Context 17.1.3. Indeed, the $\mathcal{F}$-finiteness condition ensures that $\text{tors-}\mathcal{O}_{\mathcal{F}}^f\text{-mod} \longrightarrow \mathcal{O}_{\mathcal{F}}^f\text{-mod}$ is only a collection of copies of the example of torsion $k[c]$-modules treated in Section 17.2.

We thus have the inclusions

$$\text{tors-}\mathcal{O}_{\mathcal{F}}^f\text{-mod} \xrightarrow{i} \mathcal{O}_{\mathcal{F}}^f\text{-mod} \xrightarrow{j} \mathcal{O}_{\mathcal{F}}\text{-mod}$$

of full subcategories, and our first business is to describe their right adjoints. In what follows we shall use the algebra of Euler classes

$$\mathcal{E} = \{c^v \mid v : \mathcal{F} \longrightarrow \mathbb{Z}_{\geq 0} \text{ with finite support}\}.$$

introduced in Sections 4.5 and 4.6.

DEFINITION 17.3.1. For an arbitrary module M we let

$$\phi M = \bigoplus_H e_H M. \quad \square$$

The following statements are readily verified.

LEMMA 17.3.2. *(i) For any $\mathcal{O}_{\mathcal{F}}$-module M we have the equality*

$$\phi\Gamma_c M = \Gamma_c \phi M.$$

(ii) If L is $\mathcal{F}$-finite then

$$\text{Hom}(L, M) = \text{Hom}(L, \phi M)$$

(iii) If T is a torsion module then

$$\text{Hom}(T, M) = \text{Hom}(T, \Gamma_c M). \quad \square$$

Restricting domains appropriately we reach the required conclusion.

COROLLARY 17.3.3. *The functors ϕ and Γ_c are right adjoints to inclusions in the diagram*

$$\textit{tors-}\mathcal{O}_{\mathcal{F}}^f\textit{-mod} \underset{\Gamma_c}{\overset{i}{\rightleftarrows}} \mathcal{O}_{\mathcal{F}}^f\textit{-mod} \underset{\phi}{\overset{j}{\rightleftarrows}} \mathcal{O}_{\mathcal{F}}\textit{-mod}$$

$\square$

DEFINITION 17.3.4. We let

$$\Gamma_{\mathcal{E}} M = \{x \in M \mid ex = 0 \text{ for some Euler class } e\},$$

denote the associated torsion functor.

We may paraphrase 4.6.6 as follows.

LEMMA 17.3.5. *There are equalities*

$$\Gamma_{\mathcal{E}} M = \Gamma_c \phi M = \phi \Gamma_c M. \quad \square$$

In practice it will be the right derived functors that are of more importance to us.

LEMMA 17.3.6. *For all $i \geq 0$*

$$R^i\Gamma_{\mathcal{E}} = (R^i\Gamma_c) \circ \phi.$$

Proof: Since ϕ is exact $R^*\phi = \phi$. Since ϕ is right adjoint to an exact functor it preserves injectives. □

LEMMA 17.3.7. *(i) For all $i \geq 0$ we have*

$$R^i\Gamma_{\mathcal{E}}(\mathcal{E}^{-1}M) = 0.$$

(ii) The functor $\Gamma_{\mathcal{E}}$ is left exact, and only has one right derived functor; this is given by

$$R^1\Gamma_{\mathcal{E}}M = \mathcal{E}^{-1}M/M.$$

Proof: (i) Evidently $\mathcal{E}^{-1}M$ has no Euler torsion, so the result is clear when $i = 0$. It follows in general since an Euler-local module admits an injective resolution by Euler-local modules.

(ii) First note that Euler-torsion modules may be resolved by Euler-torsion injectives, so that $R^i\Gamma_{\mathcal{E}}T = 0$ for $i > 0$, if T is Euler-torsion. Thus the short exact sequence

$$0 \longrightarrow \Gamma_{\mathcal{E}}M \longrightarrow M \longrightarrow M/\Gamma_{\mathcal{E}}M \longrightarrow 0$$

shows that for $i \geq 1$ we have $R^i\Gamma_{\mathcal{E}}M \cong R^i\Gamma_{\mathcal{E}}(M/\Gamma_{\mathcal{E}}M)$. Now, by definition

$$0 \longrightarrow \Gamma_{\mathcal{E}}M \longrightarrow M \longrightarrow \mathcal{E}^{-1}M$$

is exact, and this gives the short exact sequence

$$0 \longrightarrow M/\Gamma_{\mathcal{E}}M \longrightarrow \mathcal{E}^{-1}M \longrightarrow \mathcal{E}^{-1}M/M \longrightarrow 0.$$

Since $\mathcal{E}^{-1}M/M$ is an $\mathcal{F}$-finite torsion module, the lemma follows by Part (i). □

Since this is the first example of the phenomenon, we comment in more detail on the need to consider both $\hat{\mathbb{A}}$ and $\hat{\hat{\mathbb{A}}}$. One might expect it was worth including the inclusions

$$\text{tors-}\mathcal{O}_{\mathcal{F}}^f\text{-mod} \longrightarrow \text{tors-}\mathcal{O}_{\mathcal{F}}\text{-mod} \longrightarrow \mathcal{O}_{\mathcal{F}}\text{-mod}$$

in the other order. The point is that neither tors-$\mathcal{O}_{\mathcal{F}}^f$-mod $\longrightarrow$ tors-$\mathcal{O}_{\mathcal{F}}$-mod nor the composite tors-$\mathcal{O}_{\mathcal{F}}^f$-mod $\longrightarrow$ $\mathcal{O}_{\mathcal{F}}$-mod preserve products or injectives.

EXAMPLE 17.3.8. The module $\mathbb{I}$ is an injective $\mathcal{F}$-finite torsion module, and we show it is not injective as a $\mathcal{O}_{\mathcal{F}}$-module.

As usual, injectivity may be tested by extending maps $\theta : J \longrightarrow \mathbb{I}$ defined on an ideal J over the whole ring $\mathcal{O}_{\mathcal{F}}$. For example we take $J = (e_1, e_2, \ldots)$ define, and define the function θ by $\theta(e_i) = (0, \ldots, 0, 1, 0 \ldots)$. This takes values in $\bigoplus_H \mathbb{Q}_H$, but does not extend over $\mathbb{Q}^{\mathcal{F}}$ because there is no common bound to the length of the vectors $\theta(v)$ for v in the ideal. □

17.4. The torsion model.

For the torsion model $\mathbb{A} = \mathcal{A}_t$ we take $\hat{\mathbb{A}}$ to be the category of all maps $t_*^{\mathcal{F}} \otimes V \longrightarrow T$ with T an $\mathcal{F}$-finite module and $\hat{\hat{\mathbb{A}}}$ to be the category of these maps in which T is arbitrary. It is easy to simplify the discussion to apply to the semifree torsion model.

In this section we show this gives an example of Context 17.1.3, where $\mathbb{A}$ and $\hat{\mathbb{A}}$ have injective dimension 2. It is convenient to work throughout with the variant $\mathcal{A}'_t$ from 6.2.4, consisting of objects $V \longrightarrow \mathrm{Hom}(t_*^{\mathcal{F}}, T)$ with T an $\mathcal{F}$-finite torsion module, and analogously $\hat{\mathcal{A}}'_t$ and $\hat{\hat{\mathcal{A}}}'_t$, in which T may be any $\mathcal{F}$-finite module or any module at all.

We saw in Section 6.3 that sufficiently many injectives in $\mathcal{A}'_t$ are of form $e'_t(V) = (V \longrightarrow 0)$ for a graded vector space V or $f'_t(I) = (\mathrm{Hom}(t_*^{\mathcal{F}}, I) \longrightarrow \mathrm{Hom}(t_*^{\mathcal{F}}, I))$ for an $\mathcal{F}$-finite torsion injective I. We also gave injective resolutions of length 2. Evidently these constructions apply equally well to $\hat{\mathcal{A}}'_t$, and i preserves injectives since this is true for the inclusion of $\mathcal{F}$-finite torsion $\mathcal{O}_{\mathcal{F}}$-modules in all $\mathcal{F}$-finite $\mathcal{O}_{\mathcal{F}}$-modules.

We thus have inclusions

$$\Gamma_{\mathcal{E}} T = \Gamma_c \phi T \xrightarrow{k} \phi T \xrightarrow{l} T.$$

We may then pullback along k and l as displayed in the following diagram:

$$\begin{array}{ccccc}
k^* l^* V & \longrightarrow & l^* V & \longrightarrow & V \\
\downarrow & & \downarrow & & \downarrow \\
\mathrm{Hom}(t_*^{\mathcal{F}}, \Gamma_c \phi T) & \xrightarrow{k} & \mathrm{Hom}(t_*^{\mathcal{F}}, \phi T) & \xrightarrow{l} & \mathrm{Hom}(t_*^{\mathcal{F}}, T).
\end{array}$$

We extend terminology from the case of $\mathcal{O}_{\mathcal{F}}$-modules as follows.

DEFINITION 17.4.1. Given an object $B = (V \longrightarrow \mathrm{Hom}(t_*^{\mathcal{F}}, T))$ we write

$$\phi(B) = (l^* V \longrightarrow \mathrm{Hom}(t_*^{\mathcal{F}}, \phi T)),$$

$$\Gamma_c(B) = (k^* V \longrightarrow \mathrm{Hom}(t_*^{\mathcal{F}}, \Gamma_c T))$$

and

$$\Gamma_{\mathcal{E}}(B) = (k^* l^* V \longrightarrow \mathrm{Hom}(t_*^{\mathcal{F}}, \Gamma_{\mathcal{E}} T)). \quad \square$$

It is immediate from the universal property of the pullback that these construct the right adjoints we need.

LEMMA 17.4.2. *The functors ϕ and Γ_c are right adjoints to inclusions in the diagram*

$$\mathcal{A}'_t \underset{\Gamma_c}{\overset{i}{\rightleftarrows}} \hat{\mathcal{A}}'_t \underset{\phi}{\overset{j}{\rightleftarrows}} \hat{\hat{\mathcal{A}}}'_t$$

$\square$

By construction and the fact that $e'_t(V)$ is injective, it is easy to see that the derived functors of both ϕ and Γ_c have torsion parts which are the corresponding derived functors for $\mathcal{O}_{\mathcal{F}}$-modules as described in Section 17.3.

It is also quite routine to calculate the vertices from resolutions. For example, if $B = (t_*^{\mathcal{F}} \otimes V \longrightarrow T)$ has has the property that T is $\mathcal{F}$-finite and $0 \longrightarrow T \longrightarrow I \longrightarrow J \longrightarrow 0$ is an injective resolution then the vertices of $R^*\Gamma_c B$ are lie in exact sequences

$$0 \longrightarrow \mathrm{Hom}(t_*^{\mathcal{F}}, T) \longrightarrow A \longrightarrow \mathrm{Hom}(t_*^{\mathcal{F}}, \Gamma_c T) \longrightarrow \mathrm{Ext}(t_*^{\mathcal{F}}, T) \longrightarrow Vert(R^2\Gamma_c B) \longrightarrow 0$$

and

$$0 \longrightarrow Vert(\Gamma_c B) \longrightarrow V \oplus \mathrm{Hom}(t_*^{\mathcal{F}}, \Gamma_c I) \longrightarrow A \longrightarrow Vert(R^1\Gamma_c B) \longrightarrow 0.$$

Since both ϕ and Γ_c are right adjoints to inclusions, they both preserve injectives, and so there is a composite functor spectral sequence for calculating right derived functors of $\Gamma_{\mathcal{E}}$ from those of Γ_c and ϕ.

CHAPTER 18

Torsion functors for the semifree standard model.

In this chapter we take $t = k[c, c^{-1}]$, and $\mathbb{A}$ to be the semifree standard model, with objects the $k[c]$-maps $N \longrightarrow t \otimes V$ which become an isomorphism when c is inverted. The category $\hat{\mathbb{A}}$ consists of all $k[c]$-maps $N \longrightarrow t \otimes V$.
we prove

THEOREM 18.0.3. *The semifree standard model $\mathbb{A}$ together and the category $\hat{\mathbb{A}}$ gives an example of Context 17.1.1.*

This example is significantly more complicated than the previous cases, and the identification of $\hat{\Gamma}$ requires some thought. However, it is clear that i is a full, faithful and exact functor. Furthermore in Sections 5.4 and 5.5 we have given an explicit construction of sufficiently many injectives in $\mathbb{A}$, namely $e(V)$ for a rational vector space V and $f(I)$ for an injective $k[c]$-module I. We also gave explicit injective resolutions of length 1. These constructions apply without change to $\hat{\mathbb{A}}$, and the fact that i preserves injectives follows since a torsion $k[c]$-injective I is an injective.

It therefore only remains to construct the torsion functor. A convenient construction proceeds by identifying its vertex and then using a pullback construction. Accordingly Section 18.1 introduces an ingredient in identifying the vertex, preparing us to define the torsion functor in Section 18.2. This is followed by Section 18.3, in which a number of important examples are given. Two alternative constructions of this functor are given in Chapter 19.

18.1. Maps out of spheres.

It remains to discuss $\hat{\Gamma}$. We will identify the vertex of $\hat{\Gamma}C$ by assuming the functor exists with the defining property

$$\mathrm{Hom}(A, C) = \mathrm{Hom}(A, \hat{\Gamma}C)$$

whenever A lies in $\mathbb{A}$. For this we need to consider objects good as domains. Inspired by topology, we consider spheres analogous to the the representation spheres S^V. These are closely associated with Euler classes, as made explicit in 5.8.2. More to the point, we can calculate the vertex in the topological case by

$$\pi_*(\Phi^{\mathbb{T}}X) = \pi_*^{\mathbb{T}}(X \wedge \tilde{E}\mathcal{F}) = \lim_{\rightarrow\, V^{\mathbb{T}}=0} \pi_*^{\mathbb{T}}(X \wedge S^V) = \lim_{\rightarrow\, V^{\mathbb{T}}=0} [S^{-V}, X]_*^{\mathbb{T}}.$$

We shall see in 18.2.1 below that the algebraic counterpart of this works at the level of the abelian category $\mathbb{A}$.

The algebraic analogue of the sphere for n times the the natural representation is the object

$$S^{n\lambda} := (\Sigma^{2n}\mathcal{O} \stackrel{c^{-n}}{\longrightarrow} t).$$

Of course we also permit the notation $S^{m+n\lambda} = \Sigma^m S^{n\lambda}$. When we consider infinitely many subgroups, degree 0 is obviously a special degree in $t^{\mathcal{F}}_*$; even though $t = k[c, c^{-1}]$ is k in each even dimension we must emphasize that degree 0 is still special. Because morphisms in the standard category are required to be the identity on the t part of the vertex factor, $t \otimes \Sigma^m k$ must be distinguished for different values of m; nonetheless we occasionally use t as an abbreviation for $t \otimes k$. For example, $\Sigma^m S^{n\lambda} = (\Sigma^{m+2n}\mathcal{O} \stackrel{c^{-n}}{\longrightarrow} t \otimes \Sigma^m k)$.

We begin by investigating the functor $S^{n\lambda}$ corepresents.

LEMMA 18.1.1. *For an object $C = (P \stackrel{\gamma}{\longrightarrow} t \otimes W)$ of $\hat{\mathbb{A}}$, we have*

$$\mathrm{Hom}(S^{n\lambda}, C) = \Sigma^{-2n} P(c^{-n}) \subseteq \Sigma^{-2n} P,$$

where

$$P(c^{-n}) = \Sigma^{2n}\gamma^{-1}(c^{-n} \otimes W) \subseteq P.$$

Thus,
- *$S^{n\lambda}$ corepresents $\Sigma^{-2n}P(c^{-n})$ and*
- *$P(c^n)$ is corepresented by $S^{2n-n\lambda}$*

Proof: A map

$$\begin{array}{ccc} \Sigma^{2n}\mathcal{O} & \stackrel{\theta}{\longrightarrow} & P \\ c^{-n}\downarrow & & \downarrow\gamma \\ t\otimes k & \stackrel{1\otimes\phi}{\longrightarrow} & t\otimes W \end{array}$$

is obviously determined by $\theta(\Sigma^{2n}1)$. However it is subject to the constraint that $\gamma(\theta(\Sigma^{2n}1)) = (1 \otimes \phi)(c^{-n} \otimes 1) = c^{-n} \otimes \phi(1)$.

The reader should not proceed until he is content about the suspensions in the statement of the lemma. □

18.2. The torsion functor.

In view of the fact that $P(c^n)$ is corepresented by $S^{2n-n\lambda}$, we expect the vertex of an element of $\mathbb{A}$ to be a direct limit of the spaces $P(c^n)$ under multiplication by the Euler class c. We use the notation

$$P(c^0)[1/c] = \varinjlim(P(c^0) \stackrel{c}{\longrightarrow} \Sigma^2 P(c^1) \stackrel{c}{\longrightarrow} \Sigma^4 P(c^2) \stackrel{c}{\longrightarrow} \cdots)$$

despite the fact $P(c^0)$ is not a $k[c]$-module. To avoid confusion if an element of P lies in several different submodules $P(c^n)$, we introduce a dummy variable χ of degree 0 and write $p\chi^n$ if p is viewed as an element of $P(c^n)$. It is then natural to write $pc^{-n}\chi^{-n}$ for an element of $P(c^0)[1/c]$ which is represented in the direct limit by $p \in P(c^n)$. From the map $P(c^0) \longrightarrow P$ we may form the maps $\nu' : P(c^0)[1/c] \longrightarrow P[1/c]$ and the composite

$$\nu : P(c^0)[1/c] \stackrel{\nu'}{\longrightarrow} P[1/c] \stackrel{\gamma[1/c]}{\longrightarrow} t \otimes W,$$

which is defined by $\nu(pc^{-n}\chi^{-n}) = c^{-n}\gamma(p)$.

PROPOSITION 18.2.1. *If C is an object of the semifree standard model $\mathbb{A}$ then the natural maps*

$$\nu : P(c^0)[1/c] \xrightarrow{\cong} c^0 \otimes W$$

and

$$\tilde{\nu} : t \otimes P(c^0)[1/c] \xrightarrow{\cong} P[1/c].$$

are isomophisms.

Before giving the proof we remark that it means that if $C = (P \longrightarrow t \otimes W)$ is an arbitrary object of $\hat{\mathbb{A}}$ it is easy to identify the vertex of $\hat{\Gamma}C = (P' \longrightarrow t \otimes W')$:

$$W' = P'(c^0)[1/c] = P(c^0)[1/c].$$

It is possible to motivate the definition of P' in the same spirit, as is done in Section 19.3, but for the present we adopt a more efficient approach.

Proof: Since γ induces an isomorphism $P[1/c] \cong t \otimes W$, it suffices to prove that ν is an isomorphism.

To see ν is surjective choose an element $c^0 \otimes w$ in the codomain; since C is an object of $\mathbb{A}$, the cokernel of the structure map is c-power torsion, and for some $n \geq 0$ and $p \in P$ we have $c^n \otimes w = \gamma(p)$. But then $p \in P(c^n)$ represents an element $pc^{-n}\chi^{-n} \in P(c^0)[1/c]$, and we have $\nu(pc^{-n}\chi^{-n}) = c^{-n} \cdot \gamma(p) = c^0 \otimes w$.

For injectivity suppose given an element $p\chi^{-n}c^{-n} \in P(c^0)[1/c]$ with $\nu(p\chi^{-n}c^{-n}) = 0$. This means that $\gamma(p) = 0$. But then, since C is in $\mathcal{A}$, the kernel of γ is c-power torsion and there is an $m \geq 0$ with $c^m p = 0$. By definition of $P(c^0)[1/c]$ this means $p\chi^{-n}c^{-n} = 0$. $\square$

Finally we may define the torsion functor.

DEFINITION 18.2.2. If $C = (P \longrightarrow t \otimes W)$ is an arbitrary object of $\hat{\mathbb{A}}$ we define $\hat{\Gamma}C = (P' \longrightarrow t \otimes W')$ as follows

(i) the vertex is defined by

$$W' = P(c^0)[1/c]$$

(ii) the nub P' is defined by the pullback diagram

$$\begin{array}{ccc} P' & \longrightarrow & P \\ {\scriptstyle \gamma'}\downarrow & & \downarrow \\ t \otimes (P(c^0)[1/c]) & \xrightarrow{\tilde{\nu}} & P[1/c] \end{array}$$

(iii) The structure map is the left hand vertical γ' in the pullback square.

PROPOSITION 18.2.3. *Definition 18.2.2 gives a functor $\hat{\Gamma} : \hat{\mathbb{A}} \longrightarrow \mathbb{A}$, and this is right adjoint to inclusion.*

Proof: First note that $\hat{\Gamma}C$ is an object of $\mathbb{A}$. Indeed the right hand vertical $P \longrightarrow P[1/c]$ from the definition becomes an isomorphism when c is inverted, and isomorphisms are preserved under pulling back. The construction is obviously natural.

By construction we have a map $P' \longrightarrow P \longrightarrow t \otimes W$, and since c is invertible on the codomain this extends uniquely over $P'[1/c]$ to give a natural transformation $\eta : \hat{\Gamma}C \longrightarrow C$. To see that this is the counit of an adjunction $i \dashv \hat{\Gamma}$ it remains to check that η is an isomorphism if C lies in $\mathbb{A}$. However if C lies in $\mathbb{A}$ then $\tilde{\nu} : t \otimes P(c^0)[1/c] \longrightarrow P[1/c]$ is an isomorphism by 18.2.1, so the map $P' \longrightarrow P$ obtained by pulling this back along $P \longrightarrow P[1/c]$ is also an isomorphism. □

We end the section with an alternative view.

DEFINITION 18.2.4. If $C = (P \longrightarrow t \otimes W)$ is an arbitrary object of $\hat{\mathbb{A}}$ we let

$$P(c^*) = \bigoplus_n P(c^n),$$

and we let

$$s : P(c^*) \longrightarrow P$$

denote the map with the inclusions $P(c^n) \longrightarrow P$ as components. To refer to elements of $P(c^*)$ without ambiguity, we use the dummy variable χ of degree 0, and write $p\chi^n$ for an element $p \in P$ if we consider it as an element of $P(c^n)$. Thus the module structure on $P(c^*)$ is given by $c \cdot (p\chi^i) = (cp)\chi^{i+1}$, and s is a map of $\mathcal{O}$-modules. □

REMARK 18.2.5. The natural map $s : P(c^*) \longrightarrow P$ is not usually surjective. For example, if $P = t$ is included as the diagonal subspace of $t \otimes (k \oplus \Sigma^{2a}k)$ with $a \neq 0$ then $P(c^*) = 0$.

LEMMA 18.2.6. *For an arbitrary element C of $\hat{\mathbb{A}}$, the natural map including the direct system*

$$P(c^0) \xrightarrow{c} \Sigma^2 P(c^1) \xrightarrow{c} \Sigma^4 P(c^2) \xrightarrow{c} \cdots$$

into the direct system

$$\bigoplus_n P(c^n) \xrightarrow{c} \Sigma^2 \bigoplus_n P(c^n) \xrightarrow{c} \Sigma^4 \bigoplus_n P(c^n) \xrightarrow{c} \cdots$$

induces an isomorphism

$$t \otimes \left(P(c^0)[1/c]\right) \xrightarrow{\cong} P(c^*)[1/c].$$

Proof: From the form of the direct limit we have an isomorphism $\bigoplus_n P(c^n)[1/c] \xrightarrow{\cong} P(c^*)[1/c]$, and by cofinality multiplication by c^n gives an isomorphism $P(c^0)[1/c] \xrightarrow{\cong} P(c^n)[1/c]$.

The resulting isomorphism $\bigoplus_n P(c^0)[1/c] \longrightarrow P(c^*)[1/c]$, is given by $(n, p) \longmapsto c^n p$. □

The connection with the first approach is completed with the observation, which is easy to prove directly, but follows from 18.2.1.

LEMMA 18.2.7. *If C lies in $\mathbb{A}$ then the map*

$$s[1/c] : P(c^*)[1/c] \longrightarrow P[1/c]$$

is an isomorphism. □

18.3. Calculations of the torsion functor.

For an arbitrary object $B = (N \longrightarrow t \otimes V)$, we have a map $B \longrightarrow e(V)$, which is an isomorphism of vertices. Its kernel and cokernel are of the form $f(P)$ for a $k[c]$-module P, and its image is of the form $I \longrightarrow t \otimes V$, with monomorphic structure map. This suggests the importance of the following examples.

Although we have chosen the notation $\hat{\Gamma}$ to suggest the analogy with the torsion functor, the functor $\hat{\Gamma}$ can behave in quite startling ways.

EXAMPLE 18.3.1. *($\hat{\Gamma}$ on monomorphic objects).* The functor $\hat{\Gamma}$ annihilates any object with zero nub:

$$\hat{\Gamma}(0 \longrightarrow t \otimes V) = 0.$$

It also annihilates the object C in which $P = t$ is included by the diagonal in $t \otimes (k \oplus \Sigma^{2a} k)$ with $a \neq 0$.

More generally if $I \longrightarrow t \otimes V$ is a monomorphism, then $I(c^*) \longrightarrow I$ is also a monomorphism, and

$$\hat{\Gamma}(I \longrightarrow t \otimes V) = (I(c^*)[1/c] \cap I \longrightarrow t \otimes I(c^0)[1/c]). \quad \square$$

Next we calculate $\hat{\Gamma} f(P)$ for an arbitrary $k[c]$-module P.

LEMMA 18.3.2. *For an arbitrary even $k[c]$-module P we have*

$$\hat{\Gamma} f(P) = (P \oplus K(P[1/c]) \longrightarrow t \otimes P[1/c])$$

where $K(P[1/c]) = \ker(t \otimes P[1/c] \longrightarrow P[1/c])$. The P component of the structure map is

$$p_{2i} \longmapsto c^{-i} \otimes c^i \overline{p}_{2i}$$

and the $K(P[1/c])$ component of the structure map is the inclusion. Furthermore the structure map $P' \longrightarrow t \otimes W'$ of $\hat{\Gamma} f(P)$ has kernel $\Gamma_c P$ and cokernel $P[1/c]/P$, so that the torsion part of $\hat{\Gamma} C$ is $R^* \Gamma_c P$.

Proof: First note that if $C = (P \longrightarrow 0)$, then $P(c^n) = P$ for all n. It follows that $P(c^0)[1/c] = P[1/c]$.

By definition the nub is the pullback

$$\begin{array}{ccc} P' & \longrightarrow & P \\ \downarrow & & \downarrow \\ t \otimes P[1/c] & \longrightarrow & P[1/c]. \end{array}$$

Since the lower horizontal is obviously surjective, it follows directly that there is a short exact sequence

$$0 \longrightarrow K(P[1/c]) \longrightarrow P' \longrightarrow P \longrightarrow 0,$$

and furthermore $K(P[1/c])$ is c-divisible and hence injective, so that

$$P' \cong P \oplus K(P[1/c]).$$

To determine the P component of the structure map we must specify a splitting, and we have used

$$p_{2i} \longmapsto (p_{2i}, c^{-i} \otimes c^i \overline{p}_{2i}). \quad \square$$

One particular case deserves to be highlighted.

COROLLARY 18.3.3. *If c is invertible on P then*

$$\hat{\Gamma} f(P) = e(P). \quad \square$$

It remains to discuss the right derived functor of $\hat{\Gamma}$. We give an example to see it does not vanish in general, but prove it does vanish on any object $f(P)$.

EXAMPLE 18.3.4. *(Non-vanishing of $R^1\hat{\Gamma}$.)*
(a) The object C with $P = 0$, and $V = k$ admits an injective resolution

$$0 \longrightarrow C \longrightarrow e(k) \longrightarrow f(t) \longrightarrow 0.$$

We have already remarked in 18.3.1 that $\hat{\Gamma} C = 0$. Since $e(k)$ lies in $\mathbb{A}$, $\hat{\Gamma}$ does not change it, and $\hat{\Gamma} f(t) = e(t)$ by 18.3.2. The object $R^1\hat{\Gamma} C$ is the cokernel of the map

$$e(k) \longrightarrow e(t),$$

namely $e(t/k)$.
(b) Now consider the object C with $P = t$ included as the diagonal in $t \otimes (k \oplus \Sigma^{2a} k)$ with $a \neq 0$. This has an injective resolution

$$0 \longrightarrow C \longrightarrow e(k \oplus \Sigma^{2a} k) \longrightarrow f(t) \longrightarrow 0.$$

We have already remarked that $\hat{\Gamma} C = 0$. Since $e(k \oplus \Sigma^{2a} k)$ lies in $\mathbb{A}$, the functor $\hat{\Gamma}$ does not change it, and $\hat{\Gamma} f(t) = e(t)$ by 18.3.2. The object $R^1\hat{\Gamma} C$ is the cokernel of the map

$$e(k \oplus \Sigma^{2a} k) \longrightarrow e(t),$$

namely $e(t/(k \oplus \Sigma^{2a} k))$. $\square$

The following result on vanishing of higher derived functors is reminiscent of the corresponding property of flabby sheaves. It turns out that it shows that all the objects which arise in our constructions are acyclic, and hence it explains why our applications do not force us to discuss $R\hat{\Gamma}$.

PROPOSITION 18.3.5. *If the structure map of $C = (P \longrightarrow t \otimes W)$ is surjective modulo c-power torsion then $R^1\hat{\Gamma} C = 0$.*

Proof: We fit an arbitrary object C into short exact sequences $0 \longrightarrow f(A) \longrightarrow C \longrightarrow I \longrightarrow 0$ and $0 \longrightarrow I \longrightarrow e(W) \longrightarrow f(B) \longrightarrow 0$. Under the given hypothesis, $f(B)$ lies in $\mathbb{A}$, so that $\hat{\Gamma}$ preserves the epimorphism $e(W) \longrightarrow f(B)$ and hence $R^1\hat{\Gamma}I = 0$. The proposition follows from a special case.

LEMMA 18.3.6. *For any $k[c]$-module A, we have $R^1\hat{\Gamma}f(A) = 0$.*

Proof: It suffices to show that the nub A' of $R^1\hat{\Gamma}f(A)$ is zero. However, if $0 \longrightarrow A \longrightarrow I_0 \longrightarrow I_1 \longrightarrow 0$ is an injective resolution then $0 \longrightarrow K(A[1/c]) \longrightarrow K(I_0[1/c]) \longrightarrow K(I_1[1/c]) \longrightarrow 0$ is also exact. Thus by 18.3.2, the sequence $0 \longrightarrow A' \longrightarrow I_0' \longrightarrow I_1' \longrightarrow 0$ is also exact. □

CHAPTER 19

Wide spheres and representing the semifree torsion functor.

In Chapter 18 we constructed the torsion functor for the semifree standard model. In this chapter, we present two alternative approaches to the torsion functor. These are less useful in calculation, but both are illuminating. The first step is to identify some useful small objects: in fact we classify them in Section 19.1, and this is of independent interest. There are an enormous number of small indecomposable objects in the semifree standard model: in most cases there are an infinite number of isomorphism types with the same numerical invariants. In Section 19.2 we identify the functors corepresented by these objects, and thus pick out the 'wide spheres' for particular attention. In Section 19.3 these wide spheres are used to give an approach to the torsion functor analogous to the description $\Gamma_I M = \lim_{\to k} \mathrm{Hom}(A/I^k, M)$ of the I-power torsion functor. However the standard model is better behaved than commutative algebra, so we can dualize this in a very satisfying way: the duality is explained in Section 19.4, and the dual approach to the torsion functor is presented in Section 19.5.

19.1. Some indecomposable objects.

In this section we make explicit a large number of objects of $\mathbb{A}$, showing the richness of even the semifree standard model. In the next section we use them to give an alternative description of the torsion functor $\hat{\Gamma}$.

We already understand objects of the form $f(T)$ in $\mathbb{A}$: they correspond to the torsion $k[c]$-modules T, which are classified in the finitely generated case. The indecomposables are suspensions of $\mathbb{I}$ and the finite cyclic modules $T = k[c]/c^n$. For an arbitrary object $A = (M \longrightarrow t \otimes U)$ the kernel of the structure map consists of the torsion submodule of M. Thus A contains the subobject $f(TM)$, which we may factor out. It is therefore natural to consider objects A in which the structure map $M \longrightarrow t\otimes U$ is a monomorphism. In 19.1.6 at the end of this section we explain the topological interest in this class.

Suppose then that $A = (M \longrightarrow t\otimes U)$ is in even degree and has monomorphic structure map. The dimension of U is the crudest feature.

EXAMPLE 19.1.1. ($\dim(U) = 1$). In this case, the object is determined by the degree $2a$ in which U is nonzero together with the degree $2n$ in which the generator of M lies (we allow $n = \infty$ for the case $e(U)$): this is the sphere $S^{2a+(n-a)\lambda}$ that proved useful above. □

The basic arena for discussion is the ungraded vector space

$$\overline{U} = (t \otimes U)_0 = \bigoplus_n U_{2n}$$

with the specified direct sum decomposition as part of the structure. We let

$$\pi(1) \geq \pi(2) \geq \cdots \geq \pi(s) > 0$$

be the dimensions of the nonzero spaces U_{2n}, and we refer to $d = dim(U) = \pi(1) + \pi(2) + \cdots + \pi(s)$ as the *dimension* of A. We identify each even degree in $t \otimes U$ with $\overline{U}$ using an appropriate power of c. We thus have the subspaces

$$\cdots \subseteq M_2 \subseteq M_0 \subseteq M_{-2} \subseteq M_{-4} \subseteq M_{-6} \subseteq \cdots \subseteq \overline{U},$$

which increase as the degree decreases. If d is finite then $M_{2i} = \overline{U}$ provided i is sufficiently negative, because $(t \otimes U)/M$ is torsion. The significant features are thus the subspaces

$$\overline{U}_0 \subset \overline{U}_1 \subset \overline{U}_2 \subset \cdots \subset \overline{U}_r = \overline{U}$$

which occur in the sequence of spaces M_{2i}, their dimensions

$$d(0) < d(1) < d(2) < \cdots < d(r) = dim(\overline{U}),$$

and the degrees

$$\infty = n(0) > n(1) > n(2) > \cdots > n(r)$$

in which the dimension of M_{2i} jumps. Thus $dim(M_{2i}) = d(s)$ when $2n(s) \geq 2i > 2n(s+1)$. This data determines C. Conversely a graded vector space U together with a flag $\overline{U}_0 \subset \overline{U}_1 \subset \overline{U}_2 \subset \cdots \subset \overline{U}_r$ in $\overline{U}$ and a sequence $\mathbf{n}$ of jump dimensions allows us to construct an object C. Any morphism between objects with monomorphic structure map is determined by the map of vertices. In particular, isomorphic objects have isomorphic vertices, the same length flag and the same dimension and degree vectors $\mathbf{d}$ and $\mathbf{n}$. Furthermore since scalar multiplication has no effect on the flag, the moduli space of isomorphism types of monomorphic objects with vertex U, flag length r, dimension vector $\mathbf{d}$ and jump degrees $\mathbf{n}$ is

$$Flag(U, r, \mathbf{d})/PGL(U),$$

independent of $\mathbf{n}$, indeed, it only depends on the *type*

$$d(0) < d(1) < d(2) < \cdots < d(r-1) < \pi(1) \oplus \pi(2) \oplus \cdots \oplus \pi(s)$$

of the object. We discuss the cases where U is of dimension ≤ 3 in some detail, and observe that if U is of dimension 3 or more and k is infinite, there are infinitely many isomorphism types of indecomposables with the same discrete invariants.

In the following examples we suppose $d(0) = 0$ (ie that the nub is bounded above). The case with $d(0) \neq 0$, flag length r and type $\mathbf{T}$ is essentially equivalent to the case with $d(0) = 0$, flag length $r + 1$ and type $0 < \mathbf{T}$.

EXAMPLE 19.1.2. (dim(U) = 2). If U is of dimension 2 then either $r = 1$ or 2. If U is concentrated in a single degree, or if $\overline{U}_1$ coincides with U_{2a} for some a we may start a basis for U with one for $\overline{U}_1$ and again obtain a decomposition of C as a sum of spheres. This leaves the case in which U is in two distinct degrees $2a < 2b$, and in which $\overline{U} = U_{2a} \oplus U_{2b}$.

If $r = 1$ then

$$M_{2i} = \begin{cases} 0 & \text{if } i > n(1) \\ \overline{U} & \text{if } n(1) \geq i \end{cases},$$

and the type is $0 < 1 \oplus 1$. In this case C is a sum of two spheres: we may choose a basis of U, and this gives a decomposition of C. If $r = 2$ we have $n(1) > n(2)$ and

$$M_{2i} = \begin{cases} 0 & \text{if } i > n(1) \\ \overline{U}_1 & \text{if } n(1) \geq i > n(2) \\ \overline{U} & \text{if } n(2) \geq i \end{cases}$$

where $\overline{U}_1$ is one dimensional; the type is $0 < 1 < 1 \oplus 1$. We have already remarked that A splits as a product if $\overline{U}_1$ coincides with U_{2a} or U_{2b}. The remaining case is that $\overline{U}_1 \subseteq \overline{U} = U_{2a} \oplus U_{2b}$ does not coincide with either factor. However, since we have $GL_1(U_{2a}) \times GL_1(U_{2b})$ acting by automorphisms, any two such objects are isomorphic. In other words, for each pair $a < b$ and each pair $n(1) > n(2)$ there is up to isomorphism a unique object in which $\overline{U}_1 \cap U_{2a} = 0 = \overline{U}_1 \cap U_{2b}$, and it is easy to see it is not decomposable as a direct sum. □

EXAMPLE 19.1.3. ($\dim(U) = 3$). When U is of dimension 3 then $r = 1, 2$ or 3. If U is concentrated in a single degree or if $r = 1$ it is easy to see the object is a sum of lower dimensional objects.

Next suppose $U = U_{2a} \oplus U_{2b}$ and U_{2a} is 2-dimensional. If $r = 2$ the type is $0 < 2 < 2 \oplus 1$ or $0 < 1 < 2 \oplus 1$. In the first case $\overline{U}_1$ is of dimension 2, the intersection $\overline{U}_1 \cap U_{2a}$ is non-trivial and the object is again a sum. In the second case, $\overline{U}_1$ is one dimensional, and there is a unique isomorphism class of indecomposables, where $\overline{U}_1$ meets U_{2a} and U_{2b} trivially. If $r = 3$ the type is $0 < 1 < 2 < 2 \oplus 1$. If $\overline{U}_2 = U_{2a}$ the object is decomposable as a sum, so we suppose $\overline{U}_2 \cap U_{2a}$ is one dimensional and spanned by a vector e. Now $\overline{U}_1$ is spanned by $\lambda e + \mu f$ where f is a nonzero vector in U_{2b}. If $\overline{U}_1$ lies in either U_{2a} or U_{2b} the object is decomposable, so we may suppose $\lambda \neq 0$. There are then two cases: either $\mu = 0$ or $\mu \neq 0$, and in the latter case, we may rescale so that $\mu = 1$. Both cases are indecomposable.

Finally, suppose $U = U_{2a} \oplus U_{2b} \oplus U_{2c}$ where the three subspaces are one dimensional. We may argue immediately that the group $PGL(U)$ is two dimensional, whilst the space of flags $0 < 1 < 2$ is three dimensional so that there will be infinitely many isomorphism classes of type $0 < 1 < 2 < 1 \oplus 1 \oplus 1$ if k is infinite dimensional. We proceed with a more detailed analysis.

If $r = 2$ the type is $0 < 1 < 1 \oplus 1 \oplus 1$ or $0 < 2 < 1 \oplus 1 \oplus 1$. In each case the result is a direct sum if $\overline{U}_1$ meets one of the three one dimensional subspaces, but there is a unique isomorphism class in which $\overline{U}_1 \cap U_{2i} = 0$, and this is indecomposable. Finally there is the case $r = 3$ with type $0 < 1 < 2 < 1 \oplus 1 \oplus 1$. If $\overline{U}_2$ contains two of the one dimensional subspaces, A decomposes as a direct sum. There are then three cases in which $\overline{U}_2$ meets one subspace, and we consider the one in which it has basis $e = (\lambda, 0, 0), f = (0, \mu, \nu)$ with $\lambda\mu\nu \neq 0$; rescaling we may suppose $\lambda = \mu = \nu = 1$. If $\overline{U}_1$ is one of the three subspaces the object decomposes as a direct sum, so we may suppose $\alpha e + f$ is a basis for $\overline{U}_1$. Finally, we may rescale so that $\alpha = 1$. This leaves the case in which $\overline{U}_2$ meets none of the one dimensional subspaces. It is therefore a sum of the subspaces $\overline{U}_2 \cap (1 \oplus 1 \oplus 0)$ and $\overline{U}_2 \cap (0 \oplus 1 \oplus 1)$, and so it has basis $e = (\lambda, \mu, 0), f = (0, \nu, \sigma)$. Rescaling we may suppose $e = (1, 1, 0), f = (0, \nu, \nu)$, and then replacing f by a scalar multiple we may suppose $f = (0,1,1)$. Now $\overline{U}_1$ is spanned by $\alpha e + \beta f$. There are the two degenerate cases with $\alpha\beta = 0$, and otherwise we may suppose $\beta = 1$, and $\alpha \neq 0$. Thus $\overline{U}_1$ is spanned by $(\alpha, 1+\alpha, 1)$. We claim that no two examples with distinct α are isomorphic. Indeed, a matrix $diag(r,s,t) \in GL(U)$, only preserves $\overline{U}_2$ if $r = s$ and $s = t$.

To summarize the isomorphism classes of indecomposables

Type	Number of classes of indecomposables
$0 < 1 < 2 \oplus 1$	1
$0 < 1 < 2 < 2 \oplus 1$	2
$0 < 1 < 1 \oplus 1 \oplus 1$	1
$0 < 2 < 1 \oplus 1 \oplus 1$	1
$0 < 1 < 2 < 1 \oplus 1 \oplus 1$	$3 + (\lvert k \rvert - 1)$ □

EXAMPLE 19.1.4. ($\dim(U) = 4$). When U is 4-dimensional a complete classification would be rather complicated. We consider the special case when U is concentrated in four distinct degrees, and $r = 4$, and we claim that if k is infinite there are infinitely many non-isomorphic indecomposables. We shall be content to give an argument by dimension. The conclusion can be reached by more elementary methods, and therefore applies to all infinite fields. The space of complete flags in a 4-dimensional space U is of dimension $6 = 3+2+1$. When U is decomposed as a sum of four lines, almost all flags intersect each of the four lines trivially. We thus have a 6-dimensional space of indecomposable objects, with the group $PGL(U)$ of dimension 3 acting on it. The quotient space of isomorphism classes is thus of dimension ≥ 3. □

We pause to remark that none of the monomorphic objects in which U has dimension ≥ 2 are actually simple, even when we require the quotient to also be monomorphic.

LEMMA 19.1.5. *An object $M \longrightarrow t \otimes U$ with monomorphic structure map and with U of dimension d admits a filtration of length d with spheres as subquotients.*

Proof: We argue by induction on d, the case $d = 0$ being trivial. In general we may choose a nonzero vector $u \in U_{2a}$ and find the largest n with $c^{-n}u \in M$. There is then a map from $S^{2a+n\lambda}$ with the generator mapping to $c^{-n}u$. The resulting map $S^{n\lambda} \longrightarrow A$ is injective, and the cokernel still has monomorphic structure map, and is lower dimensional. □

For example if A is the two dimensional indecomposable discussed in 19.1.2 above, then there are short exact sequences

$$0 \longrightarrow S^{2a+(n(2)-a)\lambda} \longrightarrow A \longrightarrow S^{2b+(n(1)-b)\lambda} \longrightarrow 0$$

and

$$0 \longrightarrow S^{2b+(n(2)-b)\lambda} \longrightarrow A \longrightarrow S^{2a+(n(1)-a)\lambda} \longrightarrow 0.$$

Note that this shows there is no Jodan-Hölder refinement theorem for filtrations for the class of monomorphic objects.

Finally, we remark that the above classification also classifies a class of semifree spectra that arise quite naturally.

LEMMA 19.1.6. *The class of finite semifree complexes X with $\pi_*(\Phi^{\mathbb{T}}X)$ and $\pi_*^{\mathbb{T}}(\Sigma E\mathbb{T}_+\wedge X)$ in even degrees corresponds to the class of even objects $A = (M \longrightarrow t\otimes U)$ with $k = \mathbb{Q}$, U finite dimensional and M bounded above.*

REMARK 19.1.7. If T is a non-zero torsion module in even degrees then $f(T)$ has non-zero even degree homotopy, and it also has odd degree homotopy unless T is divisible by 3.1.1. It therefore follows that the class of finite semifree complexes X may also be described as those with $\Phi^{\mathbb{T}}X$ a wedge of even spheres and $E\mathbb{T}_+ \wedge X$ a wedge of even suspensions of $E\mathbb{T}_+$.

Proof: Suppose X is a finite semifree complex classified by $M \longrightarrow t\otimes U$ with $k = \mathbb{Q}$. Since X is finite, M is bounded above, since this is true for S^0 and $\mathbb{T}_+$. If $M \longrightarrow t \otimes U$ is not a monomorphism, the kernel will contribute odd dimensional homotopy to $\pi_*^{\mathbb{T}}(\Sigma E\mathbb{T}_+ \wedge X)$, and $U = \pi_*(\Phi^{\mathbb{T}}X)$ is in even degrees by hypothesis.

Conversely, if $M \longrightarrow t \otimes U$ is monomorphic and even with U finite and M bounded above then we need only check that the semifree spectrum X that it classifies is in fact finite. However A admits a finite filtration with spheres as subquotients by 19.1.5. Since spheres are finite, this shows X may be built from finitely many cells. □

19.2. Wide spheres.

First we record the functor corepresented by the objects of Section 19.1.

LEMMA 19.2.1. *Suppose A is an object of $\mathbb{A}$ with finite dimensional vertex U and injective structure map, and suppose that its nub is bounded above (ie $d(0) = 0$). For an arbitrary object C of $\hat{\mathbb{A}}$, then with the notation of Section 19.1,* $\mathrm{Hom}(A, C)$ *is naturally identified with the subspace of*

$$\mathrm{Hom}(U, W) \times \mathrm{Hom}(\overline{U}_r, P_{2n(r)}) \times \mathrm{Hom}(\overline{U}_{r-1}, P_{2n(r-1)}) \times \mathrm{Hom}(\overline{U}_1, P_{2n(1)})$$

consisting of $(r+1)$-tuples $(\phi, (\theta_r, \theta_{r-1}, \dots, \theta_1))$ in which
(i) θ_r lifts $1 \otimes \phi$ in the sense that

$$\gamma \circ \theta_r = 1 \otimes \phi \text{ on } \overline{U}_r = M_{2n(r)},$$

and, for $i = r-1, r-2, \dots, 1$,
(ii) θ_i is the $c^{n(i+1)-n(i)}$-division of θ_{i+1} in the sense that

$$c^{n(i+1)-n(i)}\theta_i = \theta_{i+1} \text{ on } \overline{U}_i. \quad \square$$

Once a basis of U is chosen, ϕ specifies d elements $w_1, w_2, \dots, w_d$ of W, in degrees

$$2a(1) \leq 2a(2) \leq \cdots \leq 2a(d).$$

The map θ_r then gives an element $(q_1, q_2, \dots, q_d)$ of

$$P(c^{a(1)-n(r)}) \times P(c^{a(2)-n(r)}) \times \cdots \times P(c^{a(r)-n(r)})$$

so that $\gamma(q_i) = c^{a(i)-n(r)} \otimes w_i$. Finally, if we continue to think in terms of the elements given by the image of a basis, given θ_{i+1}, the map θ_i specifies $c^{n(i)-n(i+1)}$-divisions of elements corresponding to a basis of the subspace $\overline{U}_i$.

We now want to choose an economical collection of objects to map onto elements of objects of $\mathbb{A}$. To direct our choice we examine an object C of $\mathbb{A}$. Every element p of P lies in the image of a map $A \longrightarrow C$ where A has a two step flag $\overline{U}_1 \subset \overline{U}_2 = \overline{U}$ with $dim\overline{U}_1 = 1$. More precisely, suppose that $p \in P_{2s}$, and $\gamma(p) = \sum_{j=1}^{d} c^{m(j)} \otimes w_j$ with $w_j \neq 0$ and of degree $2s + 2m(j)$. Then since γ is surjective modulo c-power torsion, there exists an N_0 so that for any $N \geq N_0$ we have $c^N p = \Sigma_j p_j$ with $\gamma(p_j) = c^{m(j)+N} \otimes w_j$. We may therefore take U to be concentrated in the distinct degrees $2s + 2m(1), 2s + 2m(2), \ldots, 2s + 2m(d)$, taking $\overline{U}_1$ to be in degree $2s$ and $\overline{U}_2$ to be in degree $n(2) = 2s - 2N$, with $\overline{U}_1$ the diagonal subspace of $\overline{U}_2$. This suggests we give special attention to these objects.

DEFINITION 19.2.2. Given $d \geq 2$, $n(1) \geq n(2)$ and $a(1) \leq a(2) \leq \cdots \leq a(d)$ we define the wide sphere

$$S^{n(1)}_{n(2)}(\mathbf{a}) = S^{n(1)}_{n(2)}(a(1), a(2), \ldots, a(d))$$

to be the object in which U has dimension d and basis in dimensions $2a(1), 2a(2), \ldots, 2a(d)$, with the submodule M of $t \otimes U$ generated by the diagonal element in degree $2n(1)$ and the entire subspace in degree $2n(2)$. The diagonal element is called the *fundamental class*.

REMARK 19.2.3. (i) It is natural to extend the notation to $d = 1$ when $n(1) = n(2)$ by taking

$$S^n_n(a) = S^{2a+(n-a)\lambda}.$$

(ii) In general the number $n(1)$ is the complex dimension of the sphere, d is the *width*, $n(1) - n(2)$ is its *spike length*, and $\mathbf{a}$ is its *basing dimension*.
(iii) The effect of suspension is

$$\Sigma^{2k} S^{n(1)}_{n(2)}(a(1), a(2), \ldots, a(d)) = S^{n(1)+k}_{n(2)+k}(a(1)+k, a(2)+k, \ldots, a(d)+k).$$

and more generally

$$\Sigma^{2k+l\lambda} S^{n(1)}_{n(2)}(a(1), a(2), \ldots, a(d)) = S^{n(1)+k+l}_{n(2)+k+l}(a(1)+k, a(2)+k, \ldots, a(d)+k). \quad \square$$

Specializing 19.2.1, we obtain.

COROLLARY 19.2.4. *Suppose $\infty > n(1) > n(2)$, and that a basis of U is given. Any element of*

$$\mathrm{Hom}(S^{n(1)}_{n(2)}(\mathbf{a}), C)$$

is specified by
(i) elements $q_1, q_2, \ldots, q_d \in Q$ of degree $2n(2)$ with $q_i \in P(c^{a(i)-n(2)})$
(ii) an element p of degree $2n(1)$ with $c^{n(1)-n(2)}p = q_1 + q_2 + \cdots + q_d$. $\square$

19.3. Corepresenting the torsion functor.

We use the wide spheres from Section 19.2 to see that the functor $\hat{\Gamma}$ also admits a description analogous to the description $\Gamma_I(\cdot) = \lim_{\rightarrow n} \mathrm{Hom}(A/I^n, \cdot)$ of the I-power torsion functor of an ideal I in a commutative ring A. Indeed, the right adjoint $\hat{\Gamma}$ to an inclusion of categories $i : \mathbb{A} \longrightarrow \hat{\mathbb{A}}$ can always be described using direct limits as follows, provided there are no set theoretic difficulties. For any object C in $\hat{\mathbb{A}}$ we may consider the category of objects $x : T \longrightarrow C$ over C with T in $\mathbb{A}$. Assuming it has a sufficiently large small subcategory we may take the direct limit $\hat{\Gamma}(C) = \lim_{\rightarrow}(x, T)$. In the present case case we may simply restrict to the sub-category of objects with cardinality at most that of t. The collection of morphisms $T \longrightarrow C$ obviously forms a set. However, to obtain an interesting description we want to use as small an indexing category as possible.

In fact we shall construct a diagram $\mathbf{D}_n$ corepresenting the nub in the sense that

$$Nub(\hat{\Gamma}C) = \lim_{\rightarrow T \in \mathbf{D}_n} \mathrm{Hom}(T, C),$$

a diagram $\mathbf{D}_{tv}$ corepresenting the extended vertex in the sense that

$$t \otimes Vertex(\hat{\Gamma}C) = \lim_{\rightarrow T \in \mathbf{D}_{tv}} \mathrm{Hom}(T, C)$$

and a morphism

$$b : \mathbf{D}_{tv} \longrightarrow \mathbf{D}_n$$

inducing the basing map. The first property this functor must have is that it is the identity on objects C of $\mathbb{A}$.

Combining 18.1.1 and 18.2.1 we may corepresent the vertex of the torsion functor.

LEMMA 19.3.1. *If C is an object of $\mathbb{A}$ and we take $\mathbf{D}_v$ to be*

$$\cdots \longrightarrow S^{-2\lambda} \longrightarrow S^{-\lambda} \longrightarrow S^0,$$

we may recover the vertex by applying $\mathrm{Hom}(\cdot, C)$ *and taking direct limits:*

$$W = \lim_{\rightarrow n} \mathrm{Hom}(S^{-n\lambda}, C). \quad \square$$

REMARK 19.3.2. Changing to wide sphere notation, $\mathbf{D}_v$ is the diagram

$$\cdots \longrightarrow S^{-2}_{-2}(0) \longrightarrow S^{-1}_{-1}(0) \longrightarrow S^0_0(0).$$

As a first step towards the nub diagram, fix a sequence $\mathbf{a}$ of distinct integers and an integer $n(1)$ and observe that we have the inverse system

$$\cdots \longrightarrow S^{n(1)}_{n(1)-3}(\mathbf{a}) \longrightarrow S^{n(1)}_{n(1)-2}(\mathbf{a}) \longrightarrow S^{n(1)}_{n(1)-1}(\mathbf{a}) \longrightarrow S^{n(1)}_{n(1)}(\mathbf{a})$$

in which the maps are inclusions. Thus in particular the fundamental classes correspond under the maps in the inverse system.

COROLLARY 19.3.3. *If C is in $\mathbb{A}$ then elements of*

$$\lim_{\rightarrow k} \mathrm{Hom}(S^{n(1)}_{n(2)-k}(\mathbf{a}), C)$$

correspond to elements $p \in P$ *of degree* $2n(1)$ *with*

$$\gamma(p) = c^{a(1)-n(1)} \otimes w_1 + c^{a(2)-n(1)} \otimes w_2 + \cdots + c^{a(d)-n(1)} \otimes w_d,$$

for some elements $w_i \in W$ *of degree* $2a(i)$.

Proof: This follows from 19.2.4 and the fact that $P(c^i)[1/c] \cong W$ for all i by 18.2.1. □

Using graded Homs it suffices to consider spheres with fundamental class in degree 0 (ie with $n(1) = 0$). The remaining difficulty is that we can always add terms $c^{a-n(1)} \otimes 0$. To compensate, we need to compare the diagrams for different values of $\mathbf{a}$. Indeed, if the sequence $\mathbf{a}$ is a subsequence of $\mathbf{b}$ there is a comparison map

$$\mathrm{res}^{\mathbf{b}}_{\mathbf{a}} : S^0_{-j}(\mathbf{b}) \longrightarrow S^0_{-j}(\mathbf{a}).$$

This is uniquely specified by its effect on vertices, which we take to be the identity on $U(\mathbf{a})$ (and which is necessarily zero in other degrees). We could consider the entire lattice of finite subsets $\mathbf{a}$ of $\mathbb{Z}$, but it suffices to choose the convenient cofinal sequence

$$[-1, +1] \subseteq [-2, +2] \subseteq [-3, +3] \subseteq \cdots$$

of integer intervals. This gives us the nub diagram.

DEFINITION 19.3.4. The nub diagram $\mathbf{D}_n$ is the lattice

$$\begin{array}{ccccccccc}
 & & \downarrow & & \downarrow & & \downarrow & & \downarrow \\
\cdots & \longrightarrow & S^0_{-3}([-3,3]) & \longrightarrow & S^0_{-2}([-3,3]) & \longrightarrow & S^0_{-1}([-3,3]) & \longrightarrow & S^0_0([-3,3]) \\
 & & \downarrow & & \downarrow & & \downarrow & & \downarrow \\
\cdots & \longrightarrow & S^0_{-3}([-2,2]) & \longrightarrow & S^0_{-2}([-2,2]) & \longrightarrow & S^0_{-1}([-2,2]) & \longrightarrow & S^0_0([-2,2]) \\
 & & \downarrow & & \downarrow & & \downarrow & & \downarrow \\
\cdots & \longrightarrow & S^0_{-3}([-1,1]) & \longrightarrow & S^0_{-2}([-1,1]) & \longrightarrow & S^0_{-1}([-1,1]) & \longrightarrow & S^0_0([-1,1])
\end{array}$$

of wide spheres.

From 19.3.3 we deduce this is correct on objects on $\mathbb{A}$.

LEMMA 19.3.5. *If* C *is an object of* $\mathbb{A}$ *and we take* $\mathbf{D}_n$ *as above, we may recover the vertex by applying* $\mathrm{Hom}(\cdot, C)$ *and taking direct limits:*

$$P = \lim_{\substack{\rightarrow \\ j,n}} \mathrm{Hom}(S^0_{-j}([-n,n], C). \quad \square$$

If $\lim_{\substack{\rightarrow \\ j,n}} \mathrm{Hom}(S^0_{-j}([-n,n]), C)$ is to be a reasonable definition we must observe that it is a $k[c]$-module. Indeed, c is induced by the map $c : \mathbf{D}_n \longrightarrow \Sigma^\lambda \mathbf{D}_n$ given by multiplication by c on the terms.

Finally, we need to corepresent the comparison map $P' \longrightarrow t \otimes W'$. Now, we can certainly corepresent $t \otimes W'$ by $\coprod_n \Sigma^{2n} \mathbf{D}_v$, but it is convenient to enlarge this slightly.

LEMMA 19.3.6. *The diagram* $\coprod_n \Sigma^{2n}\mathbf{D}_v$ *is pro-isomorphic to the diagram* $\mathbf{D}_{tv}$ *displayed below*

$$\begin{array}{ccccccccc}
 & & \downarrow & & \downarrow & & \downarrow & & \downarrow \\
\cdots & \longrightarrow & S^{-3}_{-3}([-3,3]) & \longrightarrow & S^{-2}_{-2}([-3,3]) & \longrightarrow & S^{-1}_{-1}([-3,3]) & \longrightarrow & S^{0}_{0}([-3,3]) \\
 & & \downarrow & & \downarrow & & \downarrow & & \downarrow \\
\cdots & \longrightarrow & S^{0}_{-3}([-2,2]) & \longrightarrow & S^{-2}_{-2}([-2,2]) & \longrightarrow & S^{-1}_{-1}([-2,2]) & \longrightarrow & S^{0}_{0}([-2,2]) \\
 & & \downarrow & & \downarrow & & \downarrow & & \downarrow \\
\cdots & \longrightarrow & S^{0}_{-3}([-1,1]) & \longrightarrow & S^{-2}_{-2}([-1,1]) & \longrightarrow & S^{-1}_{-1}([-1,1]) & \longrightarrow & S^{0}_{0}([-1,1]).
\end{array}$$

Proof: We need a map

$$e : \coprod_n \Sigma^{2n}\mathbf{D}_v \longrightarrow \mathbf{D}_{tv},$$

which may be specified by giving its components. The diagram $\Sigma^{2n}\mathbf{D}_v$ has jth term $S^{n-j}_{n-j}(n)$, and provided n occurs in $\mathbf{a}$ and $m \geq n-j$, there is a unique inclusion

$$S^{n-j}_{n-j}(n) \longrightarrow S^{m}_{n-j}(\mathbf{a})$$

corresponding to the inclusion of n in $\mathbf{a}$. We take the map $\Sigma^{2n}\mathbf{D}_v \longrightarrow \mathbf{D}_n$ to have jth component zero for $j \leq |n|$ and the natural inclusion

$$S^{n-j}_{n-j}(n) \longrightarrow S^{n-j}_{n-j}([-(|n|+j), |n|+j])$$

when $j > |n|$.

Since the components are injective for $j > n$ the kernel of e is pro-zero. To see the cokernel is pro-zero, note that for any object $S^{-j}_{-j}([-n,n])$ there is a map in the system from $S^{-j-r}_{-j-r}([-n,n])$, but for $-j-r < -n$ (ie for $r > n-j$) the latter is in the image of e. $\square$

Now we can represent the map $P' \longrightarrow t \otimes W'$ by

$$\mathbf{D}_n \xleftarrow{b} \mathbf{D}_{tv} \cong \coprod_n \Sigma^{2n}\mathbf{D}_v.$$

Note that replacing $\coprod_n \Sigma^{2n}\mathbf{D}_v$ with $\mathbf{D}_{tv}$ makes b injective, with the cokernel obviously equal to the diagram

$$\begin{array}{ccccccccc}
 & & \downarrow & & \downarrow & & \downarrow & & \downarrow \\
\cdots & \longrightarrow & f(\mathcal{O}/c^3) & \longrightarrow & f(\mathcal{O}/c^2) & \longrightarrow & f(\mathcal{O}/c^1) & \longrightarrow & f(\mathcal{O}/c^0) \\
 & & \downarrow & & \downarrow & & \downarrow & & \downarrow \\
\cdots & \longrightarrow & f(\mathcal{O}/c^3) & \longrightarrow & f(\mathcal{O}/c^2) & \longrightarrow & f(\mathcal{O}/c^1) & \longrightarrow & f(\mathcal{O}/c^0) \\
 & & \downarrow & & \downarrow & & \downarrow & & \downarrow \\
\cdots & \longrightarrow & f(\mathcal{O}/c^3) & \longrightarrow & f(\mathcal{O}/c^2) & \longrightarrow & f(\mathcal{O}/c^1) & \longrightarrow & f(\mathcal{O}/c^0)
\end{array}$$

We are now ready to consider the effect of inverting c. To corepresent the inversion of c in the module arising from one of the diagrams $\mathbf{D}$, we form the inverse system

$$\underleftarrow{\lim}(\mathbf{D}, c) = \left(\cdots \xrightarrow{c} \Sigma^{-2\lambda}\mathbf{D} \xrightarrow{c} \Sigma^{-\lambda}\mathbf{D} \xrightarrow{c} \mathbf{D}\right)$$

of diagrams under multiplication by c.

LEMMA 19.3.7. *The natural map*

$$\lim_{\leftarrow}(\mathbf{D}_v, c) \longrightarrow \mathbf{D}_v$$

is an isomorphism.

Proof: the diagram $\lim_{\leftarrow}(\mathbf{D}_v, c)$ has $\mathbf{D}_v$ as a diagonal, which is cofinal. □

PROPOSITION 19.3.8. *The natural map*

$$b_* : \lim_{\leftarrow}(\mathbf{D}_{tv}, c) \longrightarrow \lim_{\leftarrow}(\mathbf{D}_n, c)$$

is a pro-isomorphism.

Proof: Note that $\lim_{\leftarrow}(\cdot, c)$ is an exact functor on diagrams. Since b is a monomorphism, b_* is also a monomorphism, and $cok(b_*) = \lim_{\leftarrow}(cok(b))$. However we calculated the cokernel of b, and it has bounded c torsion at each place, so $\lim_{\leftarrow}(cok(b), c)$ is pro-zero. □

THEOREM 19.3.9. *The functor $\hat{\Gamma}' : \hat{\mathbb{A}} \longrightarrow \mathbb{A}$ defined by*

$$\Gamma' C = \left(\lim_{\rightarrow} \mathrm{Hom}(\mathbf{D}_n, C) \xrightarrow{b^*} \lim_{\rightarrow} \mathrm{Hom}(\mathbf{D}_{tv}, C)\right)$$

is right adjoint to $i : \mathbb{A} \longrightarrow \hat{\mathbb{A}}$. Hence in particular $\hat{\Gamma}' = \hat{\Gamma}$.

Proof: The construction is evidently functorial, and it follows from 19.3.8 that $\hat{\Gamma}'C$ is an object of $\mathbb{A}$. Evaluation on the fundamental class of each wide sphere gives a natural transformation $\hat{\Gamma}'C \longrightarrow C$, and it remains to check this is an isomorphism when C lies in $\mathbb{A}$.

The fact that it is isomorphic on nubs is 19.3.5; this implies it is also an isomorphism on vertices (or we could use 19.3.1 and 19.3.6). □

19.4. Duals of wide spheres.

It would be perverse not to dualize the contents of Section 19.3. First we formalize the duality statement. For this we need to specify a new object.

DEFINITION 19.4.1. Given $n(1) \geq n(2)$, $d \geq 2$ and numbers $a(1) \leq a(2) \leq \cdots \leq a(d)$ the dual sphere

$$T^{n(1)}_{n(2)}(\mathbf{a}) = T^{n(1)}_{n(2)}(a(1), a(2), \ldots, a(2))$$

is the object with vertex U having dimension d and basis in degrees $2a(1), 2a(2), \ldots, 2a(d)$, monomorphic structure map, and nub M generated by the codiagonal codimension 1 subspace in degree $2n(1)$ and all elements in degree $2n(2)$.

REMARK 19.4.2. (i) It is natural to extend the notation to $d = 1$ in the case $n(1) = n(2)$ by taking

$$T^n_n(a) = S^{2a+(n-a)\lambda} = S^n_n(a).$$

(ii) If $d = 2$ the codiagonal of codimension 1 is the diagonal of dimension 1, so we have

$$T^{n(1)}_{n(2)}(a(1), a(2)) = S^{n(1)}_{n(2)}(a(1), a(2)).$$

(iii) In general the number $n(1)$ is the complex dimension of the dual sphere, d is the *width*, $n(1) - n(2)$ is the *cospike length*, and $\mathbf{a}$ is its *basing dimension*.

(iv) The effect of suspension is

$$\Sigma^{2k} T^{n(1)}_{n(2)}(a(1), a(2), \dots, a(d)) = T^{n(1)+k}_{n(2)+k}(a(1) + k, a(2) + k, \dots, a(d) + k).$$

and more generally

$$\Sigma^{2k+l\lambda} T^{n(1)}_{n(2)}(a(1), a(2), \dots, a(d)) = T^{n(1)+k+l}_{n(2)+k+l}(a(1) + k, a(2) + k, \dots, a(d) + k). \quad \square$$

LEMMA 19.4.3. *If $n(1) < \infty$ then there is a natural isomorphism*

$$\mathrm{Hom}(A \otimes S^{n(1)}_{n(2)}(\mathbf{a}), B) = \mathrm{Hom}(A, T^{-n(2)}_{-n(1)}(-\mathbf{a}) \otimes B).$$

Proof: Categorical duality shows it is sufficient to construct suitable maps

$$\eta : S^0 \longrightarrow S^{n(1)}_{n(2)}(\mathbf{a}) \otimes T^{-n(2)}_{-n(1)}(-\mathbf{a})$$

and

$$\epsilon : S^{n(1)}_{n(2)}(\mathbf{a}) \otimes T^{-n(2)}_{-n(1)}(-\mathbf{a}) \longrightarrow S^0.$$

Concentrating on vertices, we choose an identification $U(\mathbf{a})^* = U(-\mathbf{a})$, this means that the codiagonal in $U(-\mathbf{a})$ is identified with the annihilator Δ^0 of the diagonal Δ in $U(\mathbf{a})$. Now we aim to ensure that η and ϵ give the standard maps

$$\eta_v : k \longrightarrow U(\mathbf{a}) \otimes U(\mathbf{a})^*$$

and

$$\epsilon_v : U(\mathbf{a}) \otimes U(\mathbf{a})^* \longrightarrow k$$

on vertices. It suffices to show these maps preserve nubs. In that case, η_v and ϵ_v satisfy the required categorical identities, so the identities for η and ϵ follow because the structure maps are monomorphic.

Consider ϵ_v. There is nothing to check in degrees ≤ 0 since the nub of S^0 exhausts $t \otimes k$. The nub of the tensor product is zero in degrees $> n(1) - n(2)$, so there is again nothing to check. In each remaining positive degree the only non-zero contributions are $\Delta \otimes \Delta^0 \longrightarrow k$. Since ϵ_v is evaluation, it is zero on this subspace of $U(\mathbf{a}) \otimes U(\mathbf{a})^*$.

Consider η_v. There is nothing to check in degrees ≥ 0 since the nub of S^0 is zero. In negative degrees $< n(2) - n(1)$ the nub is exhausts $t \otimes U(\mathbf{a}) \otimes U(\mathbf{a})^*$ so there is again nothing to check. Since η_v is the trace map, we need to verify that in the remaining degrees the nub contains the trace element. The nub certainly contains $\Delta \otimes U(\mathbf{a})^* + U(\mathbf{a}) \otimes \Delta^0$. If we start the basis of $U(\mathbf{a})$ with an element of the diagonal, it is visible that the trace element lies in this subspace. $\square$

19.5. Representing the torsion functor.

After Section 19.4, we are in a position to dualize our inverse systems of wide spheres to obtain direct systems. Indeed, we may take $s = 0$ or j and apply 19.4.3 with $A = S^0$ to find

$$\lim_{\rightarrow j,n} \mathrm{Hom}(S^{-s}_{-j}([-n,n]), B) = \mathrm{Hom}(S^0, \lim_{\rightarrow j,n} T^j_s([-n,n]) \otimes B).$$

Furthermore the tensor product commutes with direct limits, so that we may simply restrict attention to the two objects

$$\Delta_n := \lim_{\rightarrow j,n} T^j_0([-n,n])$$

and

$$\Delta_{tv} := \lim_{\rightarrow j,n} T^j_j([-n,n]).$$

We also have a map

$$b^* : \Delta_n \longrightarrow \Delta_{tv}.$$

We make these explicit. The object Δ_{tv} is rather familiar, and no surprise.

PROPOSITION 19.5.1. *The object representing the extended vertex is*

$$\Delta_{tv} = \bigoplus_n \Sigma^{2n} S^{\infty\lambda} = e(t).$$

The map

$$b^* : \Delta_n \longrightarrow \Delta_{tv} = e(t)$$

is an isomorphism on vertices. On nubs it is an isomorphism in degrees $0, -2, -4, \dots$, and in positive degrees $2, 4, 6, \dots$ is the inclusion of the kernel of the multiplication map

$$\mu : t \otimes t \longrightarrow t.$$

Explicitly, in degree $2n \geq 2$ it is the subspace

$$\{ \sum_{i+j=-n} \lambda_i c^i \otimes c^j \mid \sum_i \lambda_i = 0\}.$$

Proof: The identification of Δ_{tv} is straightforward, since since $T^j_j(\mathbf{a}) = S^j_j(\mathbf{a})$. The fact that b^* is a monomorphism follows since the terms $T^j_j([-n,n]) \longrightarrow T^j_0([-n,n])$ are monomorphic. The identification of the image is immediate from the definition of $T^j_0(\mathbf{a})$. □

COROLLARY 19.5.2. *The functor $\hat{\Gamma}'' : \hat{\mathbb{A}} \longrightarrow \mathbb{A}$ defined by*

$$\hat{\Gamma}'' C = \big(\mathrm{Hom}(S^0, \Delta_n \otimes C) \longrightarrow \mathrm{Hom}(S^0, \Delta_{tv} \otimes C)\big)$$

is right adjoint to the inclusion. Hence $\hat{\Gamma}'' = \hat{\Gamma}' = \hat{\Gamma}$. □

Proof 1: The result is equivalent to 19.3.9, using duality 19.4.3. □

Proof 2: It is preferable to give a direct proof which does not depend on the earlier sections of this chapter.

To see that $\hat{\Gamma}''C$ lies in $\mathbb{A}$ note that the map $b^* : \Delta_n \longrightarrow \Delta_{tv}$ becomes an isomorphism when c is inverted, as is obvious from our identification 19.5.1.

Next, we need a natural transformation $\hat{\Gamma}'' \longrightarrow 1$. For the vertex part we use the passage to vertex map

$$\mathrm{Hom}(S^0, \Delta_{tv} \otimes C) \longrightarrow \mathrm{Hom}(k, t \otimes W) = t \otimes W.$$

For the nub part, let $N\Delta_n$ be the nub of Δ_n. By definition, the multiplication of t factors to give a map ϵ in the diagram

$$\begin{array}{ccc} N\Delta_n & \overset{\epsilon}{\dashrightarrow} & \mathcal{O} \\ \downarrow & & \downarrow \\ t \otimes t & \xrightarrow{\mu} & t. \end{array}$$

This gives a map

$$\mathrm{Hom}(S^0, \Delta_n \otimes C) \longrightarrow \mathrm{Hom}(\mathcal{O}, N\Delta_n \otimes_{\mathcal{O}} P) \xrightarrow{\epsilon_*} \mathrm{Hom}(\mathcal{O}, \mathcal{O} \otimes_{\mathcal{O}} P) = P,$$

where the first map is restriction to the nub component. It may be verified that the nub and vertex parts are compatible. If $(\theta, 1 \otimes \phi)$ gives a map $S^0 \longrightarrow \Delta_n \otimes C$, with $\theta(1) = \sum_{i,j,k} c^i \otimes c^j \otimes p_k$ and $\gamma(p_k) = \sum_l c^l \otimes w_{k,l}$, then compatibility gives $\sum_{j,k} \sum_{i+l=s} c^j \otimes w_{k,l} = 0$ if $s \neq 0$, and $\phi(1) = \sum_{i,j,k} c^j \otimes w_{k,-i}$. We then calculate

$$\gamma \epsilon_* \theta(1) = \gamma(\sum_{i,j,k} c^{i+j} p_k) = \sum_{i,j,k,l} c^{i+j+l} \otimes w_{k,l},$$

which is $\phi(1)$ by the compatibility.

Finally, we need to check that the natural transformation $\hat{\Gamma}''C \longrightarrow C$ is the identity for objects C of $\mathbb{A}$. Since Δ_n is flat, and since $\mathrm{Hom}(S^0, \cdot)$ is left exact, the functor $Nub(\hat{\Gamma}''C)$ is left exact. In view of the short exact sequences $0 \longrightarrow f(A) \longrightarrow C \longrightarrow I \longrightarrow 0$ and $0 \longrightarrow I \longrightarrow e(W) \longrightarrow f(B) \longrightarrow 0$, the Five Lemma shows that it suffices to prove the isomorphism for the objects $e(W)$, $f(A)$ and $f(B)$.

When $C = e(W)$, we have $\Delta_n \otimes e(W) = \Delta_{tv} \otimes e(W)$, so it suffices to observe that the natural transformation is obviously the identity on vertices.

When $C = f(T)$ we have

$$\mathrm{Hom}(S^0, \Delta_n \otimes f(T)) = \mathrm{Hom}(S^0, f(N\Delta_n \otimes_{\mathcal{O}} T)) = N\Delta_n \otimes_{\mathcal{O}} T.$$

However, we have the diagram

$$\begin{array}{ccccccccc} 0 & \longrightarrow & K' & \longrightarrow & N\Delta_n & \longrightarrow & \mathcal{O} & \longrightarrow & 0 \\ & & \downarrow & & \downarrow & & \downarrow & & \\ 0 & \longrightarrow & K & \longrightarrow & t \otimes t & \longrightarrow & t & \longrightarrow & 0. \end{array}$$

This shows that $K' = K$ and hence, since T is torsion, that $K \otimes_{\mathcal{O}} T = 0$. It follows that $N\Delta_n \otimes_{\mathcal{O}} T \cong \mathcal{O} \otimes_{\mathcal{O}} T \cong T$ as required. □

CHAPTER 20

Torsion functors for the full standard model.

The purpose of this chapter is to construct a torsion functor for the standard model $\mathcal{A}$. We will eventually show that it is not necessary for our understanding of the Hom functor to establish that $\mathcal{A}$ fits into Context 17.1.3, because it has finite flat dimension. The corresponding phenomenon in the semifree case can also be explained by the flabbiness result 18.3.5.

The idea is to translate the methods which worked for the semifree case; this can be done very smoothly. Thus Section 20.1 records the functors corepresented by algebraic spheres and Section 20.2 translates the construction from the semifree case in a naive way, which is adequate provided we do not need homological control. In Section 20.3 we make the calculations that are needed for topological applications in Chapter 24.

20.1. Maps out of spheres.

Once again the first step in identifying the torsion functor is to identify its vertex, and this requires us to look at the functors corepresented by the algebraic analogue of topological spheres. Recall from 5.8.2 that these correspond to functions $v : \mathcal{F} \longrightarrow \mathbb{Z}$ with finite support, and are given by

$$S^v = (\mathcal{O}_{\mathcal{F}}(v) \longrightarrow t_*^{\mathcal{F}}),$$

where the submodule

$$\mathcal{O}_{\mathcal{F}}(v) := \{x \in t_*^{\mathcal{F}} \mid c^v x \in \mathcal{O}_{\mathcal{F}}\}$$

of $t_*^{\mathcal{F}}$ consists in degree $2n$ of functions on $\mathcal{F}$ which are only non-zero at H if $n \leq v(H)$. If λ is the dimension function of the natural representation (restricted to the trivial subgroup) this notation is consistent with the notation used for the semifree case in Section 18.2.

Next we describe the functor S^v corepresents: the idea is that $\mathrm{Hom}(S^v, C)$ is specified by elements of the nub of C which are "c^v-divisions of elements of $\gamma^{-1}(c^0 \otimes W)$". To be more precise, suppose $C = (P \stackrel{\gamma}{\longrightarrow} t_*^{\mathcal{F}} \otimes W)$. Then, for example S^0 co-represents $\gamma^{-1}(c^0 \otimes W)$. If v is constant at $n > 0$ on a finite set ϕ of subgroups and zero on its complement, then $c^v = c_\phi^n = e_\phi c^n + (1 - e_\phi)$. In this case $P = e_\phi P \times (1 - e_\phi)P$, and

$$\mathrm{Hom}(S^v, C) = \gamma^{-1}(e_\phi c^{-n} \otimes W) \times \gamma^{-1}((1 - e_\phi) \otimes W)) \subseteq \Sigma^{-2n} e_\phi P \times (1 - e_\phi)P.$$

More generally, if ϕ_i is the set of subgroups on which v takes the value i,

$$\mathrm{Hom}(S^v, C) = \prod_i \gamma^{-1}(e_{\phi_i} c^{-i} \otimes W) \subseteq \Sigma^{-v} P.$$

We let

$$P(c^v) := \Sigma^v \mathrm{Hom}(S^{-v}, C) \subseteq P.$$

Explicitly, an element $p \in P$ of degree d lies in $P(c^v)$ if for each i, $e_{\phi_i}\gamma(p) = c^i \otimes w_i$ for some homogeneous element $w_i \in W$ of degree $|w_i| = d + 2i$.

20.2. The torsion functor.

The torsion functor $\hat{\Gamma}$ is the right adjoint to the inclusion

$$k : \mathcal{A} \longrightarrow \hat{\mathcal{A}}$$

where $\hat{\mathcal{A}}$ is the category with objects $P \longrightarrow t_*^{\mathcal{F}} \otimes W$, without restriction on the structure map. This is adequate for our purposes, but note that $\hat{\mathcal{A}}$ does not have finite injective dimension, so we have not justified the construction of the derived category of $\hat{\mathcal{A}}$.

We are now ready to proceed with the construction of $\hat{\Gamma}$. For an object of $\mathcal{A}$ we expect the analogue of

$$\pi_*(\Phi^{\mathbb{T}} X) = \pi_*^{\mathbb{T}}(X \wedge \tilde{E}\mathcal{F}) = \lim_{\rightarrow\, V^{\mathbb{T}}=0} \pi_*^{\mathbb{T}}(X \wedge S^V) = \lim_{\rightarrow\, V^{\mathbb{T}}=0} [S^{-V}, X]_*^{\mathbb{T}}.$$

We thus expect the vertex of an element of $\mathcal{A}$ to be a direct limit of the spaces $P(c^v)$ under multiplication by the Euler classes. We use the notation

$$\mathcal{E}^{-1}P(c^0) = \lim_{\rightarrow} \Sigma^v P(c^v) = \lim_{\rightarrow\, v} \mathrm{Hom}(S^{-v}, P),$$

despite the fact $P(c^0)$ is not a $\mathcal{O}_{\mathcal{F}}$-module. We use the notation $pc^{-v}\chi^{-v}$ for the element of $\mathcal{E}^{-1}P(c^0)$ represented by $p \in P(c^v)$.

Now consider the map $P(c^0) \longrightarrow P$ and form the map $\nu' : \mathcal{E}^{-1}P(c^0) \longrightarrow \mathcal{E}^{-1}P$ and the composite

$$\nu : \mathcal{E}^{-1}P(c^0) \xrightarrow{\nu'} \mathcal{E}^{-1}P \xrightarrow{\mathcal{E}^{-1}\gamma} t_*^{\mathcal{F}} \otimes W,$$

noting that the image lies in $c^0 \otimes W$ by definition of $P(c^0)$. In terms of formulae, note that if $p \in P(c^v)$

$$\nu(pc^{-v}\chi^{-v}) = c^{-v} \cdot \gamma(p).$$

PROPOSITION 20.2.1. *If C is an object of the standard model $\mathcal{A}$ then the natural maps*

$$\nu : \mathcal{E}^{-1}P(c^0) \xrightarrow{\cong} c^0 \otimes W$$

and

$$\tilde{\nu} : t_*^{\mathcal{F}} \otimes \mathcal{E}^{-1}P(c^0) \xrightarrow{\cong} \mathcal{E}^{-1}P$$

are isomorphisms.

Proof: It suffices to show that ν is an isomorphism.

To see ν is surjective choose an element $c^0 \otimes w$ in the codomain; since C is an object of $\mathcal{A}$, the cokernel of the structure map is Euler-torsion, and for some $v \geq 0$ and $p \in P$ we have $c^v \otimes w = \gamma(p)$. But then $p \in P(c^v)$ represents an element $pc^{-v}\chi^{-v} \in \mathcal{E}^{-1}P(c^0)$, and we have $\nu(pc^{-v}\chi^{-v}) = c^{-v} \cdot \gamma(p) = c^0 \otimes w$.

For injectivity suppose given an element $p\chi^{-v}c^{-v} \in \mathcal{E}^{-1}P(c^0)$ with $\nu(p\chi^{-v}c^{-v}) = 0$. This means that $\gamma(p) = 0$, and so, since C is in $\mathcal{A}$, the kernel of γ is Euler-torsion and there is a $w \geq 0$ with $c^w p = 0$. By definition of $\mathcal{E}^{-1}P(c^0)$ this means $p\chi^{-v}c^{-v} = 0$. □

We may now define the torsion functor.

DEFINITION 20.2.2. If $C = (P \longrightarrow t_*^{\mathcal{F}} \otimes W)$ is an arbitrary object of $\hat{\mathcal{A}}$ we define $\hat{\Gamma} C = (P' \longrightarrow t_*^{\mathcal{F}} \otimes W')$ as follows
(i) the vertex is defined by

$$W' = \mathcal{E}^{-1}P(c^0)$$

(ii) the nub P' is defined by the pullback diagram

$$\begin{array}{ccc} P' & \longrightarrow & P \\ {\scriptstyle \gamma'}\downarrow & & \downarrow \\ t_*^{\mathcal{F}} \otimes \mathcal{E}^{-1}P(c^0) & \xrightarrow{\tilde{\nu}} & \mathcal{E}^{-1}P. \end{array}$$

(iii) The structure map γ' is the left hand vertical in the pullback square.

PROPOSITION 20.2.3. *Definition 20.2.2 gives a functor* $\hat{\Gamma} : \hat{\mathcal{A}} \longrightarrow \mathcal{A}$, *and this is right adjoint to the inclusion* $k : \mathcal{A} \longrightarrow \hat{\mathcal{A}}$.

Proof: First note that $\hat{\Gamma} C$ is an object of $\mathcal{A}$. Indeed the right hand vertical $P \longrightarrow \mathcal{E}^{-1}P$ of the definition becomes an isomorphism when $\mathcal{E}$ is inverted, and isomorphisms are preserved under pulling back. The construction is obviously natural.

By construction we have a map $P' \longrightarrow P \longrightarrow t_*^{\mathcal{F}} \otimes W$, and since $\mathcal{E}$ is invertible on the codomain this extends uniquely over $\mathcal{E}^{-1}P'$ to give a natural transformation $\eta : \hat{\Gamma} C \longrightarrow C$. To see that this is the counit of an adjunction $i \dashv \hat{\Gamma}$ it remains to check that η is an isomorphism if C lies in $\mathcal{A}$. However if C lies in $\mathcal{A}$ then $t_*^{\mathcal{F}} \otimes \mathcal{E}^{-1}P(c^0) \longrightarrow \mathcal{E}^{-1}P$ is an isomorphism by 20.2.1, so the map $P' \longrightarrow P$ obtained by pulling this back along $P \longrightarrow \mathcal{E}^{-1}P$ is also an isomorphism. $\square$

20.3. Calculations of the torsion functor.

The general behaviour of the torsion functor is the same as that of the semifree standard model as presented in Section 18.3, so we restrict ourselves to the cases that are particularly relevant to topology.

LEMMA 20.3.1. *If P occurs as a nub then*

$$\hat{\Gamma} f(P) = \left(P \oplus K(\mathcal{E}^{-1}P) \longrightarrow t_*^{\mathcal{F}} \otimes \mathcal{E}^{-1}P\right)$$

where $K(\mathcal{E}^{-1}P) = \ker(t_*^{\mathcal{F}} \otimes \mathcal{E}^{-1}P \longrightarrow \mathcal{E}^{-1}P)$. *Furthermore the structure map* $P' \longrightarrow t_*^{\mathcal{F}} \otimes W'$ *of* $\hat{\Gamma} f(P)$ *has kernel* $\Gamma_{\mathcal{E}} P$ *and cokernel* $\mathcal{E}^{-1}P/P$, *so that the torsion part of* $\hat{\Gamma} f(P)$ *is* $R^*\Gamma_{\mathcal{E}} P$.

Proof: Note that $P(c^v) = P$ for all v, and hence $\mathcal{E}^{-1}P(c^0) = \mathcal{E}^{-1}P$.

By definition, the nub is the pullback

$$\begin{array}{ccc} P' & \longrightarrow & P \\ \downarrow & & \downarrow \\ t_*^{\mathcal{F}} \otimes \mathcal{E}^{-1}P & \longrightarrow & \mathcal{E}^{-1}P \end{array}$$

It follows directly that there is a short exact sequence

$$0 \longrightarrow K(\mathcal{E}^{-1}P) \longrightarrow P' \longrightarrow P \longrightarrow 0.$$

Finally we claim that $K(\mathcal{E}^{-1}P)$ is injective. Indeed, since P is a nub we have an isomorphism $\mathcal{E}^{-1}P \cong t_*^{\mathcal{F}} \otimes V$, for some V, so $K(\mathcal{E}^{-1}P) = K(t_*^{\mathcal{F}} \otimes V) = K(t_*^{\mathcal{F}}) \otimes V$. However multiplication $t_*^{\mathcal{F}} \otimes t_*^{\mathcal{F}} \longrightarrow t_*^{\mathcal{F}}$ is split by $\tau \longmapsto \tau \otimes 1$. This shows $K(\mathcal{E}^{-1}P)$ is a summand of $t_*^{\mathcal{F}} \otimes t_*^{\mathcal{F}} \otimes V$, and the latter is injective for baseable modules by 5.3.1. □

It is worth drawing attention to two immediate consequences.

COROLLARY 20.3.2. *If $\mathcal{E}$ is invertible on P then*

$$\hat{\Gamma} f(P) = e(\mathcal{E}^{-1}P). \quad \square$$

COROLLARY 20.3.3. *Given any $\mathcal{O}_{\mathcal{F}}$-module P there is a cofibre sequence*

$$f(\Gamma_{\mathcal{E}}P) \vee \Sigma^{-1} f(R^1\Gamma_{\mathcal{E}}P) \longrightarrow \hat{\Gamma} f(P) \longrightarrow e(\mathcal{E}^{-1}P)$$

in the standard model. □

It would be desirable to be able to deduce this by applying a functor $R\hat{\Gamma} f(\cdot)$ to a fibre sequence $F \longrightarrow P \longrightarrow \mathcal{E}^{-1}P$. Unfortunately we have not justified the existence of $R\hat{\Gamma}$, but an ad hoc argument based on the analogues of the flabbiness result 18.3.5 from the semifree standard model is also possible. We record the following result on vanishing of higher derived functors It turns out that it shows that all the objects which arise in our constructions are acyclic, and hence it explains why our applications do not force us to discuss $R\hat{\Gamma}$.

PROPOSITION 20.3.4. *If the structure map of $C = (P \longrightarrow t_*^{\mathcal{F}} \otimes W)$ is surjective modulo Euler torsion then $R^i\hat{\Gamma} C = 0$ for $i \geq 1$.*

Proof: Formally identical to 18.3.5. □

CHAPTER 21

Product functors.

It is well known that, although it is easy to show that products exist in categories of spectra, it can be hard to understand anything about them that the categorical properties of products do not guarantee. However the category of $\mathbb{T}$-spectra is simple enough that we can give a complete analysis, and the author has found the process of understanding this extremely instructive. In any case, the analysis of products is important for two other reasons: firstly the case of the standard model allows us to give an analysis of function spectra (on objects), and secondly it allows one to understand inverse limits over a sequence of spectra. The algebraic and categorical issues which are important here occur again in a more complicated form in our discussion of the tensor and Hom functors. This chapter depends on Chapters 17 to 20.

21.1. General discussion.

Suppose then that A is an indexing set, and that we wish to understand the product $\Pi_{\alpha \in A} Y_\alpha$ of $\mathbb{T}$-spectra Y_α in terms of the algebraic models of the various spectra Y_α. The reason products are amenable to analysis is that the the functor $\Pi^{\mathbb{C}} : \mathbb{C}^A \longrightarrow \mathbb{C}$ is right adjoint to the very well behaved diagonal functor $\Delta : \mathbb{C} \longrightarrow \mathbb{C}^A$. Since we know that the model of the topological diagonal is the algebraic diagonal, we know that their right adjoints correspond, and hence we need only identify the product in the algebraic context. To avoid confusion we shall only use the undecorated symbol Π for a functor which is the product on underlying sets.

Now suppose that $\mathbb{C} = dg\mathbb{A}$ where $\mathbb{A}$ is one of our finite dimensional abelian categories 17.1.4-17.1.8, such as the category of torsion $k[c]$-modules. We shall see that there is a product $\prod^{\mathbb{A}} : \mathbb{A}^A \longrightarrow \mathbb{A}$; this therefore induces a functor $\prod^{dg\mathbb{A}} : dg\mathbb{A}^A \longrightarrow dg\mathbb{A}$, which is also the product functor. Since $\prod^{dg\mathbb{A}}$ preserves the homotopy relation, it follows that we can define a functor $\prod^{D\mathbb{A}} : D\mathcal{A}^A \longrightarrow D\mathcal{A}$ by applying $\prod^{dg\mathbb{A}}$ to a fibrant approximation of an object (ie $\prod^{D\mathbb{A}} = R\prod^{dg\mathbb{A}}$). Furthermore, since $\prod^{dg\mathbb{A}}$ is right adjoint to an exact functor, it preserves injectives and pullbacks, and hence it is still right adjoint to the diagonal functor.

SUMMARY 21.1.1. If M_α is a model of Y_α with fibrant approximation $\hat{M}_\alpha$, then a model of $\Pi_\alpha Y_\alpha$ is

$$\Pi_\alpha^{D\mathbb{A}} M_\alpha = \Pi_\alpha^{\mathbb{A}} \hat{M}_\alpha. \quad \square$$

To make the description useful we must also explain how to calculate the homology of the model.

Now suppose we are in Context 17.1.1. The purpose of considering $\hat{\mathbb{A}}$ is its convenience for calculation. We may therefore hope that the product in $\hat{\mathbb{A}}$ is easily identified and exact, so that $\Pi^{D\hat{\mathbb{A}}} = \Pi^{\hat{\mathbb{A}}}$. We thus have

$$\Pi^{\mathbb{A}} = \hat{\Gamma}\Pi^{\hat{\mathbb{A}}}$$

and

$$\Pi^{D\mathbb{A}} = R\Pi^{\mathbb{A}} = R\hat{\Gamma} \circ \Pi^{\hat{\mathbb{A}}}.$$

In cases where Context 17.1.3 is relevant

$$\Pi^{\mathbb{A}} = \hat{\Gamma}\phi\Pi^{\hat{\hat{\mathbb{A}}}}$$

and

$$\Pi^{D\mathbb{A}} = R\Pi^{\mathbb{A}} = R\hat{\Gamma} \circ R(\phi\Pi^{\hat{\mathbb{A}}}).$$

In the following brief sections we discuss the various cases.

21.2. Torsion modules.

We begin with the category $\mathcal{F}$-finite torsion modules, leaving the easier category of torsion $k[c]$-modules to the reader. The present case is an ingredient in the discussion of the torsion model, and also illustrates the points which arise.

Recall the context and notation from Section 17.3, especially the functor ϕ defined by $\phi(M) = \bigoplus_H M(H)$ and the Euler torsion functor $\Gamma_{\mathcal{E}}$. First we note that the set theoretic product of $\mathcal{F}$-finite modules need not be $\mathcal{F}$-finite. However, since ϕ is right adjoint to inclusion by 17.3.2, it is immediate that the the categorical product of $\mathcal{F}$-finite modules M_α is $\phi\Pi_\alpha M_\alpha$. Hence for $\mathcal{F}$-finite torsion modules M_α the product is $\Gamma_c\phi\Pi_\alpha M_\alpha$.

It may help some readers to note that this is directly analogous to the situation in topology. If Y_α is a $\mathcal{F}$-spectrum for each α then the product *in the category of $\mathcal{F}$-spectra* is $E\mathcal{F}_+ \wedge \Pi_\alpha Y_\alpha$: indeed, if T is a $\mathcal{F}$-spectrum,

$$[T, \Pi_\alpha Y_\alpha]^{\mathbb{T}} = [T, E\mathcal{F}_+ \wedge \Pi_\alpha Y_\alpha]^{\mathbb{T}}.$$

PROPOSITION 21.2.1. *If $\mathbb{A}$ is the category of $\mathcal{F}$-finite torsion $\mathcal{O}_{\mathcal{F}}$-modules, then*

$$\Pi_\alpha^{\mathbb{A}} M_\alpha = \Gamma_{\mathcal{E}}\Pi_\alpha M_\alpha$$

and

$$\Pi_\alpha^{D\mathbb{A}} M_\alpha = R\Gamma_{\mathcal{E}}\Pi_\alpha M_\alpha.$$

Proof: It remains only to show that right derived functors respect composition, which follows if the relevant functors preserve injectives. This is a little delicate, because Π (as a functor on families of $\mathcal{F}$-finite torsion modules) is not a right adjoint. However, we saw in Section 17.3 that

$$\Gamma_{\mathcal{E}}\Pi = \Gamma_c\phi\Pi.$$

Now $\phi\Pi$ is the product in the category of $\mathcal{F}$-finite modules, and thus a right adjoint. It is exact because it is the composite of exact functors. Thus we have $R(\Gamma_{\mathcal{E}}\Pi) = R\Gamma_c \circ R(\phi\Pi) = R\Gamma_c \circ (\phi\Pi)$, and the result follows. □

Since objects of dg$\mathbb{A}$ are formal in this case, it is obviously useful to record the homology of the product.

COROLLARY 21.2.2. *The homology of the product in the derived category is calculated by a split short exact sequence*

$$0 \longrightarrow R^1\Gamma_{\mathcal{E}}\Pi_\alpha H_*(\Sigma^{-1}M_\alpha) \longrightarrow H_*(\Pi_\alpha^{D\mathbb{A}}M_\alpha) \longrightarrow \Gamma_{\mathcal{E}}\Pi_\alpha H_*(M_\alpha) \longrightarrow 0. \quad \square$$

Thus, for example, if M_α is a dg object in even degrees, then the product has model $\Gamma_{\mathcal{E}}\Pi_\alpha M_\alpha \oplus \Sigma^{-1}R^1\Gamma_{\mathcal{E}}\Pi_\alpha M_\alpha$.

21.3. The torsion model.

Next consider the torsion model $\mathbb{A} = \mathcal{A}_t$. We may reflect that we already know the torsion part of the model for the product $\Pi_\alpha Y_\alpha$, and we almost know the vertex. Indeed

$$\Pi_\alpha(Y_\alpha \wedge E\mathcal{F}_+) \longrightarrow \Pi_\alpha Y_\alpha$$

is an $\mathcal{F}$-equivalence, and therefore the torsion part of the product is given by the product of the torsion parts in the category of $\mathcal{F}$-finite torsion modules. The vertex can almost be calculated from the exact sequence

$$\cdots \longrightarrow \pi_*^{\mathbb{T}}(E\mathcal{F}_+ \wedge \Pi_\alpha Y_\alpha) \longrightarrow \pi_*^{\mathbb{T}}(\Pi_\alpha Y_\alpha) \longrightarrow \pi_*^{\mathbb{T}}(\tilde{E}\mathcal{F} \wedge \Pi_\alpha Y_\alpha) \longrightarrow \cdots .$$

However, the categorical procedure makes this a redundant prop to our confidence.

For the rest of the section it is convenient to work with the 6.2.4 version $\mathcal{A}_t'$ of the torsion model, with objects $V \longrightarrow \mathrm{Hom}(t_*^{\mathcal{F}}, T)$ with T $\mathcal{F}$-finite and torsion. As in Section 17.4, we have inclusions

$$\mathcal{A}_t' \xrightarrow{\ i\ } \hat{\mathcal{A}}_t' \xrightarrow{\ j\ } \hat{\hat{\mathcal{A}}}_t',$$

where T is arbitrary in $\hat{\hat{\mathcal{A}}}_t'$ and $\mathcal{F}$-finite in $\hat{\mathcal{A}}_t'$.

It is immediate that the product of objects $M_\alpha = (V_\alpha \longrightarrow \mathrm{Hom}(t_*^{\mathcal{F}}, T_\alpha))$ in $\hat{\hat{\mathcal{A}}}_t'$ is $\Pi_\alpha M_\alpha = (\Pi_\alpha V_\alpha \longrightarrow \mathrm{Hom}(t_*^{\mathcal{F}}, \Pi_\alpha T_\alpha))$.

PROPOSITION 21.3.1. *The product of objects $M_\alpha = (V_\alpha \longrightarrow \mathrm{Hom}(t_*^{\mathcal{F}}, T_\alpha))$ in the torsion model category is*

$$\Pi_\alpha^{\mathcal{A}_t'} M_\alpha = \Gamma_c\phi\Pi_\alpha M_\alpha.$$

It therefore has
(i) torsion part $T_\Pi = \Gamma_{\mathcal{E}}\Pi_\alpha T_\alpha$,
(ii) vertex V_Π and structure map are described by the pullback square

$$\begin{array}{ccc} V_\Pi & \longrightarrow & \Pi_\alpha V_\alpha \\ \downarrow & & \downarrow \\ \mathrm{Hom}(t_*^{\mathcal{F}}, \Gamma_{\mathcal{E}}\Pi_\alpha T_\alpha) & \longrightarrow & \mathrm{Hom}(t_*^{\mathcal{F}}, \Pi_\alpha T_\alpha) \end{array}$$

where the lower horizontal is induced by the inclusion $\Gamma_{\mathcal{E}}\Pi_\alpha T_\alpha \longrightarrow \Pi_\alpha T_\alpha$. The behaviour on morphisms is as implied by this categorical description. $\square$

COROLLARY 21.3.2. *The product on the derived torsion model is*

$$\Pi^{D\mathcal{A}_t} = R(\Gamma_c \phi \Pi) = R(\Gamma_c) \circ R(\phi \Pi).$$

Proof: The only part requiring further comment is the asserted composite. First note that $\phi\Pi : (\mathcal{A}'_t)^A \longrightarrow \hat{\mathcal{A}}'_t$ does relate two categories of finite injective dimension, so that the right derived functor is defined. We need then only comment that $\phi\Pi$ preserves injectives: to see this we note that it can be viewed as the restriction of the right adjoint $(\hat{\mathcal{A}}'_t)^A \longrightarrow \hat{\mathcal{A}}'_t$, and recall that i preserves injectives. □

As remarked in Section 21.1, this automatically gives a model for the product of $\mathbb{T}$-spectra. However we would like to know the homology of the product and, more exactly, its vertex and torsion part. For the torsion part we already have an answer, provided by Corollary 21.2.2 above.

The discussion in Section 17.4 explains how to calculate $R^*\Gamma_c$, and exactly analogous methods give $R^*(\phi\Pi)$, and a composite functor spectral sequence helps calculate $R^*(\Gamma_c\phi\Pi)$. Alternatively, $R^*(\Gamma_c\phi\Pi)$ can be calculated directly. One advantage of the composite approach is that if, for example, all torsion objects were supported on the same finite set of subgroups, then $\phi\Pi_\alpha M_\alpha = \Pi_\alpha M_\alpha$ and this is an exact functor.

21.4. The standard models.

The two-object models are a little more complicated. Let us begin with the case of $k[c]$-modules, with $\hat{\mathbb{A}}$ being the category of $k[c]$-morphisms $N \longrightarrow t \otimes V$. As before we swiftly calculate the product in the category $\hat{\mathbb{A}}$, because $\hat{\mathbb{A}}$ contains the object $\hat{E} = (0 \longrightarrow t)$ corepresenting the passage to vertex functor

$$\mathrm{Hom}(\hat{E}, (P \longrightarrow t \otimes V)) = V.$$

Thus if $C_\alpha = (P_\alpha \longrightarrow t \otimes W_\alpha)$, and if we assume $\prod_\alpha^{\hat{\mathbb{A}}} C_\alpha = (Q \longrightarrow t \otimes H)$ then we apply $\mathrm{Hom}(\hat{E}, \cdot)$ to deduce $H = \Pi_\alpha W_\alpha$. Applying $\mathrm{Hom}(S^{n\lambda}, \cdot)$ we deduce $Q(c^{-n}) = \Pi_\alpha P_\alpha(c^{-n})$. We give a more direct construction. We let $\delta : t \otimes \Pi_\alpha W_\alpha \longrightarrow \Pi_\alpha(t \otimes W_\alpha)$ denote the diagonal, and we let δ^* be the pullback functor:

$$\begin{array}{ccc} \delta^*\Pi_\alpha P_\alpha & \longrightarrow & \Pi_\alpha P_\alpha \\ \downarrow & & \downarrow \\ t \otimes \Pi_\alpha W_\alpha & \xrightarrow{\delta} & \Pi_\alpha(t \otimes W_\alpha) \end{array}$$

LEMMA 21.4.1. *The product on the category $\hat{\mathbb{A}}$ is $\prod^{\hat{\mathbb{A}}} = \delta^* \prod$.*

Proof: Immediate from the universal property of pullbacks. □

COROLLARY 21.4.2. *The product on the semifree category $D\mathbb{A}$ is*

$$\Pi^{D\mathbb{A}} = R(\hat{\Gamma}\delta^*\Pi) = R\hat{\Gamma}R(\delta^*\Pi).$$

Proof: The preservation of composites follows since $\delta^*\Pi$ is a right adjoint and so preserves injectives. $\square$

The full standard model is precisely analogous. The only difference is that because we have not established the appropriate context for homological control we have no means to interpret $R(\delta^*\Pi)$. However, we may still consider individual derived functors $R^i(\delta^*\Pi)$, and since $\delta^*\Pi$ is a right adjoint we still have a composite functor spectral sequence for calculation.

COROLLARY 21.4.3. *The product on the standard model category $D\mathcal{A}$ is*

$$\Pi^{D\mathcal{A}} = R(\hat{\Gamma}\delta^*\Pi). \quad \square$$

CHAPTER 22

The tensor-Hom adjunction.

In this chapter we construct the internal tensor and Hom functors for the semifree and full standard model categories. After Section 22.1 specifying the categorical context and giving familiar examples, we start in Section 22.2 by identifying the Hom functor in some special cases, assuming its existence. The semifree Hom functor is constructed in Section 22.3, and essentially the same construction is the principal input in constructing the Hom functor in the standard model in Section 22.6. The passage to derived categories is quite subtle, and treated in Chapter 23.

22.1. General discussion.

Once again we are led to consider an abstract categorical situation because it arises in several algebraic ways. The situation here extends that of Chapter 17 outlined in Section 17.1.

Thus our principal interest is in an abelian category $\mathbb{A}$ inside a larger abelian category $\hat{\mathbb{A}}$, which is useful for calculations. We shall treat four of our five examples 17.1.4 to 17.1.8, namely, the cases where $\mathbb{A}$ is the category of torsion $k[c]$-modules, Euler torsion $\mathcal{O}_{\mathcal{F}}$-modules, the category of semifree modules or the standard model category $\mathcal{A}$. The author does not know how to treat the torsion model in an analogous way.

We shall illustrate the general discussion with the two single object examples, and return to the examples with objects of the form $(N \longrightarrow t \otimes V)$ in the next section, since they are much more substantial.

We begin by summarizing the framework in which we operate. The rest of the chapter shows that all four of the above examples fit into this framework. Firstly, suppose that the inclusion $i : \mathbb{A} \longrightarrow \hat{\mathbb{A}}$ is exact, and has a right adjoint $\hat{\Gamma} : \hat{\mathbb{A}} \longrightarrow \mathbb{A}$ as in Context 17.1.1. We have given explicit constructions of $\hat{\Gamma}$ in all the relevant examples. The new data for this chapter is that for any object N of $\hat{\mathbb{A}}$ there is a tensor product functor $\otimes N : \hat{\mathbb{A}} \longrightarrow \hat{\mathbb{A}}$, and that, if N is in $\mathbb{A}$, this functor restricts to a functor on $\mathbb{A}$. Finally, we suppose that the tensor product functor is associative, and has a right adjoint $\hat{\mathbf{Hom}}(N, \cdot) : \hat{\mathbb{A}} \longrightarrow \hat{\mathbb{A}}$; we use bold face to emphasize that $\hat{\mathbf{Hom}}(N, P)$ is an object of $\hat{\mathbb{A}}$, and not just the abelian group of homomorphisms. Of course, it then follows that if N is an object of $\mathbb{A}$ the functor $\otimes N : \mathbb{A} \longrightarrow \mathbb{A}$ has right adjoint $\mathbf{IntHom}(N, \cdot) = \hat{\Gamma}\, \hat{\mathbf{Hom}}(N, \cdot)$. To summarize, we have the diagram of adjoint pairs of functors

22.1.1.

$$\mathbb{A} \underset{\hat{\Gamma}}{\overset{i}{\rightleftarrows}} \hat{\mathbb{A}} \underset{\hat{\mathbf{Hom}}(N,\cdot)}{\overset{\otimes N}{\rightleftarrows}} \hat{\mathbb{A}}$$

where the left adjoints are displayed at the top.

We shall verify below that this applies to all our Examples 17.1.4 to 17.1.8 except the torsion model. The two single object examples are familiar, and we record them here.

Example 17.1.4 continued: Here $\mathbb{A}$ is the category of torsion $k[c]$-modules, and $\hat{\mathbb{A}}$ is the category of all $k[c]$-modules. The right adjoint to inclusion is the c-power torsion functor: for an arbitrary $k[c]$-module $\hat{N}$ we have $\hat{\Gamma}N = \Gamma_c N$. Here $\hat{\mathbf{Hom}}$ is the usual Hom functor, and the internal Hom functor on the category of torsion $k[c]$-modules is

$$\mathbf{IntHom}(N,P) = \Gamma_c \operatorname{Hom}(N,P). \quad \square$$

The next example is very similar.

Example 17.1.5 continued: Here $\mathbb{A}$ is the category of $\mathcal{F}$-finite torsion modules, $\hat{\mathbb{A}}$ is the category of $\mathcal{F}$-finite modules, and $\hat{\hat{\mathbb{A}}}$ is the category of all $\mathcal{O}_{\mathcal{F}}$-modules. The right adjoint to $i : \mathbb{A} \longrightarrow \hat{\mathbb{A}}$ is the c power torsion functor, Γ_c. The right adjoint to the inclusion $j : \hat{\mathbb{A}} \longrightarrow \hat{\hat{\mathbb{A}}}$ is the exact functor $\phi(M) = \bigoplus_H M(H)$. The Euler torsion functor is related to it by $\Gamma_{\mathcal{E}} = \Gamma_c \phi$. In this case $\hat{\mathbf{Hom}}$ is the $\mathcal{F}$-finite part of the usual Hom functor:

$$\hat{\mathbf{Hom}}(N,P) = \phi \operatorname{Hom}(N,P),$$

and the internal Hom functor on the category of torsion $\mathcal{F}$-finite $\mathcal{O}_{\mathcal{F}}$-modules is

$$\mathbf{IntHom}(N,P) = \Gamma_{\mathcal{E}} \operatorname{Hom}(N,P). \quad \square$$

The point of factorizing the internal Hom functor is twofold. Firstly, as the examples illustrate, the functor $\hat{\mathbf{Hom}}$ is much closer to the usual group of homomorphisms, and thus easier to identify. Much more important is the second reason: we wish to pass to derived categories and retain the adjointness of functors. This will be explained in Chapter 23.

22.2. Calculations in the semifree standard model.

To discuss the tensor-Hom adjunction we need objects $A = (M \longrightarrow t \otimes U)$, $B = (N \longrightarrow t \otimes V)$ and $C = (P \longrightarrow t \otimes W)$. The tensor product is the obvious one:

$$A \otimes B = (M \otimes_{\mathcal{O}} N \longrightarrow (t \otimes_k U) \otimes_{\mathcal{O}} (t \otimes_k V) = t \otimes_k (U \otimes_k V)).$$

This clearly preserves $\mathbb{A}$.

In this section we give some easy calculations, assuming the existence of a right adjoint $\hat{\mathbf{Hom}}(B, \cdot)$ to $\otimes B$: we use the equality

$$\operatorname{Hom}(A, \hat{\mathbf{Hom}}(B,C)) = \operatorname{Hom}(A \otimes B, C).$$

The functor $\hat{\mathbf{Hom}}(B,C)$ will be constructed in Section 22.3.

LEMMA 22.2.1. *For any object $B = (N \longrightarrow t_*^{\mathcal{F}} \otimes V)$ of $\mathcal{A}$ and rational vector space W we have*

$$\hat{\mathbf{Hom}}(B, e(W)) = e(\operatorname{Hom}(V,W)).$$

Proof: We make a short calculation in which $A = (M \longrightarrow t_*^{\mathcal{F}} \otimes U)$ is an arbitrary object of $\hat{\mathbb{A}}$

$$\begin{aligned}\mathrm{Hom}(A, \hat{\mathbf{Hom}}(B, e(W))) &= \mathrm{Hom}(A \otimes B, e(W)) \\ &= \mathrm{Hom}(U \otimes V, W) \\ &= \mathrm{Hom}(U, \mathrm{Hom}(V, W)) \\ &= \mathrm{Hom}(A, e(\mathrm{Hom}(V, W))). \quad \square\end{aligned}$$

LEMMA 22.2.2. *For any object $B = (N \longrightarrow t \otimes V)$, and any $\mathcal{F}$-finite torsion module P*

$$\hat{\mathbf{Hom}}(B, f(P)) = f(\mathrm{Hom}(N, P)).$$

Proof: We make a short calculation in which $A = (M \longrightarrow t_*^{\mathcal{F}} \otimes U)$ is an arbitrary element of $\hat{\mathbb{A}}$

$$\begin{aligned}\mathrm{Hom}(A, \hat{\mathbf{Hom}}(B, f(P)) &= \mathrm{Hom}(A \otimes B, f(P)) \\ &= \mathrm{Hom}(M \otimes N, P) \\ &= \mathrm{Hom}(M, \mathrm{Hom}(N, P)) \\ &= \mathrm{Hom}(A, f(\mathrm{Hom}(N, P))). \quad \square\end{aligned}$$

REMARK 22.2.3. It is interesting to contrast the special case of this with $B = e(V)$ with the analogous calculation with internal Homs in $\mathbb{A}$. Whereas

$$\hat{\mathbf{Hom}}(e(V), f(P)) = f(\mathrm{Hom}(t \otimes V, P))$$

we have

$$\mathbf{IntHom}(e(V), f(P)) = e(\mathrm{Hom}(t_*^{\mathcal{F}} \otimes V, P)).$$

This follows from our calculation 18.3.2 with the torsion functor, but it is interesting to see it directly. We make a short calculation in which $A = (M \longrightarrow t_*^{\mathcal{F}} \otimes U)$ is an arbitrary element of $\mathcal{A}$

$$\begin{aligned}\mathrm{Hom}(A, \mathbf{IntHom}(e(V), f(P)) &= \mathrm{Hom}(A \otimes e(V), f(P)) \\ &= \mathrm{Hom}(M \otimes_{\mathcal{O}} t_*^{\mathcal{F}} \otimes V, P) \\ &= \mathrm{Hom}(U \otimes t_*^{\mathcal{F}} \otimes V, P) \\ &= \mathrm{Hom}(U, \mathrm{Hom}(t_*^{\mathcal{F}} \otimes V, P)) \\ &= \mathrm{Hom}(A, e(\mathrm{Hom}(t_*^{\mathcal{F}} \otimes V, P))). \quad \square\end{aligned}$$

EXAMPLE 22.2.4. *The object* $\mathbf{IntHom}(f(\mathbb{I}), f(\mathbb{I}))$.

By Example 22.2.2, we find $\hat{\mathbf{Hom}}(f(\mathbb{I}), f(\mathbb{I})) = f(\mathrm{Hom}(\mathbb{I}, \mathbb{I})) = f(\mathcal{O})$. We therefore conclude

$$\mathbf{IntHom}(f(\mathbb{I}), f(\mathbb{I})) = \hat{\Gamma} f(\mathcal{O}),$$

and this was calculated in 18.3.2. Indeed, this also gives

$$R\mathbf{IntHom}(f(\mathbb{I}), f(\mathbb{I})) = \mathbf{IntHom}(f(\mathbb{I}), f(\mathbb{I})) = \hat{\Gamma} f(\mathcal{O}) = \begin{pmatrix} \mathcal{O} \oplus K(t) \\ \downarrow \\ t \otimes_k t \end{pmatrix},$$

(with torsion part $\Sigma^2\mathbb{I}$). The case of torsion modules leads us to expect this may differ from $R\hat{\Gamma} R\hat{\mathbf{Hom}}(f(\mathbb{I}), f(\mathbb{I}))$, but, because of the flabbiness result 18.3.5, it does not. $\square$

22.3. Construction of the semifree Hom functor.

In this section we give a definition of the internal Hom functor in the semifree category $\hat{\mathbb{A}}$. In fact we define an object

$$\hat{\mathbf{Hom}}(B,C) = (Q \longrightarrow t \otimes H),$$

and then show that $\hat{\mathbf{Hom}}(B,\cdot)$ gives the right adjoint to $B\otimes$. The procedure is similar to that used for the torsion functor: we first identify the vertex and then use a suitable pullback construction to find the nub.

As in the examples we identify H and Q by assuming the defining adjunction

$$\mathrm{Hom}(A \otimes B, C) = \mathrm{Hom}(A, \hat{\mathbf{Hom}}(B,C)).$$

Taking $A = \hat{E} = (0 \longrightarrow t)$ we see that we must take $H = \mathrm{Hom}(V,W)$.

DEFINITION 22.3.1. With $B = (N \xrightarrow{\beta} t \otimes V)$ and $C = (P \xrightarrow{\gamma} t \otimes W)$ we take $\hat{\mathbf{Hom}}(B,C) = (Q \xrightarrow{\delta} t \otimes H)$ with $H = \mathrm{Hom}(V,W)$ and both Q and δ defined by the pullback square

$$\begin{array}{ccccc} Q & & \longrightarrow & & \mathrm{Hom}(N,P) \\ \downarrow{\scriptstyle \delta} & & & & \downarrow{\scriptstyle \gamma_*} \\ t \otimes \mathrm{Hom}(V,W) & \xrightarrow{\Delta} & \mathrm{Hom}(t \otimes V, t \otimes W) & \xrightarrow{\beta^*} & \mathrm{Hom}(N, t \otimes W). \end{array}$$

LEMMA 22.3.2. *Using definition 22.3.1 there is a bijection*

$$\mathrm{Hom}(A \otimes B, C) = \mathrm{Hom}(A, \hat{\mathbf{Hom}}(B,C))$$

natural in all three variables.

Proof: A morphism on the left is given by a square

$$\begin{array}{ccccc} M \otimes N & & \xrightarrow{\theta} & & P \\ \downarrow{\scriptstyle \alpha\otimes\beta} & & & & \downarrow{\scriptstyle \gamma} \\ (t \otimes U) \otimes (t \otimes V) & \xrightarrow{=} & t \otimes (U \otimes V) & \xrightarrow{1\otimes\phi} & t \otimes W \end{array}$$

whilst a morphism on the right is given by a diagram

$$\begin{array}{ccccc} M & \xrightarrow{\lambda} & Q & \longrightarrow & \mathrm{Hom}(N,P) \\ \downarrow{\scriptstyle \alpha} & & \downarrow{\scriptstyle \delta} & & \downarrow{\scriptstyle \gamma_*} \\ t \otimes U & \xrightarrow{1\otimes\mu} & t \otimes \mathrm{Hom}(V,W) & \xrightarrow{\beta^*\circ\Delta} & \mathrm{Hom}(N, t \otimes W) \end{array}$$

By the universal property of the pullback, λ is uniquely determined by a compatible pair consisting of $1 \otimes \mu$ and $\lambda' : M \longrightarrow \mathrm{Hom}(N,P)$.

The natural bijection arises by taking θ to be adjoint to λ' and ϕ to be adjoint to μ. We leave the reader to check that the compatibility conditions for (θ, ϕ) are equivalent to those for (λ', μ) under this correspondence. $\square$

22.4. Flabbiness of the Hom object.

Because of the contrast between the simplicity of the semifree standard model case and the complexity of the torsion $k[c]$-module case we should verify that $R\hat{\Gamma}R\hat{\mathbf{Hom}} = R\mathbf{IntHom}$ in this case. Since it is not true that $\hat{\mathbf{Hom}}$ preserves injectives, further explanation is necessary.

PROPOSITION 22.4.1. *For any objects B and C of the semifree standard model, the object $\hat{\mathbf{Hom}}(B,C)$ is acyclic in the sense that*

$$(R^1\hat{\Gamma})\hat{\mathbf{Hom}}(B,C) = 0.$$

COROLLARY 22.4.2. *For any objects B and C of the semifree standard model,*

$$R\hat{\Gamma}R\hat{\mathbf{Hom}}(B,C) = R\mathbf{IntHom}(B,C). \quad \Box$$

Proof of 22.4.1: Since $\hat{\Gamma}$ is a right adjoint, it preserves inverse limits. Thus

$$\hat{\Gamma}\hat{\mathbf{Hom}}(\lim_{\rightarrow \alpha} B_\alpha, C) = \lim_{\leftarrow \alpha} \hat{\Gamma}\hat{\mathbf{Hom}}(B_\alpha, C).$$

It therefore suffices to prove the result for a class of objects B large enough to construct all objects by direct limits.

LEMMA 22.4.3. *Any object B of the semifree standard model is the direct limit of objects with finitely generated nub.*

Proof: Certainly the vertex V is the direct limit of its finite dimensional subspaces V_β, and so B is the direct limit of the objects B_β obtained by pulling back along $t \otimes V_\beta \longrightarrow t \otimes V$. Furthermore, pullback preserves isomorphism so the objects B_β still lie in the semifree standard model. Now if N_β is the nub of B_β, since V_β is finite dimensional we may choose a finitely generated submodule $N_\beta^0 \subseteq N_\beta$ so that the cokernel of $N_\beta^0 \longrightarrow t \otimes V_\beta$ is c-power torsion. We now let $N_{\beta,\gamma}$ run over finitely generated $k[c]$-submodules of N_β containing N_β^0. Certainly

$$N_\beta = \bigcup_\gamma N_{\beta,\gamma},$$

and the objects $N_{\beta,\gamma} \longrightarrow t \otimes V_\beta$ lie in the standard model since $N_{\beta,\gamma}$ contains N_β^0. $\Box$

LEMMA 22.4.4. *If N is finitely generated then $\hat{\mathbf{Hom}}(B,C) = (Q \longrightarrow t \otimes H)$ is flabby in the sense that its structure map has torsion cokernel.*

Proof: Suppose given $\theta : V \longrightarrow W$. We must show that there is an n so that $c^n \otimes \theta$ lies in the image of Q. In other words that for some n there is a solution to the problem

$$\begin{array}{ccc} N & \dashrightarrow & P \\ {\scriptstyle\beta}\downarrow & & \downarrow{\scriptstyle\gamma} \\ t \otimes V & \xrightarrow{c^n \otimes \theta} & t \otimes W. \end{array}$$

Since N is finitely generated, and $cok(\gamma)$ is torsion there exists an i such that $c^i(1\otimes\theta)\beta(N)$ lies in the image of γ. We thus view $c^i(1\otimes\theta)\beta$ as a map $N \longrightarrow im(\gamma)$. The obstruction to lifting $c^i(1\otimes\theta)\beta$ to P lies in $\mathrm{Ext}(N, ker(\gamma))$. Since $ker(\gamma)$ is torsion and N is finitely generated, we may multiply by c^j to kill this obstruction, and take $n = i + j$. □

The proposition now follows from 18.3.5. □

22.5. Calculations in the standard model.

The modifications from the semifree model are superficial in this section and the next, so we simply record the results.

To discuss the tensor-Hom adjunction we need objects $A = (M \longrightarrow t_*^{\mathcal{F}} \otimes U)$, $B = (N \longrightarrow t_*^{\mathcal{F}} \otimes V)$ and $C = (P \longrightarrow t_*^{\mathcal{F}} \otimes W)$. The tensor product is the obvious one:

$$A \otimes B = (M \otimes_{\mathcal{O}_{\mathcal{F}}} N \longrightarrow (t_*^{\mathcal{F}} \otimes U) \otimes_{\mathcal{O}_{\mathcal{F}}} (t_*^{\mathcal{F}} \otimes V) = t_*^{\mathcal{F}} \otimes (U \otimes_k V)).$$

This clearly preserves $\mathcal{A}$.

In this section we give some easy calculations, assuming the existence of a right adjoint $\hat{\mathbf{Hom}}(B, \cdot)$ to $\otimes B$: we use the equality

$$\mathrm{Hom}(A, \hat{\mathbf{Hom}}(B, C)) = \mathrm{Hom}(A \otimes B, C).$$

The functor $\hat{\mathbf{Hom}}(B, C)$ will be constructed in Section 22.6.

LEMMA 22.5.1. *For any object $B = (N \longrightarrow t_*^{\mathcal{F}} \otimes V)$ of $\mathcal{A}$ and rational vector space W we have*

$$\hat{\mathbf{Hom}}(B, e(W)) = e(\mathrm{Hom}(V, W)).$$

Proof: Formally identical to 22.2.1. □

LEMMA 22.5.2. *For any object $B = (N \longrightarrow t_*^{\mathcal{F}} \otimes V)$, and any $\mathcal{F}$-finite torsion module P*

$$\hat{\mathbf{Hom}}(B, f(P)) = f(\mathrm{Hom}(N, P)).$$

Proof: Formally identical to 22.2.2. □

EXAMPLE 22.5.3. *The object* $\mathbf{IntHom}(f(\mathbb{I}), f(\mathbb{I}))$.

By Example 22.5.2, we find $\hat{\mathbf{Hom}}(f(\mathbb{I}), f(\mathbb{I})) = f(\mathrm{Hom}(\mathbb{I}, \mathbb{I})) = f(\mathcal{O}_{\mathcal{F}})$. We therefore conclude

$$\mathbf{IntHom}(f(\mathbb{I}), f(\mathbb{I})) = \hat{\Gamma} f(\mathcal{O}_{\mathcal{F}}),$$

and this was calculated in 20.3.1. Indeed, this also gives

$$R\mathbf{IntHom}(f(\mathbb{I}), f(\mathbb{I})) = \mathbf{IntHom}(f(\mathbb{I}), f(\mathbb{I})) = \hat{\Gamma} f(\mathcal{O}_{\mathcal{F}}) = \begin{pmatrix} \mathcal{O}_{\mathcal{F}} \oplus K(t_*^{\mathcal{F}}) \\ \downarrow \\ t_*^{\mathcal{F}} \otimes t_*^{\mathcal{F}} \end{pmatrix},$$

(with torsion part $\Sigma^2\mathbb{I}$). □

22.6. Construction of the standard Hom functor.

In this section we give a definition of the internal Hom functor in the standard model category $\mathcal{A}$. In fact we define an object

$$\hat{\mathbf{Hom}}(B,C) = (Q \longrightarrow t_*^{\mathcal{F}} \otimes H),$$

and then show that $\hat{\mathbf{Hom}}(B,\cdot)$ gives the right adjoint to $B\otimes$. The procedure is similar to that used for the torsion functor: we first identify the vertex and then use a suitable pullback construction to find the nub.

As in the examples we identify H and Q by assuming the defining adjunction

$$\mathrm{Hom}(A \otimes B, C) = \mathrm{Hom}(A, \hat{\mathbf{Hom}}(B,C)).$$

Taking $A = \hat{E} = (0 \longrightarrow t_*^{\mathcal{F}})$ we see that we must take $H = \mathrm{Hom}(V,W)$.

DEFINITION 22.6.1. With $B = (N \xrightarrow{\beta} t_*^{\mathcal{F}} \otimes V)$ and $C = (P \xrightarrow{\gamma} t_*^{\mathcal{F}} \otimes W)$ we take $\hat{\mathbf{Hom}}(B,C) = (Q \xrightarrow{\delta} t_*^{\mathcal{F}} \otimes H)$ with $H = \mathrm{Hom}(V,W)$ and both Q and δ defined by the pullback square

$$\begin{array}{ccc} Q & \longrightarrow & \mathrm{Hom}(N,P) \\ \downarrow{\scriptstyle\delta} & & \downarrow{\scriptstyle\gamma_*} \\ t_*^{\mathcal{F}} \otimes \mathrm{Hom}(V,W) \xrightarrow{\Delta} \mathrm{Hom}(t_*^{\mathcal{F}} \otimes V, t_*^{\mathcal{F}} \otimes W) & \xrightarrow{\beta^*} & \mathrm{Hom}(N, t_*^{\mathcal{F}} \otimes W). \end{array}$$

LEMMA 22.6.2. *Using definition 22.3.1 there is a bijection*

$$\mathrm{Hom}(A \otimes B, C) = \mathrm{Hom}(A, \hat{\mathbf{Hom}}(B,C))$$

natural in all three variables.

Proof: Formally identical to 22.3.2. □

CHAPTER 23

The derived tensor-Hom adjunction.

In this chapter we construct the internal derived tensor and derived Hom functors for all four of our main examples, building on the work of Chapter 22. Here at last is a case where the standard models are simpler than categories of torsion modules.

We begin in Section 23.1 by dealing with the formalities in the case of finite flat dimension. We show this applies to the semifree standard model in Section 23.2, and in Section 23.3 that it applies to the full standard model.

We return to formalities in Section 23.4, giving an alternative homological condition which applies to the case of torsion modules. This homological condition is then established for the case of torsion $k[c]$-modules in Section 23.5,

23.1. The case of finite flat dimension.

We continue the discussion from Section 22.1. The tensor-Hom adjunction passes to derived categories, in a straightforward way if the category $\mathbb{A}$ has finite flat dimension. This covers the standard models, as we show in the following two sections.

PROPOSITION 23.1.1. *Suppose given a category $\mathbb{A}$ with flat dimension 1, and with tensor and Hom as in 22.1. For objects M, N and P of $\mathbb{A}$ we have the natural isomorphism*

$$[M \otimes^L N, P] = [M, R\mathbf{IntHom}(N, P)],$$

where $[\cdot, \cdot]$ refers to morphisms in $D\mathbb{A}$.

First we should make sense of $\otimes^L$.

LEMMA 23.1.2. *If $\mathbb{A}$ has flat dimension 1, then for $B \in \mathbb{A}$ the left derived tensor product functor $\otimes B : D\mathbb{A} \longrightarrow D\mathbb{A}$ exists, and may be calculated by flat resolutions.*

Proof: We have seen that any epimorphism $F \longrightarrow A$ with F flat gives a flat resolution $0 \longrightarrow R \longrightarrow F \longrightarrow A \longrightarrow 0$, and if $F' \longrightarrow A$ is a second, we may compare the two using the sum $F' \oplus F'' \longrightarrow A$. This shows that $cofibre(R \otimes B \longrightarrow F \otimes B)$ gives a well defined functor on the derived category. The rest of the proof that this is the total left derived functor proceeds as with cofibrant approximations: the essential point is that tensor product with a flat object preserves weak equivalences. $\square$

Proof of 23.1.1: We need to know that the tensor-Hom adjunction passes to derived categories. The argument is standard, but we reproduce it for comparison with the more complicated case in Section 23.4.

Consider the functor of three variables from $\hat{\mathbb{A}}$

$$\mathrm{Hom}(M \otimes N, P) = \mathrm{Hom}(M, \mathbf{IntHom}(N, P))$$

to graded abelian groups. It follows from associativity of the tensor product that if M is flat, $M\otimes$ preserves flat objects, and so the total right derived functor on the left can be calculated as a composite. Similarly, if P is injective $\mathbf{IntHom}(\cdot, P)$ takes flat objects to injectives so that the functor on the right can also be calculated as a composite. This gives the equality

$$R\mathrm{Hom}(M \otimes^L N, P) = R\mathrm{Hom}(M, R\mathbf{IntHom}(N, P)),$$

where right and left derived functors refer to $\hat{\mathbb{A}}$. Taking 0th homology we find

$$D\mathcal{A}(M \otimes^L N, P) = D\mathcal{A}(M, R\mathbf{IntHom}(N, P))$$

as required. □

23.2. Flat dimension of the semifree category.

To show that 23.1.1 applies to the semifree category $\mathbb{A}$, it remains to verify $\mathbb{A}$ has flat dimension 1. We include discussion of $\hat{\mathbb{A}}$ and the inclusion i to show that $\mathbb{A}$ could equally well be treated by the more complicated method of Section 23.4.

LEMMA 23.2.1. *The categories $\mathbb{A}$ and $\hat{\mathbb{A}}$ have enough flat objects, and are of flat dimension 1. Furthermore the map i preserves the flat objects.*

Proof: Since the $k[c]$-module $t \otimes U$ is always flat, an object of $\mathbb{A}$ or $\hat{\mathbb{A}}$ is flat if and only if its nub is flat. Accordingly, the functor i preserves flat objects. Furthermore, since any torsion free module is flat, it follows that the submodule of a flat module is flat. It therefore suffices to show there are enough flat objects.

We have seen in 19.2.4 that any element of the nub of an object of $\mathbb{A}$ lies in the image of a map from a wide sphere. It is obvious that wide spheres are flat. We may therefore construct a map from a sum of wide spheres which is an epimorphism on nubs. It is therefore also an epimorphism on vertices. This shows there are enough flat objects.

For $\hat{\mathbb{A}}$, we may map onto the vertex since the object $\hat{E} = (0 \longrightarrow t)$ co-representing the vertex is flat. For the nub, if $p \in P$ is of degree $2r$ and satisfies $\gamma(p) = \sum_{i=1}^{d} c^{n_i} \otimes w_i$, we may construct the object T with vertex U having generators $u_1, u_2, \ldots, u_d$ in degrees $2r + 2n_1, 2r + 2n_2, \ldots, 2r + 2n_d$, and in which the nub is generated by $s = \sum_{i=1}^{d} c^{n_i} \otimes u_i$. There is a unique map $T \longrightarrow C$ in which s maps to p. □

23.3. Flat dimension of the standard model.

To show that 23.1.1 applies to the standard model category $\mathcal{A}$, it remains to verify it has flat dimension 1. To obtain enough flat objects we must introduce wide spheres for the full standard model. Because the context is significantly more complicated than in the semifree case we must adapt the notation accordingly.

DEFINITION 23.3.1. A *wide sphere* is an object with injective structure map described as follows. It has vertex U with basis $e_1, e_2, \ldots, e_d$ in degrees $2a_1, 2a_2, \ldots, 2a_d$, and the nub is the submodule of $t^{\mathcal{F}}_* \otimes U$ generated by $c^{v_1} \otimes e_1, c^{v_2} \otimes e_2, \ldots, c^{v_d} \otimes e_d$ and an element $g \in t^{\mathcal{F}}_* \otimes U$ so that $g = \tau_1 \otimes e_1 + \tau_2 \otimes e_2 + \cdots + \tau_d \otimes e_d$. We write

$$S^g_{c^{v_1}\otimes e_1, c^{v_2}\otimes e_2, \ldots, c^{v_d}\otimes e_d}(\mathbf{a}) = S^{\tau_1\otimes e_1+\tau_2\otimes e_2+\cdots+\tau_d\otimes e_d}_{c^{v_1}\otimes e_1, c^{v_2}\otimes e_2, \ldots, c^{v_d}\otimes e_d}(a_1, a_2, \ldots, a_d)$$

for this object. The element g is called the *fundamental class.*

REMARK 23.3.2. (i) By construction the structure map of the wide spheres becomes an isomorphism when $\mathcal{E}$ is inverted, so they are objects of the standard model.
(ii) The notation is related to that of the ordinary spheres by

$$S^{n+v} = S^{c^{-v}\otimes e}_{c^{-v}\otimes e}(n).$$

(iii) In the semifree case the elements $v_1, v_2, \ldots, v_d, \tau_1, \tau_2, \ldots, \tau_d$ are determined by their degrees, and furthermore, a cofinal collection of objects have all the elements in the subscript in a single degree. For such objects, both superscript and subscript are described by a single integer, and this explains how to obtain the present notation is related to that of 19.2.2.

To be of use to us, we must show that these wide spheres are flat. By way of motivation, we note that S^v is flat when v takes finitely many values. Indeed the module $\mathcal{O}_{\mathcal{F}}(v)$ is projective as a $\mathcal{O}_{\mathcal{F}}$-module, since it is a finite sum of suspensions of $e_\phi \mathcal{O}_{\mathcal{F}}$ where $\phi \subseteq \mathcal{F}$ is a set on which v is constant.

LEMMA 23.3.3. *The nub of a wide sphere is a projective $\mathcal{O}_{\mathcal{F}}$-module, and hence a wide sphere is flat.*

Proof: Let S be a wide sphere as in 23.3.1. We express S as a sum of objects with projective nubs. First note that for any finite subgroup H the summand $e_H S$ has projective nub since it is a torsion free module over $\mathbb{Q}[c_H]$. We use this to peel off the subgroups which behave exceptionally.

First let $H_1, H_2, \ldots, H_m$ be the subgroups on which some function v_i is non-zero, and let ϕ be the complementary set of finite subgroups. The nub of the object $e_\phi S$ is the submodule of $e_\phi t^{\mathcal{F}}_* \otimes U$ generated by $e_\phi \mathcal{O}_{\mathcal{F}} \otimes U$ and the element

$$e_\phi g = e_\phi \tau_1 \otimes e_1 + e_\phi \tau_2 \otimes e_2 + \cdots + e_\phi \tau_d \otimes e_d.$$

For each i we have $e_\phi \tau_i = e_\phi c^{-w_i} x_i$ for some Euler class c^{w_i}, and some element $x_i \in \mathcal{O}_{\mathcal{F}}$. Now let $K_1, K_2, \ldots, K_n$ be the subgroups on which some w_i is non-zero, and let ψ be the complementary set of finite subgroups. Then the nub of $e_\psi e_\phi S$ is the projective module $e_\psi e_\phi \mathcal{O}_{\mathcal{F}} \otimes U$. We have an isomorphism

$$S \cong \bigoplus_{i=1}^{m} e_{H_i} S \oplus \bigoplus_{j=j_0+1}^{n} e_{K_j} S \oplus e_\psi e_\phi S$$

expressing S as a sum of objects with projective nubs, where $K_1, K_2, \ldots, K_{j_0}$ are the subgroups K_j occuring as amongst the H_i. □

LEMMA 23.3.4. *There are enough flat objects in the standard model.*

Proof: It suffices to show that there is a map from a flat object which maps onto the nub, since inverting $\mathcal{E}$ is exact.

Given any element $p \in P$ in the nub of an object of the standard model, we have $\gamma(p) = \sum_i \sigma_i \otimes w_i$, for suitable elements $w_i \in W$. Since γ becomes an isomorphism when $\mathcal{E}$ is inverted, for each i there is an Euler class c^{v_i} so that $c^{v_i} \otimes w_i = \gamma(p_i)$ for some $p_i \in P$. Hence

$$\gamma(c^{v_1+v_2+\cdots+v_d}p) = \gamma(\sum_i \sigma_i c^{v_1+v_2+\cdots+v_d}/c^{v_i}p_i).$$

Accordingly we have

$$c^v(c^{v_1+v_2+\cdots+v_d}p) = c^v(\sum_i \sigma_i c^{v_1+v_2+\cdots+v_d}/c^{v_i}p_i),$$

for some v.

Now take $a_i = deg(w_i)$, and $\tau_i = c^v\sigma_i c^{v_1+v_2+\cdots+v_d}/c^{v_i}$, and take S to be the associated wide sphere. By construction there is a map $S \longrightarrow C$ taking the fundamental class to p. □

LEMMA 23.3.5. *The standard model has flat dimension 1.*

Proof: By 23.3.4, for any object C of $\mathcal{A}$ we may construct an epimorphism $\pi : \Sigma \longrightarrow C$ where Σ is a sum of wide spheres. We claim $K = ker(\pi)$ is flat: indeed, the nub L of K is a submodule of a projective module, and is baseable. By 5.3.1 we have $\mathrm{Ext}^1(L, N) = 0$ whenever N is baseable. Now construct an epimorphism $\pi' : \Sigma' \longrightarrow K$, with Σ' a sum of spheres. It follows that π' splits, so that L is projective. □

23.4. The case without enough flat objects.

We continue the discussion from Section 22.1, explaining how the tensor-Hom adjunction passes to derived categories when $\mathbb{A}$ does not have enough flat objects. To control the passage to derived categories, we shall need the following condition to be satisfied: it applies to the category of torsion $k[c]$-modules and the category of $\mathcal{F}$-finite torsion modules.

CONDITION 23.4.1. The adjoint pair of functors

$$\mathbb{A} \underset{\hat{\Gamma}}{\overset{i}{\rightleftarrows}} \hat{\mathbb{A}}$$

gives Context 17.1.1, and in addition

(i) The category $\hat{\mathbb{A}}$ has flat dimension 1, and

(ii) $I \otimes M = 0$ for all M in $\mathbb{A}$ whenever I is $\mathbb{A}$-injective.

We begin by stating the result and then discuss what is involved in making sense of it.

PROPOSITION 23.4.2. *Suppose given adjoint pairs of functors as in 22.1.1, and suppose in addition that Condition 23.4.1 is satisfied. For objects M, N and P of $\mathbb{A}$ we have the natural isomorphism*

$$[M \otimes^L N, P] = [M, R\hat{\Gamma}\; R\hat{\mathbf{Hom}}(N, P)],$$

where $[\cdot, \cdot]$ refers to morphisms in $D\mathbb{A}$.

The two points to notice here concern the way the composite functors have been treated. The first is that it is not immediately clear how to interpret $M \otimes^L N$ as an object of $\mathbb{A}$, since $\mathbb{A}$ may not have enough projective or flat objects. The second point is that on the right hand side

$$R\hat{\Gamma}\; R\hat{\mathbf{Hom}}(N, P) \neq R\mathbf{IntHom}(N, P)$$

in general. This can happen because the functor $\hat{\mathbf{Hom}}(N, \cdot)$ does not preserve injectives, as we saw in Example 16.2.1.

Proof: Both pairs $l : \mathbb{D} \rightleftarrows \mathbb{E} : r$ of adjoint functors from Section 22.1 do indeed pass to adjoint functors $\mathbf{L}l : D\mathbb{D} \rightleftarrows D\mathbb{E} : \mathbf{R}r$ on derived categories. For the pair $i, \hat{\Gamma}$ this is because i is exact, and therefore the functor on dg categories preserves both cofibrations and acyclic cofibrations. For the tensor-Hom pair the reason is different. First note that $R\hat{\mathbf{Hom}}$ exists since $\hat{\mathbb{A}}$ has enough injectives. If $\hat{\mathbb{A}}$ has enough projectives we know that $\otimes^L$ exists, but it also suffices that $\hat{\mathbb{A}}$ has enough flat objects by 23.1.2. Now consider the functor of three variables from $\hat{\mathbb{A}}$

$$\mathrm{Hom}(M \otimes N, P) = \mathrm{Hom}(M, \hat{\mathbf{Hom}}(N, P))$$

to graded abelian groups. It follows from associativity of the tensor product that if M is flat, $M\otimes$ preserves flat objects, and so the total right derived functor on the left can be calculated as a composite. Similarly, if P is injective $\hat{\mathbf{Hom}}(\cdot, P)$ takes flat objects to injectives so that the functor on the right can also be calculated as a composite. This gives the equality

$$R\mathrm{Hom}(M \otimes^L N, P) = R\mathrm{Hom}(M, R\hat{\mathbf{Hom}}(N, P)),$$

where right and left derived functors refer to $\hat{\mathbb{A}}$. Taking 0th homology we find

$$D\hat{\mathbb{A}}(M \otimes^L N, P) = D\hat{\mathbb{A}}(M, R\hat{\mathbf{Hom}}(N, P)).$$

We may now deduce the statement in the proposition. Firstly, we have already observed that $i = Li$ is left adjoint to $R\hat{\Gamma}$, so that if M is in $\mathbb{A}$,

$$D\hat{\mathbb{A}}(M, R\hat{\mathbf{Hom}}(N, P)) = D\mathbb{A}(M, R\hat{\Gamma}\; R\hat{\mathbf{Hom}}(N, P)).$$

Finally, if M, N and P are in $\mathbb{A}$ then

$$D\mathbb{A}(M \otimes^L N, P) = D\hat{\mathbb{A}}(M \otimes^L N, P)$$

provided $M \otimes^L N$ can be interpreted unambiguously as an object of $D\mathbb{A}$, since $D\hat{\mathbb{A}}(L, P) = D\mathbb{A}(L, P)$ when L and P are in $\mathbb{A}$, by 17.1.2.

Now consider the meaning of $M \otimes^L N$ on the left. If $\mathbb{A}$ had enough flat objects, it would need no comment by 23.1.2. However, if $\mathbb{A}$ does not have enough flat objects, more needs to be said. In $\hat{\mathbb{A}}$, the tensor product functor is right exact and has only one left derived functor, the torsion product functor $\mathrm{Tor}(\cdot, \cdot)$, which is left exact. Furthermore, $\mathrm{Tor}(\cdot, \cdot)$

restricts to a functor on the category $\mathbb{A}$, where we may consider its total right derived functor. In the cases that $\mathbb{A}$ does not have enough flat objects we have the equality

$$M \otimes^L N = \Sigma R\mathrm{Tor}(M, N)$$

in $\mathrm{D}\hat{\mathbb{A}}$ provided M or N is in $\mathbb{A}$, by Condition 23.4.1 (ii) (the argument is given in more detail in a special case in 23.5.2 below). □

23.5. Torsion $k[c]$-modules.

In this section we complete the justification for the category of torsion $k[c]$-modules as in 17.1.4. The modifications for Example 17.1.5 are minimal.

It remains to show that Conditions 23.4.1 are satisfied in this case. We verified 17.1.1 in Section 17.2, and Conditions 23.4.1 (i) is well known. It remains to establish Condition 23.4.1 (ii).

The short exact sequence

$$0 \longrightarrow k[c] \longrightarrow k[c, c^{-1}] \longrightarrow \Sigma^2\mathbb{I} \longrightarrow 0$$

immediately gives the following.

LEMMA 23.5.1. *For any $k[c]$-module M there is an exact sequence*

$$0 \longrightarrow \mathrm{Tor}(M, \Sigma^2\mathbb{I}) \longrightarrow M \longrightarrow M[c^{-1}] \longrightarrow M \otimes \Sigma^2\mathbb{I} \longrightarrow 0,$$

and hence, for any torsion module T, we have $T \otimes \Sigma^2\mathbb{I} = 0$ and $\mathrm{Tor}(T, \Sigma^2\mathbb{I}) = T$. □

This completes the justification, but it may reassure some readers to see the arguments of 23.4.2 repeated in a familiar context.

Note that this means in particular that we may calculate $\mathrm{Tor}_*(S, T)$ for torsion modules S and T using an injective resolution of S. Indeed if $0 \longrightarrow S \longrightarrow I \longrightarrow J \longrightarrow 0$ is an injective resolution using sums of suspensions of $\mathbb{I}$ we obtain the exact sequence

$$0 \longrightarrow \mathrm{Tor}(S, T) \longrightarrow \mathrm{Tor}(I, T) \longrightarrow \mathrm{Tor}(J, T) \longrightarrow S \otimes T \longrightarrow 0.$$

Our problem is that there are not enough projectives amongst torsion modules, so that it is not clear that the left derived torsion product exists. However the category of $k[c]$-modules is one dimensional, and thus the torsion product functor is left exact, so we may use the right derived torsion product instead.

LEMMA 23.5.2. *On the category of $k[c]$-modules, the left derived torsion product is equivalent to the suspension of the right derived torsion product on torsion modules. More precisely, if M or N is a torsion module*

$$M \otimes^L N = \Sigma R\mathrm{Tor}(M, N).$$

Proof: It is enough to establish that $k[c] \otimes^L N = \Sigma R\mathrm{Tor}(k[c], N)$. Consider the exact sequence

$$0 \longrightarrow k[c] \longrightarrow k[c, c^{-1}] \longrightarrow \Sigma^2\mathbb{I} \longrightarrow 0.$$

Since the first two terms are flat, it gives the calculation

$$\mathrm{Tor}(\Sigma^2\mathbb{I}, N) = ker(N \longrightarrow N[c^{-1}]).$$

On the other hand, $k[c, c^{-1}]$ is both flat and injective, so that $R\mathrm{Tor}(k[c, c^{-1}], N) = 0$. Thus $R\mathrm{Tor}(k[c], N) = \Sigma^{-1}R\mathrm{Tor}(\Sigma^2\mathbb{I}, N)$. Moreover, since $\mathbb{I}$ is injective $R\mathrm{Tor}(\Sigma^2\mathbb{I}, N) = \mathrm{Tor}(\Sigma^2\mathbb{I}, N)$. □

We write $M \circledast N$ for this functor, and refer to it as the left derived tensor product. If M and N are torsion modules it should be interpreted as the suspension of the right derived torsion product: by the lemma this is not ambiguous.

LEMMA 23.5.3. *The right adjoint of* $\circledast B : D(tors\ k[c]) \longrightarrow D(tors\ k[c])$ *is*

$$R'\mathrm{Hom}(B, \cdot) := R\Gamma_c\, R\mathrm{Hom}(B, \cdot).$$

Proof: The adjoint pair $i : k[c]\text{-mod} \leftrightarrows \text{tors } k[c]\text{-mod} : \Gamma_c$ passes to derived categories since i is exact. The adjoint pair $\otimes B : k[c]\text{-mod} \leftrightarrows k[c]\text{-mod} : \mathrm{Hom}(B, \cdot)$ passes to derived categories by use of projectives. This shows the required adjunction, or more properly, the first equality

$$D(k[c])(A \circledast B, C) = D(k[c])(A, R\mathrm{Hom}(B, C)) = D(\text{tors-}k[c])(A, R\Gamma_c\, R\mathrm{Hom}(B, C)) :$$

the second equality is the fact that $R\Gamma_c$ is right adjoint to $Li = i$. Finally, we note that

$$D(\text{tors-}k[c])(A \circledast B, C) = D(k[c])(A \circledast B, C)$$

when A, B and C are torsion. This is because we may calculate $A \circledast B = \Sigma R\mathrm{Tor}(A, B)$ on the right using torsion injectives. □

CHAPTER 24

Smash products, function spectra and Lewis-May fixed points.

Our identification of the derived level tensor and Hom functors suggests algebraic models of smash products and function objects in the category of $\mathbb{T}$-spectra. Both in the algebraic and in the topological world the constructions are homotopy invariant and exact. We use this together with our models of products (Chapter 21) and knowledge of maps between injective spectra (Section 5.7) to model these constructions on objects. In Section 24.1 we show that the smash product is modelled by the left derived tensor product, In Section 24.2 we state the theorem giving a model for function spectra, give some examples and prove a special case. The proof that the function is modelled by the right adjoint of the left derived tensor product is given in the Section 24.3.

Finally, in Section 24.4 we return to the Lewis-May K-fixed point functor, introducing an appropriate relative internal Hom object, and showing that on the standard model the algebraic model is $\Psi^K C = R\mathbf{IntHom}_{\overline{\mathcal{O}}_{\mathcal{F}}}(S^0, C)$.

24.1. Models of smash products.

In this section we show that the left derived tensor product as discussed in Chapter 23 is indeed the model of the smash product, in the category of free, semifree, almost free or arbitrary $\mathbb{T}$-spectra, as appropriate. Unfortunately, we are only able to describe the model on objects; neither the smash product nor its adjoint preserve pure parity objects in general, so one cannot get a firm grip on morphisms. A satisfactory resolution would be to prove the equivalence of homotopy categories from a string of equivalences at the level of model categories, following the example of Quillen [**23**].

The first observation is that the smash product preserves $\mathcal{F}$-spectra and $\mathcal{F}$-contractible spectra. More precisely $X \wedge Y \wedge E\mathcal{F}_+ \simeq (X \wedge E\mathcal{F}_+) \wedge (Y \wedge E\mathcal{F}_+)$ and $\Phi^{\mathbb{T}}(X \wedge Y) \simeq \Phi^{\mathbb{T}} X \wedge \Phi^{\mathbb{T}} Y$. Since $\Phi^{\mathbb{T}}(X \wedge Y) = \Phi^{\mathbb{T}}(X) \wedge \Phi^{\mathbb{T}}(Y)$, the Künneth theorem gives

$$\pi_*(\Phi^{\mathbb{T}}(X \wedge Y)) = \pi_*(\Phi^{\mathbb{T}} X) \otimes \pi_*(\Phi^{\mathbb{T}} Y),$$

and we shall show that $\pi_*^{\mathbb{T}}(E\mathcal{F}_+ \wedge X \wedge Y)$ is the left derived tensor product of $\pi_*^{\mathbb{T}}(E\mathcal{F}_+ \wedge X)$ and $\pi_*^{\mathbb{T}}(E\mathcal{F}_+ \wedge Y)$. This might suggest the use of the torsion model, but we warn that there is a suspension which makes the gluing map inaccessible.

The connection with topology arises from the fact that $\Sigma\mathbb{I} \circledast M = M$ for any $\mathcal{F}$-finite torsion module M, where $\circledast$ denotes the right derived torsion product functor. Indeed, we have $E\langle H \rangle \wedge Y = Y$, and hence

$$\pi_*^{\mathbb{T}}(E\langle H \rangle \wedge Y) = \Sigma \mathrm{Tor}(\pi_*^{\mathbb{T}}(E\langle H \rangle), \pi_*^{\mathbb{T}}(Y)) = \pi_*^{\mathbb{T}}(E\langle H \rangle) \circledast \pi_*^{\mathbb{T}}(Y).$$

We may now prove the special case of the theorem in which both spectra are $\mathcal{F}$-spectra.

PROPOSITION 24.1.1. *For $\mathcal{F}$-spectra X and Y, their smash product is modelled by the dg $\mathcal{O}_{\mathcal{F}}$-module $\pi_*^{\mathbb{T}}(X) \circledast \pi_*^{\mathbb{T}}(Y)$. In particular there is a short exact sequence*

$$0 \longrightarrow \pi_*^{\mathbb{T}}(X) \otimes \pi_*^{\mathbb{T}}(Y) \longrightarrow \pi_*^{\mathbb{T}}(X \wedge Y) \longrightarrow \Sigma\mathrm{Tor}(\pi_*^{\mathbb{T}}(X), \pi_*^{\mathbb{T}}(Y)) \longrightarrow 0.$$

Proof: We have observed that the case when Y is injective is immediate. For the general case use an Adams resolution $Y \longrightarrow I(Y) \longrightarrow J(Y)$, realizing $\pi_*^{\mathbb{T}}(Y) \longrightarrow I \xrightarrow{d} J$ in homotopy, and obtain a cofibre sequence $X \wedge Y \longrightarrow X \wedge I(Y) \longrightarrow X \wedge J(Y)$. The model of $X \wedge Y$ is thus the fibre of $(S \circledast I \longrightarrow S \circledast J) = \Sigma(\mathrm{Tor}(S, I) \longrightarrow \mathrm{Tor}(S, J))$. Providing we identify the map as $1 \circledast d = \Sigma\mathrm{Tor}(1, d)$ the proof is complete. However both $I(Y)$ and $J(Y)$ are wedges of suspensions of $E\langle H\rangle$, so it suffices to show for an arbitrary map $f : \Sigma^n E\langle H\rangle \longrightarrow E\langle H\rangle$ that the map $1 \wedge f : \Sigma^n X \wedge E\langle H\rangle \longrightarrow X \wedge E\langle H\rangle$ induces $1 \circledast p(f)$. But in 2.4.1 we classified all such maps f: they are multiplications by c_H^m for some m. Identifying $X \wedge E\langle H\rangle$ with X, the map $1 \wedge f$ is identified with c_H^m by definition of the $\mathbb{Q}[c_H]$-module structure. □

THEOREM 24.1.2. *For arbitrary rational $\mathbb{T}$-spectra X and Y, their smash product is modelled by $\pi_*^{\mathcal{A}}(X) \circledast \pi_*^{\mathcal{A}}(Y)$. In particular there is a short exact sequence*

$$0 \longrightarrow \pi_*^{\mathcal{A}}(X) \otimes \pi_*^{\mathcal{A}}(Y) \longrightarrow \pi_*^{\mathcal{A}}(X \wedge Y) \longrightarrow \Sigma\mathrm{Tor}(\pi_*^{\mathcal{A}}(X), \pi_*^{\mathcal{A}}(Y)) \longrightarrow 0.$$

Proof: We have observed that the result is immediate for sufficiently many injective spectra Y (ie when Y is $\mathcal{F}$-contractible or an injective $\mathcal{F}$-spectrum). For the general case use an Adams resolution $Y \longrightarrow I(Y) \longrightarrow J(Y)$, realizing $\pi_*^{\mathcal{A}}(Y) \longrightarrow I \xrightarrow{d} J$ in homotopy, and obtain a cofibre sequence $X \wedge Y \longrightarrow X \wedge I(Y) \longrightarrow X \wedge J(Y)$. The model of $X \wedge Y$ is thus the fibre of $(\pi_*^{\mathcal{A}}(X) \circledast I \longrightarrow \pi_*^{\mathcal{A}}(X) \circledast J)$. It remains to identify the map as $1 \circledast d$.

However $I(Y)$ is a wedge of suspensions of spectra $E\langle H\rangle$, so it suffices to deal with one factor $\delta : E\langle H\rangle \longrightarrow J(Y)$ where $J(Y)$ is a $\mathcal{F}$-free injective. In Section 5.7 we identified such maps as composites of certain standard maps and retractions. The smash product and its model respect retracts, so it suffices to check that the standard maps are accurately modelled. The standard maps are of two types. Those of the first typ are the maps $c^n : E\langle H\rangle \longrightarrow \Sigma^{2n}E\langle H\rangle$, and these are correctly modelled since the model is built with $\mathcal{O}_{\mathcal{F}}$-modules. Those of the second type are the maps $g : \tilde{E}\mathcal{F} \longrightarrow \Sigma E\mathcal{F}_+ \wedge S^{-nV(H)}$. This map g is the cofibre of the map $f : S^{-nV(H)} \longrightarrow \tilde{E}\mathcal{F}$ which is the identity on geometric fixed points. The codomain of f is injective, so f is determined by its induced map. To describe the model let v be the function with $v(K) = n$ when $K \subseteq H$, and $v(K) = 0$ otherwise, and let $\mathcal{O}_{\mathcal{F}}(-v)$ be the submodule of $t_*^{\mathcal{F}}$, which in degree m consists of the product of the factors with $m \leq -v(K)$. Thus g is modelled by the map depicted thus

$$\begin{array}{ccc} t_*^{\mathcal{F}} & \longrightarrow & t_*^{\mathcal{F}}/\mathcal{O}_{\mathcal{F}}(-v) \\ \downarrow & & \downarrow \\ t_*^{\mathcal{F}} & \longrightarrow & 0. \end{array}$$

In other words, this is the cokernel of the inclusion $\phi : S^{-v} \longrightarrow e(\mathbb{Q})$, in the notation of Section 22.1. The modelling process preserves cofibre sequences, so it suffices to show that $1 \wedge \phi : \pi_*^{\mathcal{A}}(X) \circledast S^{-v} \longrightarrow \pi_*^{\mathcal{A}}(X) \circledast e(\mathbb{Q})$ models $1 \wedge f : X \wedge S^{-nV(H)} \longrightarrow X \wedge \tilde{E}\mathcal{F}$. It suffices to deal with the case when X is even. However in this case $X \wedge S^{-nV(H)}$ and $X \wedge \tilde{E}\mathcal{F}$ are also both even, so the map $1 \wedge f$ is determined by its induced map. Furthermore, since $X \wedge \tilde{E}\mathcal{F}$ is $\mathcal{F}$-contractible, this is determined by its effect on vertices, which is the identity. This is faithfully modelled by $1 \circledast \phi$. □

24.2. Models of function spectra.

The obvious conjecture is that the function spectrum corresponds to the internal Hom functor before passage to homotopy, and hence that in the homotopy category the function spectrum is modelled by something closely related to RHom. In the case of $\mathcal{F}$-spectra the model will not be RHom itself, but in any case the functor should be recognized as the right adjoint $R'\hat{\mathbf{Hom}}$ of the total left derived tensor product. Of course, this would follow immediately if the identification of the smash product with the total left derived functor of tensor product was functorial. In Chapter 23 we have studied the candidate model of function spectra, and established various formal properties. In particular, it is a homotopy functor that preserves cofibre sequences, and we show in this section and the next that the function spectrum is object-accessible in the sense of 7.3 and modelled by $R'\hat{\mathbf{Hom}}$ on generators.

THEOREM 24.2.1. *If Y and Z are rational $\mathbb{T}$-spectra modelled by B and C in the standard model, then $F(Y, Z)$ is modelled by $R\mathbf{IntHom}(B, C)$.*

The analogous result holds for semifree spectra. We motivate the proof by treating the case of free spectra.

PROPOSITION 24.2.2. *If Y and Z are free spectra modelled by torsion $\mathbb{Q}[c]$-modules S and T then the internal function spectrum $F(Y, Z) \wedge E\mathbb{T}_+$ is modelled by the dg torsion module $R\Gamma_c\, R\mathrm{Hom}(S, T)$.*

Proof: We note that this result follows from the general case using a little algebra, since the theorems on function spectra and smash products assert that $F(Y, Z) \wedge E\mathbb{T}_+$ is modelled by $R\mathbf{IntHom}(f(S), f(T)) \circledast \Sigma\mathbb{I}(1)$. Nonetheless we give a more direct proof, by way of preparation for the case of arbitrary spectra.

Considerations of accessibility show that we ought to concentrate on the case when $\otimes S$ preserves cofibrations. Whilst this only happens in degenerate cases, the torsion product functor preserves cofibrations when S is injective. This suggests that it is expedient to identify $F(Y, Z) \wedge E\mathbb{T}_+$ as the cofibre of

$$F(I(Y), Z) \wedge E\mathbb{T}_+ \longleftarrow F(J(Y), Z) \wedge E\mathbb{T}_+$$

as might otherwise seem rather unnatural: we refer to this as the dual resolution method. In our case we know that both $I(Y)$ and $I(Z)$ are wedges of suspensions of $E\mathbb{T}_+$, and so $F(I(Y), Z) \wedge E\mathbb{T}_+$ and $F(J(Y), Z) \wedge E\mathbb{T}_+$ are internal products of suspensions of spectra $F(E\mathbb{T}_+, Z) \wedge E\mathbb{T}_+ \simeq Z \wedge E\mathbb{T}_+$. Since we have modelled products, it suffices to identify the map between the products. However the map $I(Y) \longrightarrow J(Y)$ is specified by its components, and any map $E\mathbb{T}_+ \longrightarrow J(Z)$ factors as $E\mathbb{T}_+ \longrightarrow \Sigma^{2n} E\mathbb{T}_+ \longrightarrow J(Z)$ for some

n where the first map is multiplication by c^n and the second is the inclusion of a direct summand. The remaining verifications will be treated in more detail in the proof of the general case. □

However this gives very little indication of the general answer. The vertices are especially hard to describe since they involve both torsion and torsion free parts. For example if X and Y are both $\mathcal{F}$-spectra we may calculate $[X, Y]_*^{\mathbb{T}}$ using the Adams short exact sequence, and $F(X, Y) \wedge \tilde{E}\mathcal{F}$ has homotopy groups $\mathcal{E}^{-1}[X, Y]_*^{\mathbb{T}}$. Roughly speaking, this suggests a contribution $\mathcal{E}^{-1}R\mathrm{Hom}(S, T)$ from the torsion parts of a general spectrum. It is not immediately clear how this should be mixed with the vertex contribution $\mathrm{Hom}(V, W)$.

REMARK 24.2.3. If Y and Z are $\mathcal{F}$-spectra, the function spectrum $F(Y, Z)$ has Thom isomorphisms, so its vertex can be calculated by inverting Euler classes. Thus if Y and Z are modelled by the $\mathcal{F}$-finite torsion modules S and T, then 2.2.3 and 3.1.1 show that vertex of $F(Y, Z)$ is modelled by $\mathcal{E}^{-1}R\mathrm{Hom}(S, T)$. We claim that it is also modelled by $R\mathrm{Hom}(t_*^{\mathcal{F}}, \Sigma R\Gamma_{\mathcal{E}} R\mathrm{Hom}(S, T))$. This is a manifestation of Warwick duality.

To be precise, if $\mathcal{E}$ is a multiplicatively closed set in a commutative ring A we define the chain complex $K(\mathcal{E})$ by the fibre sequence

$$K(\mathcal{E}) \longrightarrow A \longrightarrow \mathcal{E}^{-1}A.$$

Denoting projective approximation by the letter P, Warwick duality **[9]** is the statement that there is a quasi-isomorphism

$$\mathrm{Hom}(P\mathcal{E}^{-1}A, \Sigma K(\mathcal{E}) \otimes X) \simeq \mathrm{Hom}(PK(\mathcal{E}), X) \otimes \mathcal{E}^{-1}A.$$

The derived analogue of this is

$$R\mathrm{Hom}(t_*^{\mathcal{F}}, \Sigma R\Gamma_{\mathcal{E}} X) \simeq \mathcal{E}^{-1}R\mathrm{Hom}(\Sigma\mathbb{I}, X).$$

The required equivalence thus follows provided $X = \mathrm{Hom}(S, T)$ is complete in the sense that the natural map

$$X = R\mathrm{Hom}(\mathcal{O}_{\mathcal{F}}, X) \longrightarrow R\mathrm{Hom}(\Sigma\mathbb{I}, X)$$

is an equivalence. However for $X = \mathrm{Hom}(S, T)$ we may use the tensor-Hom adjunction to establish completeness:

$$R\mathrm{Hom}(\mathcal{O}_{\mathcal{F}}, R\mathrm{Hom}(S, T)) \longrightarrow R\mathrm{Hom}(\Sigma\mathbb{I}, R\mathrm{Hom}(S, T)) \xrightarrow{\simeq} R\mathrm{Hom}(\Sigma\mathbb{I} \circledast S, T) = R\mathrm{Hom}(S, T). \quad \square$$

We briefly consider one general example.

EXAMPLE 24.2.4. *(Functional duals.)* Suppose the $\mathbb{T}$-spectrum X has standard model $B = (N \longrightarrow t_*^{\mathcal{F}} \otimes V)$, and we let $Q \longrightarrow t_*^{\mathcal{F}} \otimes H$ denote the torsion model of the functional dual $DX = F(X, S^0)$ if we already understand X in the standard model.

Recall from 5.8.1 that S^0 has model $(\mathcal{O}_{\mathcal{F}} \longrightarrow t_*^{\mathcal{F}})$. This has torsion part $\Sigma\mathbb{I}$, and the natural fibrant approximation is the fibre of the map $e(\mathbb{Q}) \longrightarrow f(\Sigma^2\mathbb{I})$. It follows from 24.2.2 that if S is the torsion part of X, the torsion part of DX is

$$R\Gamma_{\mathcal{E}}\, \mathrm{Hom}(S, \Sigma\mathbb{I}).$$

The vertex is described by the fibre sequence

$$H \longrightarrow V^* \longrightarrow \mathcal{E}^{-1}\mathrm{Hom}(t_*^{\mathcal{F}} \otimes V, \Sigma^2\mathbb{I}).$$

However the real aim here is to give a description of the complete model. We show that the obvious cofibre sequence $DX \longrightarrow DX \wedge \tilde{E}\mathcal{F} \longrightarrow DX \wedge \Sigma E\mathcal{F}_+$ is also natural from the algebraic point of view.

LEMMA 24.2.5. *If* $B = (N \longrightarrow t_*^{\mathcal{F}} \otimes V)$, *then the functional dual of* B *is described by*

$$R\mathbf{IntHom}(B, S^0) = \mathit{fibre}\left(e(V^*) \longrightarrow \hat{\Gamma} f(\mathrm{Hom}(N, \Sigma^2\mathbb{I}))\right).$$

Proof: We use the injective resolution $S^0 \longrightarrow e(\mathbb{Q}) \longrightarrow f(\Sigma^2\mathbb{I})$, and deduce that

$$R\mathbf{IntHom}(B, S^0) = \mathit{fibre}\left(\hat{\Gamma}\hat{\mathbf{Hom}}(B, e(\mathbb{Q})) \longrightarrow \hat{\Gamma}\hat{\mathbf{Hom}}(B, \Sigma^2 f(\mathbb{I}))\right).$$

Now by 22.5.1, we find $\hat{\mathbf{Hom}}(B, e(\mathbb{Q}))$ is the injective $e(V^*)$. On the other hand, by 22.5.2 $\hat{\mathbf{Hom}}(B, \Sigma^2 f(\mathbb{I})) = f(\mathrm{Hom}(N, \Sigma^2\mathbb{I}))$. Applying $\hat{\Gamma}$ we obtain the result. □

We can say a little more about the term $\hat{\Gamma} f(\mathrm{Hom}(N, \Sigma^2\mathbb{I}))$ in the description. For an arbitrary baseable $\mathcal{O}_{\mathcal{F}}$-module M we proved in Example 20.3.3 that there is a fibre sequence

$$\hat{\Gamma} f(P) \longrightarrow e(\mathcal{E}^{-1}P) \longrightarrow f(\mathcal{E}^{-1}P/P) \vee f(\Gamma_{\mathcal{E}}P). \quad \square$$

24.3. Proof of Theorem 24.2.1.

Since the proof is slightly long and indirect, it has a section to itself.

One might expect the proof to use the cofibration $F(Y, Z) \longrightarrow F(Y, I(Z)) \longrightarrow F(Y, J(Z))$, where $Z \longrightarrow I(Z) \longrightarrow J(Z)$ is an Adams resolution. However it is not clear how to eliminate indeterminacy in the description of the map $F(Y, I(Z)) \longrightarrow F(Y, J(Z))$. Instead, we therefore use our analysis of products to apply the dual resolution method as follows.

In the usual way, we are going to prove that the functor $F(Y, Z)$ is object-accessible, by constructing it from particular cases $F(Y', Z')$, by taking mapping cones of maps which can be identified with no indeterminacy at any stage. Of course the algebraic model, $R\mathbf{IntHom}$ is a functor on the homotopy category and preserves cofibres, so it suffices to show that it has the correct value on the building blocks $F(X', Y')$. We begin by outlining the argument.

Step 1: For an arbitrary Z we give a description of $F(\tilde{E}\mathcal{F}, Z)$ and $F(E\langle H\rangle, Z)$ for all H. This is reasonably straightforward. The description is not claimed to be natural in Z; however, it is automatically natural in the first variable, since all maps between spectra of the form $\tilde{E}\mathcal{F}$ or $E\langle H\rangle$ are essentially given by multiplication by powers of c or Euler classes as we saw in Section 5.7.

Step 2: Consider an Adams resolution $Y \longrightarrow I(Y) \longrightarrow J(Y)$ of Y. Thus both $I(Y) = \bigvee_\alpha I_\alpha$ and $J(Y) = \bigvee_\beta J_\beta$ are wedges of suspensions of $\tilde{E}\mathcal{F}$ and $E\langle H\rangle$ for various H. Furthermore $J(Y)$ does not involve any copies of $\tilde{E}\mathcal{F}$.

Step 3: Using the description of products we can describe the models of $F(I(Y), Z) = \prod_\alpha F(I_\alpha, Z)$ and $F(J(Y), Z) = \prod_\beta F(J_\beta, Z)$. We therefore have a cofibre sequence

$$F(Y,Z) \longleftarrow \prod_\alpha F(I_\alpha, Z) \stackrel{\delta^*}{\longleftarrow} \prod_\beta F(J_\beta, Z).$$

Step 4: Describe the map δ^* between the products. For this, note that the map $\delta : I(Y) \longrightarrow J(Y)$ is categorically described by its components $\delta_\alpha : I_\alpha \longrightarrow J(Y)$, so the map of function spectra is also described by its components $F(\delta_\alpha, Z)$. It therefore suffices to describe these.

Step 5: Since I_α is either a suspension of $\tilde{E}\mathcal{F}$ or of $E\langle H \rangle$ for some H, we have shown in Section 5.7 that any map into a spectrum of the form $J(Y)$ can be described as the composite of a very explicit map followed by the inclusion of a direct summand. The description in Step 1 is natural for these maps, and the product description is categorical.

The point of this analysis is that it can be carried out in parallel in the algebraic and topological categories, and the primitive pieces and constructions necessarily correspond.

We proceed with Step 1.

LEMMA 24.3.1. *The algebraic model of $F(\tilde{E}\mathcal{F}, Z)$ is correctly described by the theorem.*

Proof: We take the model $e(\mathbb{Q})$ of $\tilde{E}\mathcal{F}$, and establish the lemma in turn for Z being $\mathcal{F}$-contractible, $\mathcal{F}$-free and arbitrary.

If Z is $\mathcal{F}$-contractible $F(\tilde{E}\mathcal{F}, Z) \simeq Z$. On the other hand $C = e(W)$, which is already injective, so that $R\mathbf{IntHom}(e(\mathbb{Q}), C) = \mathbf{IntHom}(e(\mathbb{Q}), C)$. However, $\mathbf{IntHom}(e(\mathbb{Q}), C) = e(W)$ by the case $B = e(W)$ of 22.5.1.

If Z is an $\mathcal{F}$-spectrum in even degrees then $F(\tilde{E}\mathcal{F}, Z)$ is $\mathcal{F}$-contractible, and by 5.3.2 has homotopy $\mathrm{Hom}(t_*^{\mathcal{F}}, T)$ in even degrees and $\mathrm{Ext}(\Sigma t_*^{\mathcal{F}}, T)$ in odd degrees. On the other hand $C = f(T)$, and we may choose an injective resolution $f(T) \longrightarrow f(I) \longrightarrow f(J)$. Thus

$$R\mathbf{IntHom}(e(\mathbb{Q}), f(T)) = \mathit{fibre}\left(\hat{\Gamma}\hat{\mathbf{Hom}}(e(\mathbb{Q}), f(I)) \longrightarrow \hat{\Gamma}\hat{\mathbf{Hom}}(e(\mathbb{Q}), f(J))\right).$$

The result therefore follows from the case of 22.5.2 in which $B = e(\mathbb{Q})$, since both terms are even and the map is classified by its effect in homotopy.

For the general case we must consider the structure map $q_Z : Z \wedge \tilde{E}\mathcal{F} \longrightarrow Z \wedge \Sigma E\mathcal{F}_+$. Now note that the induced map $F(\tilde{E}\mathcal{F}, q) : F(\tilde{E}\mathcal{F}, Z \wedge \tilde{E}\mathcal{F}) \longrightarrow F(\tilde{E}\mathcal{F}, Z \wedge \Sigma E\mathcal{F}_+)$ is determined by the diagram

$$\begin{array}{ccc} F(\tilde{E}\mathcal{F}, Z \wedge \tilde{E}\mathcal{F}) & \stackrel{F(\tilde{E}\mathcal{F}, q)}{\longrightarrow} & F(\tilde{E}\mathcal{F}, Z \wedge \Sigma E\mathcal{F}_+) \\ \simeq\downarrow & & \downarrow \\ F(S^0, Z \wedge \tilde{E}\mathcal{F}) & \stackrel{q_X}{\longrightarrow} & F(S^0, Z \wedge \Sigma E\mathcal{F}_+) \end{array}$$

since all maps from an $\mathcal{F}$-contractible spectrum to $F(E\mathcal{F}_+, T)$ are null, using the smash-function adjunction and the fact that $E\mathbb{T}_+ \wedge \tilde{E}\mathcal{F} \simeq *$. This argument applies equally well in the algebraic category, where it describes the map induced in the model. $\square$

LEMMA 24.3.2. *The algebraic model of $F(E\langle H\rangle, Z)$ is correctly described by the theorem.*

REMARK 24.3.3. For calculational purposes we care more about knowing the the torsion part and vertex of $F(E\langle H\rangle, Z)$, both of which are easy. Indeed, the torsion part is $R\Gamma_c\, R\mathrm{Hom}(\Sigma\mathbb{I}(H), T)$ where T is the torsion part of Z, and the vertex is $\mathcal{E}^{-1}R\mathrm{Hom}(\Sigma\mathbb{I}(H), T)$. On the other hand, this does not serve the present purpose.

Proof: Note first that both the topological object and its algebraic model are trivial when Z is $\mathcal{F}$-contractible. The geometric fact that $F(E\langle H\rangle, Z\wedge \tilde{E}\mathcal{F})\simeq *$ is familiar. On the other hand, $e(W)$ is injective, so that $R\mathbf{IntHom}(f(\Sigma\mathbb{I}(H)), e(W)) = \mathbf{IntHom}(f(\Sigma\mathbb{I}(H)), e(W))$, and this is trivial by 22.5.1. We may thus assume Z is an $\mathcal{F}$-spectrum, and $C = f(\Sigma N)$ for an even $\mathcal{F}$-finite torsion module N.

Take an injective resolution $0 \longrightarrow N \longrightarrow I \longrightarrow J \longrightarrow 0$ of N by $\mathcal{F}$-finite torsion modules We thus find

$$R\mathbf{IntHom}(f(\mathbb{I}(H)), f(N)) = \mathit{fibre}\left(\hat{\Gamma}\hat{\mathbf{H}}\mathbf{om}(f(\mathbb{I}(H)), f(I)) \longrightarrow \hat{\Gamma}\hat{\mathbf{H}}\mathbf{om}(f(\mathbb{I}(H)), f(J))\right).$$

We shall prove shortly that the two terms are even. It follows that the map between them is classified by its effect in homotopy. By the discussion in Section 5.7, the relevant maps are composites of retractions and collapse maps $c^n : E\langle H\rangle \longrightarrow \Sigma^{2n}E\langle H\rangle$, so it is immediate that they are modelled by the corresponding algebraic maps.

It is possible to establish the evenness by direct calculation generalizing Example 13.5.2, but the argument is less cluttered by intricacies if we choose an $\mathcal{F}$-spectrum Y with $\pi_*^{\mathbb{T}}(Y) = \Sigma J$ and consider the cofibre sequence

$$F(E\langle H\rangle, Y) \longrightarrow F(E\langle H\rangle, Y)\wedge \tilde{E}\mathcal{F} \xrightarrow{q} F(E\langle H\rangle, Y)\wedge \Sigma E\mathcal{F}_+$$

and its algebraic analogue. We identify the second and third terms and the map between them.

Firstly, $F(E\langle H\rangle, Y)\wedge \tilde{E}\mathcal{F}$ is $\mathcal{F}$-contractible and determined by its homotopy, and hence modelled by $e(\mathcal{E}^{-1}\mathrm{Hom}(\mathbb{I}(H), J))$. Next, $F(E\langle H\rangle, Y)\wedge \Sigma E\mathcal{F}_+ \simeq Y \wedge \Sigma E\langle H\rangle$, which is modelled by $\Sigma^2 J(H)$. The corresponding algebraic equivalence is given by the fact that $R\mathbf{IntHom}(e(\mathbb{Q}), J)$ is torsion free, so that $\circledast f(\Sigma\mathbb{I})$ annihilates it. Since $J(H)$ is injective, q is classified by a map

$$q : e(\mathcal{E}^{-1}\mathrm{Hom}(\mathbb{I}(H), J)) \longrightarrow \Sigma^2 J(H)$$

in the abelian category $\mathcal{A}$. The significant fact is that it is surjective, so $F(E\langle H\rangle, J)$ is modelled in the standard model by its fibre,

$$\ker(q) \longrightarrow t_*^{\mathcal{F}} \otimes \mathcal{E}^{-1}\mathrm{Hom}(\mathbb{I}(H), J). \quad \square$$

This completes Step 1. Steps 2, 3 and 4 require no further comment. We proceed to Step 5.

LEMMA 24.3.4. *The identifications in 24.3.1 and 24.3.2 are natural for the maps in 5.7.3.*

Proof: The identifications are certainly natural for retractions. This leaves the question of naturality for the maps $c^n : E\langle H\rangle \longrightarrow \Sigma^{2n}E\langle H\rangle$ and $\tilde{E}\mathcal{F} \longrightarrow \Sigma E\mathcal{F}_+ \wedge S^{-nV(H)}$. Naturality for those of the first type is clear since the modelling is all done in the category of $\mathcal{O}_{\mathcal{F}}$-modules.

For those of the second type we may compose with projection onto the Kth factor, and it is sufficient to prove naturality for these. There are two types, depending on whether K is contained in H or not. If K is not contained in H, the map is equal to the composite $\tilde{E}\mathcal{F} \longrightarrow \Sigma E\mathcal{F}_+ \longrightarrow \Sigma E\langle K\rangle$, and thus a special case of the second type.

We may therefore suppose that $f : \tilde{E}\mathcal{F} \longrightarrow \Sigma^{-2n+1}E\langle H\rangle$ is given, and that we want to understand the map $f^* : F(\Sigma^{-2n+1}E\langle H\rangle, Z) \longrightarrow F(\tilde{E}\mathcal{F}, Z)$. Applying $F(f, \cdot)$ to the cofibre sequence $Z \wedge E\mathcal{F}_+ \longrightarrow Z \longrightarrow Z \wedge \tilde{E}\mathcal{F}$, we obtain the diagram

$$\begin{array}{ccccc} F(\Sigma^{-2n+1}E\langle H\rangle, Z \wedge E\mathcal{F}_+) & \xrightarrow{\simeq} & F(\Sigma^{-2n+1}E\langle H\rangle, Z) & \longrightarrow & F(\Sigma^{-2n+1}E\langle H\rangle, Z \wedge \tilde{E}\mathcal{F}) \\ \downarrow & & \downarrow & & \downarrow \\ F(\tilde{E}\mathcal{F}, Z \wedge E\mathcal{F}_+) & \longrightarrow & F(\tilde{E}\mathcal{F}, Z) & \longrightarrow & F(\tilde{E}\mathcal{F}, Z \wedge \tilde{E}\mathcal{F}), \end{array}$$

in which the three verticals are induced by f. Since $F(E\langle H\rangle, Z \wedge \tilde{E}\mathcal{F}) \simeq *$, the induced map is determined by the composite $\alpha : F(\Sigma^{-2n+1}E\langle H\rangle, Z \wedge E\mathcal{F}_+) \longrightarrow F(\tilde{E}\mathcal{F}, Z \wedge \tilde{E}\mathcal{F})$. Next, the codomain of α is $\mathcal{F}$-contractible, and so α is determined by its effect in homotopy. Splitting $Z \wedge E\mathcal{F}_+$ into even and odd parts, we see that the Adams short exact sequence shows that the effect of f in homotopy determines the effect of α in homotopy. Since the model is based on the effect of f in homotopy, this completes the proof. □

24.4. The Lewis-May K-fixed point functor.

We return to the Lewis-May K-fixed point functor, using what we have learnt about internal Hom functors to construct an algebraic model on the standard model. We aim to model the adjunction

$$[\mathrm{inf}_{\mathbb{T}/K}^{\mathbb{T}} X, Y]^{\mathbb{T}} = [X, \Psi^K Y],$$

where $\mathrm{inf}_{\mathbb{T}/K}^{\mathbb{T}} : \overline{\mathbb{T}}\text{-Spec} \longrightarrow \mathbb{T}\text{-Spec}$ is the inflation functor building in non-trivial representations, and $\Psi^K : \mathbb{T}\text{-Spec} \longrightarrow \overline{\mathbb{T}}\text{-Spec}$ is the Lewis-May $\mathbb{T}$-fixed point functor [**19**, II.4.4].

We recall that the quotient map $q : \mathbb{T} \longrightarrow \overline{\mathbb{T}}$ induces a function $q : \mathcal{F} \longrightarrow \overline{\mathcal{F}}$ on subgroups, and hence a ring map $q^* : \overline{\mathcal{O}}_{\overline{\mathcal{F}}} \longrightarrow \mathcal{O}_{\mathcal{F}}$. This allows us to define 9.1.1 the inflation on $\overline{\mathcal{O}}_{\overline{\mathcal{F}}}$-modules by $\mathrm{inf}(\overline{N}) = \mathcal{O}_{\mathcal{F}} \otimes_{\overline{\mathcal{O}}_{\overline{\mathcal{F}}}} \overline{N}$.

Reformulating the definition from Section 9.2, we obtain.

DEFINITION 24.4.1. The inflation is defined on an object $\overline{A}$ of the standard model $\overline{\mathcal{A}}$ by

$$\mathrm{inf}(\overline{A}) = S^0_{\mathcal{A}} \otimes_{\overline{\mathcal{O}}_{\overline{\mathcal{F}}}} \overline{A},$$

where $S^0_{\mathcal{A}}$ is the 0-sphere in $\mathcal{A}$.

Evidently the right adjoint of $S^0_{\mathcal{A}} \otimes_{\overline{\mathcal{O}}_{\overline{\mathcal{F}}}}$ as a functor on $\overline{\mathcal{A}} \longrightarrow \mathcal{A}$ should take the form $C \longmapsto \mathbf{IntHom}_{\overline{\mathcal{O}}_{\overline{\mathcal{F}}}}(S^0_{\mathcal{A}}, C)$. It is clearer to give a slightly more general definition.

DEFINITION 24.4.2. The tensor product $B \otimes_{\overline{\mathcal{O}}_{\mathcal{F}}} \overline{A}$ of an object $\overline{A}$ of $\hat{\overline{\mathcal{A}}}$ and an object B of $\hat{\mathcal{A}}$ is given by tensoring both nub and vertex over $\overline{\mathcal{O}}_{\mathcal{F}}$.

REMARK 24.4.3. By associativity of the tensor product the functor $B \otimes_{\overline{\mathcal{O}}_{\mathcal{F}}}$ is a composite $B \otimes_{\mathcal{O}_{\mathcal{F}}} S^0_{\mathcal{A}} \otimes_{\overline{\mathcal{O}}_{\mathcal{F}}}$. This shows that the inflation is the significant case.

LEMMA 24.4.4. *(i) For all $\overline{A}$ and B, the object $B \otimes_{\overline{\mathcal{O}}_{\mathcal{F}}} \overline{A}$ is an object of $\hat{\mathcal{A}}$.*
(ii) If B lies in $\mathcal{A}$ and $\overline{A}$ lies in $\overline{\mathcal{A}}$, then the object $B \otimes_{\overline{\mathcal{O}}_{\mathcal{F}}} \overline{A}$ is an object of $\mathcal{A}$.

Proof: By 9.1.3

$$\mathcal{E}^{-1}(M \otimes_{\overline{\mathcal{O}}_{\mathcal{F}}} N) \cong (\mathcal{E}^{-1}M) \otimes_{\overline{\mathcal{O}}_{\mathcal{F}}} N) \cong M \otimes_{\overline{\mathcal{O}}_{\mathcal{F}}} (\overline{\mathcal{E}}^{-1}N). \quad \square$$

The corresponding definition for Hom is analogous.

DEFINITION 24.4.5. Let $B = (N \xrightarrow{\beta} t_*^{\mathcal{F}} \otimes V)$ and $C = (P \xrightarrow{\gamma} t_*^{\mathcal{F}} \otimes W)$ be objects of $\hat{\mathcal{A}}$.
(i) We define the object $\hat{\mathbf{Hom}}_{\overline{\mathcal{O}}_{\mathcal{F}}}(B, C) = (Q \xrightarrow{\delta} t_*^{\mathcal{F}} \otimes H)$ of $\hat{\overline{\mathcal{A}}}$ by taking $H = \mathrm{Hom}(V, W)$ and both Q and δ defined by the pullback square

$$\begin{array}{ccccc} Q & & \longrightarrow & & \mathrm{Hom}_{\overline{\mathcal{O}}_{\mathcal{F}}}(N, P) \\ \downarrow{\scriptstyle\delta} & & & & \downarrow{\scriptstyle\gamma_*} \\ t_*^{\mathcal{F}} \otimes \mathrm{Hom}(V, W) & \xrightarrow{\Delta} & \mathrm{Hom}_{\overline{\mathcal{O}}_{\mathcal{F}}}(t_*^{\mathcal{F}} \otimes V, t_*^{\mathcal{F}} \otimes W) & \xrightarrow{\beta^*} & \mathrm{Hom}_{\overline{\mathcal{O}}_{\mathcal{F}}}(N, t_*^{\mathcal{F}} \otimes W). \end{array}$$

(ii) If B and C lie in $\mathcal{A}$ we take

$$\mathbf{IntHom}_{\overline{\mathcal{O}}_{\mathcal{F}}}(B, C) = \hat{\Gamma}\hat{\mathbf{Hom}}_{\overline{\mathcal{O}}_{\mathcal{F}}}(B, C).$$

We record the resulting adjunction.

LEMMA 24.4.6. *Using definition 24.4.5 there is a bijection*

$$\mathrm{Hom}(B \otimes_{\overline{\mathcal{O}}_{\mathcal{F}}} \overline{A}, C) = \mathrm{Hom}(\overline{A}, \hat{\mathbf{Hom}}_{\overline{\mathcal{O}}_{\mathcal{F}}}(B, C))$$

natural in all three variables, where B and C are objects of $\hat{\mathcal{A}}$ and $\overline{A}$ is an object of $\hat{\overline{\mathcal{A}}}$.

Proof: Formally identical to 22.3.2. $\square$

The fact that $\hat{\Gamma}$ is right adjoint to the inclusion $\mathcal{A} \longrightarrow \hat{\mathcal{A}}$ gives us the desired result.

COROLLARY 24.4.7. *Using definition 24.4.5 there is a bijection*

$$\mathrm{Hom}(B \otimes_{\overline{\mathcal{O}}_{\mathcal{F}}} \overline{A}, C) = \mathrm{Hom}(\overline{A}, \mathbf{IntHom}_{\overline{\mathcal{O}}_{\mathcal{F}}}(B, C))$$

natural in all three variables, where B and C are objects of $\mathcal{A}$ and $\overline{A}$ is an object of $\overline{\mathcal{A}}$. $\square$

We are now ready to specialize the construction to the case at hand.

DEFINITION 24.4.8. The algebraic inflation functor

$$\mathrm{inf}_{\overline{\mathcal{A}}}^{\mathcal{A}} : \overline{\mathcal{A}} \longrightarrow \mathcal{A}$$

is defined by

$$\mathrm{inf}_{\overline{\mathcal{A}}}^{\mathcal{A}} \overline{A} = S^0 \otimes_{\overline{\mathcal{O}}_{\mathcal{F}}} \overline{A}.$$

(b) The algebraic Lewis-May K-fixed point functor

$$\Psi^K : \mathcal{A} \longrightarrow \overline{\mathcal{A}}$$

is defined by

$$\Psi^K C = \mathbf{IntHom}_{\overline{\mathcal{O}}_{\mathcal{F}}}(S^0, C).$$

REMARK 24.4.9. Note that the use of the torsion functor is redundant in this case. Indeed,

$$\mathbf{IntHom}_{\overline{\mathcal{O}}_{\mathcal{F}}}(S^0_{\mathcal{A}}, C) = \hat{\mathbf{H}}\mathbf{om}_{\overline{\mathcal{O}}_{\mathcal{F}}}(S^0_{\mathcal{A}}, C)$$

because $\hat{\mathbf{H}}\mathbf{om}_{\overline{\mathcal{O}}_{\mathcal{F}}}(S^0_{\mathcal{A}}, C)$ is already an object of $\mathcal{A}$. To see this, recall from 9.1.2 that $\mathcal{O}_{\mathcal{F}}$ is a sum of finitely many retracts of $\overline{\mathcal{O}}_{\mathcal{F}}$; since the right hand vertical γ_* in 24.4.5 (i) would be an isomorphism modulo Euler torsion for $N = \overline{\mathcal{O}}_{\mathcal{F}}$, it is also an isomorphism modulo Euler torsion for the nub $N = \mathcal{O}_{\mathcal{F}}$ of $S^0_{\mathcal{A}}$. □

The fact that this gives a model of the topological Lewis-May fixed point functor follows by passage to derived categories.

THEOREM 24.4.10. *(a) The algebraic model of* $\mathrm{inf}_{\mathbb{T}/K}^{\mathbb{T}} : \overline{\mathbb{T}}$*-Spec* $\longrightarrow \mathbb{T}$*-Spec is*

$$\mathrm{inf}_{\overline{\mathcal{A}}}^{\mathcal{A}} A = S^0 \otimes_{\overline{\mathcal{O}}_{\mathcal{F}}} A$$

in the sense that

$$\begin{array}{ccc} \overline{\mathbb{T}}\text{-}Spec & \xrightarrow{\mathrm{inf}_{\mathbb{T}/K}^{\mathbb{T}}} & \mathbb{T}\text{-}Spec \\ \simeq\downarrow & & \downarrow\simeq \\ D\overline{\mathcal{A}} & \xrightarrow{\mathrm{inf}_{\overline{\mathcal{A}}}^{\mathcal{A}}} & D\mathcal{A} \end{array}$$

commutes up to natural isomorphism.
(b) The algebraic model of the Lewis-May fixed point functor of $\Psi^K : \mathbb{T}$*-Spec* $\longrightarrow \overline{\mathbb{T}}$*-Spec is*

$$R\Psi^K C = R\mathbf{IntHom}_{\overline{\mathcal{O}}_{\mathcal{F}}}(S^0, C)$$

in the sense that

$$\begin{array}{ccc} \mathbb{T}\text{-}Spec & \xrightarrow{\Psi^K} & \overline{\mathbb{T}}\text{-}Spec \\ \simeq\downarrow & & \downarrow\simeq \\ D\mathcal{A} & \xrightarrow{R\Psi^K} & D\overline{\mathcal{A}} \end{array}$$

commutes up to natural isomorphism.

REMARK 24.4.11. Conventions common in topology would allow us to abbreviate $R\Psi^K$ to Ψ^K provided it is clear that the functor on derived categories is intended.

Proof: Since we proved in 10.3.1 that the algebraic inflation functor modelled the topological one, it follows that its right adjoint models the Lewis-May fixed point functor.

Since the algebraic inflation functor is exact, the adjunction at the level of abelian categories passes to derived categories as claimed. □

This allows us to give an algebraic explanation of 10.1.1

LEMMA 24.4.12. *There is a natural isomorphism*

$$\Psi^K(S^0_{\mathcal{A}}) \otimes_{\overline{\mathcal{O}}_{\overline{\mathcal{F}}}} \overline{A} = \mathbf{IntHom}_{\overline{\mathcal{O}}_{\overline{\mathcal{F}}}}(S^0_{\mathcal{A}}, S^0_{\mathcal{A}}) \otimes_{\overline{\mathcal{O}}_{\overline{\mathcal{F}}}} \overline{A} \cong \mathbf{IntHom}_{\overline{\mathcal{O}}_{\overline{\mathcal{F}}}}(S^0_{\mathcal{A}}, S^0_{\mathcal{A}} \otimes_{\overline{\mathcal{O}}_{\overline{\mathcal{F}}}} \overline{A}) = \Psi^K(\mathrm{inf}\overline{A}).$$

Proof: This follows by duality from the finiteness of $\mathcal{O}_{\mathcal{F}}$ as an $\overline{\mathcal{O}}_{\overline{\mathcal{F}}}$ module as proved in 9.1.2. □

APPENDIX A

Mackey functors.

The algebra of Mackey functors for the circle group is so much simpler than that of a general compact Lie group that it is worth an elementary treatment. The general case is presented elsewhere [**10**]. This appendix is of interest in its own right, but parts of this section are used in justifying an assertion in Section 1.4, and it is also used in Chapter 12.

A rational Mackey functor is a contravariant additive functor on the full subcategory $h\mathcal{SO}$ of the rational stable homotopy category with the natural cells G/H_+ as objects. We shall say M is an $h\mathcal{SO}$-module.

A Mackey functor M extends to the full subcategory $h\mathcal{SVO}$ on finite wedges of natural cells. Equally, if b is a retract of a natural cell n there is an idempotent self-map $e : n \longrightarrow n$ with $b = en$ and we may define $M(b) = eM(n)$.

Now let $h\mathcal{SB}$ denote the full subcategory on the basic cells (see Section 2.1) and note that $h\mathcal{SVO} \subseteq h\mathcal{SVB}$; we thus have functors

$$h\mathcal{SO} \longrightarrow h\mathcal{SVO} \longrightarrow h\mathcal{SVB} \longleftarrow h\mathcal{SB},$$

and from the above discussion it is easy to see that they induce equivalences of module categories.

PROPOSITION A.1. *There is an equivalence of categories*

$$\textit{Mackey functors}{=}h\mathcal{SO}\textit{-modules} \simeq h\mathcal{SB}\textit{-modules}.$$

□

As a matter of notation, if M is a Mackey functor we let $M^e : h\mathcal{SB} \longrightarrow \mathbf{Ab}$ denote the corresponding $h\mathcal{SB}$-module: the letter e suggests an idempotent. This is somewhat redundant since M^e and M are equal functors on $h\mathcal{SVB}$, but it allows us to follow standard usage and abbreviate $M(G/H_+)$ to $M(H)$ and $M(\sigma_H^0)$ to $M^e(H)$. Quite explicitly we have

A.2.

$$M(\mathbb{T}) = M^e(\mathbb{T}) \text{ and } M(H) = \bigoplus_{K \subseteq H} M^e(K)$$

$$M^e(\mathbb{T}) = M(\mathbb{T}) \text{ and } M^e(H) = e_H M(H).$$

We remark that a map $M \longrightarrow N$ is a monomorphism if and only if $M^e \longrightarrow N^e$ is a monomorphism, and similarly for epimorphisms.

The great advantage of considering M^e is that $h\mathcal{SB}$ has very few morphisms (see 2.1.1 and 2.1.6), and it is much simpler to conduct proofs and homological algebra with $h\mathcal{SB}$-modules. Nonetheless, because Mackey functors are more familiar, we have stated results in terms of Mackey functors where possible.

To emphasize the simplicity of $h\mathcal{SB}$-modules, we note that M^e is specified by a diagram

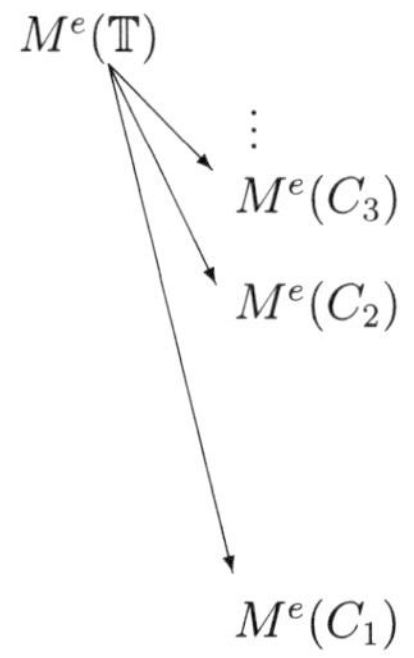

of rational vector spaces. The maps $M^e(\mathbb{T}) \longrightarrow M^e(H)$ will be called the e-restriction maps.

EXAMPLE A.3. It is immediate from the Yoneda lemma that the represented Mackey functor $\mathbb{A}_c = [\ \ , c]^{\mathbb{T}}$ is projective for any basic or natural cell c. We make the abbreviation

$$P_H = \mathbb{A}_{\sigma_H^0},$$

and, in the interests of uniformity,

$$P_{\mathbb{T}} = \mathbb{A}.$$

One readily calculates that $P^e_{\mathbb{T}}$ is $\mathbb{Q}$ at each basic cell, and the e-restriction maps are all the identity. Similarly the $h\mathcal{SB}$-module P^e_H represented by σ^0_H is $\mathbb{Q}$ at H and 0 elsewhere.

We shall refer to the projectives $P_{\mathbb{T}}$ and P_H as principal projectives. They are obviously indecomposable.

By the Yoneda lemma $\mathrm{Hom}(P_H, M) = M^e(H)$, and similarly for $\mathbb{T}$. □

One may make explicit the condition that a Mackey functor M is injective or projective, and give canonical resolutions.

LEMMA A.4. *(i) A Mackey functor M is projective if and only if each of the e-restriction maps $M^e(\mathbb{T}) \longrightarrow M^e(H)$ is a monomorphism.*
(ii) A Mackey functor M is injective if and only if the map $M^e(\mathbb{T}) \longrightarrow \prod_H M^e(H)$ is an epimorphism.

Proof: To simplify notation let $V = M^e$ throughout the proof, and work entirely with $h\mathcal{SB}$-modules.

(i) To see that any module V satisfying the specified condition is projective we suppose given a short exact sequence $0 \longrightarrow K \longrightarrow U \stackrel{f}{\longrightarrow} V \longrightarrow 0$ and construct a submodule V' of U which is a complement of K.

We begin at $\mathbb{T}$, choosing $V'(\mathbb{T})$ to be an arbitrary vector space complement to $K(\mathbb{T})$ in $U(\mathbb{T})$.

Now choose a finite subgroup H. Let $\overline{V'(\mathbb{T})}$ denote the image of $V'(\mathbb{T})$ in $U(H)$, and similarly let $\overline{V(\mathbb{T})}$ denote the image of $V(\mathbb{T})$ in $V(H)$. Since $V(\mathbb{T}) \longrightarrow V(H)$ is a monomorphism by hypothesis, and since f is a map of Mackey functors, $f(H)$ gives an isomorphism $\overline{V'(\mathbb{T})} \xrightarrow{\cong} \overline{V(\mathbb{T})}$. It follows that $\overline{V'(\mathbb{T})} \cap K(H) = 0$ and we may choose a vector space complement $V'(H)$ of $K(H)$ which contains $\overline{V'(\mathbb{T})}$.

With these choices the vector spaces $V(\mathbb{T})$ and $V'(H)$ form a sub-$h\mathcal{SB}$-module V' of U.

The necessity of the condition will follow from the fact that it holds for the principal projectives P_H above. Indeed, by the Yoneda lemma, for any module V there is an epimorphism $P_V \longrightarrow V$ with P_V a sum of principal projectives. If V is projective, the epimorphism must split, so the condition holds for V.
(ii) The proof is similar but this time we construct the splitting at finite subgroups first. Necessity again follows using a canonical monomorphism $V \longrightarrow I_V$ which we construct in the proof of A.5 below. □

Note in particular that if M is a Mackey functor with $M(\mathbb{T}) = 0$ then M is projective, and if $M(H) = 0$ for all finite subgroups H then M is injective.

LEMMA A.5. *Any Mackey functor M has a canonical projective resolution*

$$0 \longrightarrow K_M \longrightarrow P_M \longrightarrow M \longrightarrow 0$$

and a canonical injective resolution

$$0 \longrightarrow M \longrightarrow I_M \longrightarrow C_M \longrightarrow 0,$$

where the modules P_M, K_M, I_M and C_M are defined in the proof. In particular, the category of Mackey functors is of global dimension 1.

Proof: Note from the Yoneda lemma that $\mathrm{Hom}(P_H, M) = M^e(H)$. Therefore, for an arbitrary Mackey functor M we may define

$$P_M = (M^e(\mathbb{T}) \otimes P_{\mathbb{T}}) \oplus \bigoplus_H (M^e(H) \otimes P_H)$$

and a surjective map $P_M \longrightarrow M$. Since $P_H(\mathbb{T}) = 0$ the kernel K_M of this map is zero at $\mathbb{T}$, and hence projective. Evidently this resolution will not usually be minimal.

Similarly we may define an injective I_M by taking

$$I_M(\mathbb{T}) = M^e(\mathbb{T}) \oplus \prod_H M^e(H)$$

and

$$I_M(H) = M(H).$$

The e-restriction map

$$I^e_M(\mathbb{T}) = M^e(\mathbb{T}) \oplus \prod_H M^e(H) \longrightarrow M^e(H) = I^e_M(H)$$

is the e-restriction map of M^e on the $M^e(\mathbb{T})$ factor and projection on the $\prod_H M^e(H)$ factor. The module I_M is therefore injective by A.4 (ii), and there is an obvious monomorphism $M \longrightarrow I_M$, whose cokernel C_M is concentrated at $\mathbb{T}$, where we have $C_M(\mathbb{T}) = \prod_H M^e(H)$. □

LEMMA A.6. *For any projective Mackey functor Q there is a splitting*

$$Q \cong Q_{\mathcal{F}} \oplus (Q(\mathbb{T}) \otimes P_{\mathbb{T}})$$

where $Q_{\mathcal{F}}(\mathbb{T}) = 0$. Accordingly, any projective Mackey functor P is a direct sum of copies of the various indecomposable projectives $P_{\mathbb{T}}$ and P_H.

Proof: First, we note that it is obvious that any module M with $M(\mathbb{T}) = 0$ is a sum of projectives P_H.

Now, choose a map $\epsilon : Q(\mathbb{T}) \otimes P_{\mathbb{T}} \longrightarrow Q$ which is an isomorphism at $\mathbb{T}$, and define $Q_{\mathcal{F}}$ by the exact sequence

$$Q(\mathbb{T}) \otimes P_{\mathbb{T}} \xrightarrow{\epsilon} Q \longrightarrow Q_{\mathcal{F}} \longrightarrow 0.$$

By construction $Q_{\mathcal{F}}(\mathbb{T}) = 0$, and $Q_{\mathcal{F}}$ is therefore projective. It follows that $im(\epsilon)$ is a summand of Q and also projective. Hence $im(\epsilon)$ is a summand of $Q(\mathbb{T}) \otimes P_{\mathbb{T}}$; since it also contains all of the value $Q(\mathbb{T}) \otimes P_{\mathbb{T}}(\mathbb{T})$ at $\mathbb{T}$, the summand exhausts $Q(\mathbb{T}) \otimes P_{\mathbb{T}}$. This implies that ϵ is a monomorphism, and the splitting follows. □

It is appropriate to say that a Mackey functor M is finitely generated if it is a quotient of a finite sum of representable functors $\mathbb{A}_{\mathbb{T}/H}$, and similarly for an $h\mathcal{SB}$-module. We may thus consider the projective class group $K_0(\mathfrak{M}_{\mathbb{T}})$ of rational Mackey functors, which is the Grothendieck group of finitely generated projective modules.

For topological purposes, the functors $\mathbb{A}_{\mathbb{T}/H}$ and $\mathbb{A}$ represented by the natural cells are specially significant, and the quotient of $K_0(\mathfrak{M}_{\mathbb{T}})$ by the submodule they generate is the home of the Wall finiteness obstruction. Since the Mackey functors represented by basic cells can be expressed as a difference of the representable functors corresponding to the natural cells, this quotient is zero and the Wall finiteness obstruction vanishes for rational $\mathbb{T}$-spectra.

We may be more precise.

COROLLARY A.7. *The projective class group of rational Mackey functors is free abelian on the generators $P_{\mathbb{T}}$ and P_H.*

Proof: The comments above establish that the modules $P_{\mathbb{T}}$ and P_H generate the projective class group. To see they are independent we define a rank function for each subgroup. Indeed we take $r_{\mathbb{T}}(P) = dim_{\mathbb{Q}} P^e(\mathbb{T})$ and for a finite subgroup H we take $r_H(P) = dim_{\mathbb{Q}}(cok\{P^e(\mathbb{T}) \longrightarrow P^e(H)\})$; these are additive on extensions, and hence give functions on the class group.

Since $r_H(P_K) = 0$ if $K \neq H$ and $r_H(P_H) = 1$, the functions show the generators are independent. For an arbitrary projective P we have the formula

$$[P] = r_{\mathbb{T}}(P)[P_{\mathbb{T}}] + \Sigma_H r_H(P)[P_H].$$

□

We also have a structure theorem for injectives. In fact there are two special types of injectives I: the $\mathcal{F}$-injectives with $I(\mathbb{T}) = \prod_H I^e(H)$ and the e-restriction maps being projections, and the $\tilde{\mathcal{F}}$-injectives with $I(H) = 0$ for all finite subgroups H. An $\tilde{\mathcal{F}}$-injective is determined by its value at $\mathbb{T}$, and we write $L[U]$ for the $\tilde{\mathcal{F}}$-injective with value U at $\mathbb{T}$. An $\mathcal{F}$-injective I is determined by its values at the finite subgroups, and indeed

$$I \cong \prod_H I^e(H) \otimes R_H$$

where R_H is the $\mathcal{F}$-injective with $R^e_H(H) = R^e_H(\mathbb{T}) = \mathbb{Q}$ and $R^e_H(K) = 0$ for $K \neq H$.

LEMMA A.8. *For any injective I there is a splitting*

$$I \cong I_{\mathcal{F}} \oplus L[C(I)]$$

where $I_{\mathcal{F}}$ is an $\mathcal{F}$-injective and $C(I) = ker\{I^e(\mathbb{T}) \longrightarrow \prod_H I^e(H)\}$.

Proof: Clearly $L[C(I)]$ is a submodule of I, and we may define $I_{\mathcal{F}}$ by the short exact sequence

$$0 \longrightarrow L[C(I)] \longrightarrow I \longrightarrow I_{\mathcal{F}} \longrightarrow 0.$$

This splits since $L[C(I)]$ is injective. By construction, the structure map $I^e_{\mathcal{F}}(\mathbb{T}) \longrightarrow \prod_H I^e_{\mathcal{F}}(H) = \prod_H I^e(H)$ is a monomorphism, and it is also an epimorphism since $I_{\mathcal{F}}$ is injective. Thus $I_{\mathcal{F}}$ is an $\mathcal{F}$-injective as required. □

We conclude with some examples which occur geometrically. The most general class is that of spectra $X \wedge \tilde{E}\mathcal{F}$; since these are H-contractible we find the Mackey functor $\underline{\pi}^{\mathbb{T}}_n(X \wedge \tilde{E}\mathcal{F})$ is the $\tilde{\mathcal{F}}$-injective $L[\pi_n(\Phi^{\mathbb{T}} X)]$.

Perhaps more important are the homotopy functors of cells.

EXAMPLE A.9. *(Homotopy functors of basic cells.)*

By definition, $\underline{\pi}^{\mathbb{T}}_0(\sigma^0_H) = P_H$ and similarly for $\mathbb{T}$. The other homotopy functors of cells are $\underline{\pi}^{\mathbb{T}}_{2k+1}(S^0) = L[\mathbb{Q}\mathcal{F}]$ for $k \geq 0$, and $\underline{\pi}^{\mathbb{T}}_1(\sigma^0_H)$. Let $R'_H = \underline{\pi}^{\mathbb{T}}_1(\sigma^0_H)$: we shall prove that $R'_H \cong R_H$. It is easy to see there is an extension

$$0 \longrightarrow P_H \longrightarrow R'_H \longrightarrow L[\mathbb{Q}] \longrightarrow 0.$$

LEMMA A.10. *(a)* $\mathrm{Ext}(L[\mathbb{Q}], P_H) \cong \mathbb{Q}$, *and hence all non-split extensions define isomorphic modules.*
(b) The module R'_H is a non-trivial P_H extension over $L[\mathbb{Q}]$.

Proof: (a) We simply apply $\mathrm{Hom}(\cdot, P_H)$ to the projective resolution

$$0 \longrightarrow \bigoplus_K P_K \longrightarrow P_{\mathbb{T}} \longrightarrow L[\mathbb{Q}] \longrightarrow 0.$$

(b) It is enough to verify that the restriction map $R'_H(\mathbb{T}) \longrightarrow R'_H(H)$ is nonzero. For this we recall that restriction is induced by the projection $S^1 \wedge G/H_+ \longrightarrow S^1 \wedge S^0$ in $[\cdot, \sigma^0_H]^{\mathbb{T}} = [\cdot, e_H S^1]^H$. Since the projection is H-equivariantly a retraction onto a direct factor, it is nontrivial. □

APPENDIX B

Closed model categories.

We have a number of algebraic categories modelling categories of spectra: we have in mind (0) the category of torsion $\mathbb{Q}[c_H]$-modules, (1) the category of $\mathcal{F}$-finite torsion $\mathcal{O}_{\mathcal{F}}$-modules, (2) the standard model category $\mathcal{A}$, and (3) the torsion module category $\mathcal{A}_t$. Since we want to show they provide models for the homotopy theory of the categories of spectra, we need to show they admit such a structure themselves. We therefore prove that they all admit the structure of closed model categories in the sense of Quillen [**22**]. For reference, we recall that the structure of a closed model category consists of three classes of morphisms: the class **we** of weak equivalences, the class **fib** of fibrations, and the class **cof** of cofibrations. Morphisms which are weak equivalences and fibrations are known as acyclic fibrations, and similarly for cofibrations. These are required to satisfy Quillen's axioms CM1-5 [**23**].

The axioms CM1-3 are formal properties, and are easily checked in the cases which concern us, but it will be convenient to record the terminology associated with axioms CM4 and CM5. Given morphisms $i : A \longrightarrow B$ and $p : X \longrightarrow Y$, we say that we say that i has the left lifting property (LLP) with respect to p and that p has the right lifting property (RLP) with respect to i if, for any choice of f and g giving a commutative square

$$\begin{array}{ccc} A & \xrightarrow{f} & X \\ {\scriptstyle i}\downarrow & \nearrow & \downarrow{\scriptstyle p} \\ B & \xrightarrow[g]{} & Y \end{array}$$

there is a diagonal arrow making both triangles commute. Axiom CM4 states that if i is a cofibration and p is a fibration then there is such a diagonal arrow if either (i) p is a weak equivalence or (ii) i is a weak equivalence. Axiom CM5 states that any morphism $f : X \longrightarrow Y$ can be factorized both (i) (cofibrant approximation) as $X \xrightarrow{i} \hat{Y} \xrightarrow{p} Y$ where i is a cofibration and p is an acyclic fibration and (ii) (fibrant approximation) as $X \xrightarrow{i} \hat{X} \xrightarrow{p} Y$ where i is an acyclic cofibration and p is a fibration.

Our algebraic models all arise as categories of dg objects in an abelian category $\mathbb{A}$. Since we aim to form the quotient category in which homology isomorphisms are inverted we take the class **we** of weak equivalences to be the homology isomorphisms in each case. Furthermore, the abelian categories all fail to have enough projectives. We are therefore obliged to make cofibrancy automatic, and proceed as follows.

We have a family $\mathcal{BI}$ of basic injectives in $\mathbb{A}$; this is required to contain enough injectives that any object of $\mathbb{A}$ embeds in some product of elements of $\mathcal{BI}$, but it is convenient for verifications to choose $\mathcal{BI}$ to as small as possible. Regarding such a basic injective I as a differential graded object with zero differential, we obtain the Eilenberg-MacLane object (or co-sphere) $K(I)$, and the path object (or co-disc) $P(I) = fibre(1 : K(I) \longrightarrow K(I))$.

Clearly $P(I)$ is acyclic in all cases, and we use the collection of maps $P(I) \longrightarrow 0$ as the test acyclic 'fibrations'. It will transpire they are indeed fibrations. We therefore define the class **cof** to consist of maps $i : A \longrightarrow B$ having the LLP with respect to $P(I) \longrightarrow 0$ for all basic injectives I. The class **fib** then consists of all maps $p : X \longrightarrow Y$ having the RLP with respect to acyclic cofibrations. It is tautologous that the maps $P(I) \longrightarrow 0$ are indeed fibrations for basic injectives I. Furthermore, it is immediate from the definition that the class **fib** is closed under base extension, and thus in particular projections $Y \times P(I) \longrightarrow Y$ are also fibrations.

The main result is that this procedure gives dg$\mathbb{A}$ the structure of a closed model category in each case. In practice, the proof proceeds as follows; we have highlighted the steps requiring properties of the particular categories in question. These follow by an identification of the group of dg maps into $P(I)$ and $K(I)$ for basic injectives.

Step 1: Identify the cofibrations explicitly as maps $i : A \longrightarrow B$ for which i is a termwise monomorphism.
Step 2: Show that for any X there is a termwise monomorphism $\alpha : X \longrightarrow P(I)$ for some injective I.

Since we have chosen $\mathcal{BI}$ to include enough injectives, this is easy to prove, and the axiom CM5 (i) follows quickly. Indeed for any $f : X \longrightarrow Y$ we factorize it as

$$X \stackrel{\{f,\alpha\}}{\longrightarrow} Y \times P(I) \stackrel{\text{proj}}{\longrightarrow} Y.$$

Step 3: Prove that the maps $P(I) \longrightarrow K(I)$ are fibrations for all basic injectives I.

Note that any injective is a retract of a product of basic injectives, so it follows that $P(I) \longrightarrow K(I)$ is a fibration for any injective I.

Step 4: Prove the fibrant approximation axiom CM5 (ii) using only well understood fibrations.

More precisely, given $f : X \longrightarrow Y$ we form a factorization $X \longrightarrow Y' \longrightarrow Y$ with $X \longrightarrow Y'$ a homology isomorphism and $Y' \longrightarrow Y$ a fibration formed by iterated pullback of fibrations $P(I) \longrightarrow K(I)$. This is precisely dual to the usual argument attaching cells to make a map of spaces into a weak homotopy equivalence, but because the dual of the small object argument does not apply, we use the finiteness of injective dimension in $\mathbb{A}$ to see that only finitely many steps are involved in this process. The map $X \longrightarrow Y'$ can be made into a cofibration by using an acyclic as in the proof of CM5 (ii). We now see using the defining RLP, that an arbitrary fibration is a retract of such a standard one.

We will examine Case 0, the category of torsion $k[c]$-modules, in detail. It will make the structure of the argument clear, and there is an important point to be made about the use of torsion functors. We let $\mathcal{O} = k[c]$ and we are particularly concerned with the case $k = \mathbb{Q}$ and $c = c_H$.

The set $\mathcal{BI}$ of basic injectives consists of suspensions of the standard torsion injective $\mathbb{I} = \mathrm{Hom}_k(\mathcal{O}, k) = \Sigma^{-2}k[c, c^{-1}]/k[c]$. The first lemma is elementary, and is probably clearer if we allow an arbitrary graded vector space V of coefficients:

$$\mathbb{I}(V) = \mathrm{Hom}_k(\mathcal{O}, V).$$

LEMMA B.1. *For a graded vector space V, the object $\mathbb{I}(V)$ is injective, and we have equalities*
(i) $\mathrm{Hom}_{\mathcal{O}}(R, K(\mathbb{I}(V))) = \mathrm{Hom}_k(R, V)$
(ii) $\mathrm{dgHom}_{\mathcal{O}}(R, K(\mathbb{I}(V))) = \mathrm{Hom}_k(R/dR, V)$
(iii) $\mathrm{dgHom}_{\mathcal{O}}(R, P(\mathbb{I}(V))) = \mathrm{Hom}_k(\Sigma R, V)$. □

We warn that $\mathbb{I}(V)$ is only a torsion module if V is bounded below.

LEMMA B.2. *A map $i : A \longrightarrow B$ is a cofibration if and only if it is monomorphic in each degree.*

Proof: From the previous lemma we see that the problem

$$\begin{array}{ccc} A & \xrightarrow{f} & P(\Sigma^{n+1}\mathbb{I}(k)) \\ {\scriptstyle i}\downarrow & \nearrow & \downarrow{\scriptstyle p} \\ B & \longrightarrow & 0 \end{array}$$

in the category of dg $\mathcal{O}$-modules is equivalent to the problem

$$\begin{array}{ccc} A_n & \xrightarrow{f} & k \\ {\scriptstyle i}\downarrow & \nearrow & \downarrow{\scriptstyle p} \\ B_n & \longrightarrow & 0 \end{array}$$

in the category of k-vector spaces. Therefore $i : A \longrightarrow B$ has the LLP for $P(\Sigma^{n+1}\mathbb{I}(k)) \longrightarrow 0$ if and only if i is monomorphic in degree n. □

LEMMA B.3. *For any graded vector space V, the maps and $P(\mathbb{I}(V)) \longrightarrow K(\mathbb{I}(V))$ are fibrations.*

Proof: We must show that $P(\mathbb{I}(V)) \longrightarrow K(\mathbb{I}(V))$ has the RLP for all termwise monomorphisms $i : A \longrightarrow B$ which are homology isomorphisms. Using the first lemma, we translate the lifting problem about dg $\mathcal{O}$-maps into a problem about k-maps into V. We convert the problem of dg $\mathcal{O}$-modules

$$\begin{array}{ccc} A & \xrightarrow{f} & P(\mathbb{I}(V)) \\ {\scriptstyle i}\downarrow & \nearrow & \downarrow{\scriptstyle p} \\ B & \xrightarrow[g]{} & K(\mathbb{I}(V)) \end{array}$$

into the problem

$$\begin{array}{ccc} A & \xrightarrow{f'} & \Sigma^{-1}V \\ {\scriptstyle i}\downarrow & \nearrow & \uparrow{\scriptstyle \overline{g}'} \\ B & \xleftarrow[\overline{d}]{} & \Sigma^{-1}B/dB \end{array}$$

of k-maps, with the extra fact that i is a dg $\mathcal{O}$-map. Regarding A as a sub-graded vector space of B we see how to define the diagonal on A, and on dB we must take $db \longmapsto g'(b)$. The fact that i is a homology isomorphism can be used to check this is well defined. $\square$

Now, as remarked above, the modules $\mathbb{I}(V)$ will not be torsion modules if V is not bounded below. However the c-power torsion functor Γ_c is right adjoint to the full and faithful inclusion i : tors $\mathcal{O}$-mod $\longrightarrow$ $\mathcal{O}$-mod of the category of torsion $\mathcal{O}$-modules in the category of all $\mathcal{O}$-modules. Therefore, for arbitrary graded vector spaces V, the map $\Gamma_c P(\mathbb{I}(V)) \longrightarrow \Gamma_c K(\mathbb{I}(V))$ is a fibration in the category of dg torsion $\mathcal{O}$-modules. It also follows that $\Gamma_c P(\mathbb{I}(V)) \longrightarrow 0$ is a fibration, and since $\Gamma_c P(\mathbb{I}(V))$ is the fibre of the identity map of $\Gamma_c K(\mathbb{I}(V))$, it is also acyclic.

It is now clear that any X admits an embedding into an acyclic fibrant object, namely $X \longrightarrow P(\mathbb{I}(X))$, and hence CM5 (i) follows. It remains to establish fibrant approximation, and we suppose given $f : X \longrightarrow Y$, with $H_*(X) = M$, $H_*(Y) = N$. We shall construct maps $X \longrightarrow Y_3 \longrightarrow Y_2 \longrightarrow Y_1 \longrightarrow Y_0 = Y$, with $X \longrightarrow Y_3 \longrightarrow Y$ the required factorization. To construct Y_{n+1} from Y_n we form a pullback diagram

$$\begin{array}{cccc} X & & & \\ & \searrow^{f_{n+1}} & \searrow^{u_n} & \\ {\scriptstyle f_n}\searrow & Y_{n+1} & \longrightarrow & P(I_n) \\ & \downarrow & & \downarrow \\ & Y_n & \xrightarrow{\beta_n} & K(I_n). \end{array}$$

This corresponds to a short exact sequence $0 \longrightarrow Y_{n+1} \longrightarrow Y_n \oplus P(I_n) \longrightarrow K(I_n) \longrightarrow 0$ and hence, since $P(I)$ is acyclic, it gives a long exact sequence in homology

$$\cdots \longrightarrow \Sigma^{-1} I_n \longrightarrow H_*(Y_{n+1}) \longrightarrow H_*(Y_n) \xrightarrow{(\beta_n)_*} I_n \longrightarrow \cdots .$$

At each stage we have to specify maps u_n and β_n making the diagram commute, and this gives $f_{n+1} : X \longrightarrow Y_{n+1}$. Equivalently, we need to specify k-maps u'_n and β'_n so that the diagram

$$\begin{array}{ccc} X & \xrightarrow{d} & \Sigma X \\ f_n \downarrow & & \downarrow u'_n \\ Y_n & \xrightarrow{\beta'_n} & V_n \end{array}$$

commutes, where $I_n = \mathbb{I}(V_n)$. We shall usually begin by choosing β'_n so that β_n has the desired effect on homology. Provided $\beta'_n f_n$ is zero on cycles, it induces a map $X/ker(d) \longrightarrow V_n$, and we may extend this to a k-map on ΣX without obstruction. At the stage $n = 0$ we must take care about u_0, but u_1 and u_2 are arbitrary extensions.

Choose an embedding $\eta : M \longrightarrow \Sigma^{-1} I_0$ of M in an injective. First take $\beta_0 = 0$ and u_0 so that it induces η on homology. For example, we may choose a complement Z' to the cyles and let u_0 be the composite $X \longrightarrow X/(Z' + dX) \cong H_*(X) \longrightarrow \Sigma^{-1} I_0$. Thus $N_1 = N \oplus I_0$ and $f_1 : X \longrightarrow Y_1$ induces $\{f_*, \eta\} : M \longrightarrow N_1 = N \oplus I_0$. Let C_1 be the cokernel, and choose an injective resolution

$$0 \longrightarrow C_1 \longrightarrow I_1 \longrightarrow \Sigma I_2 \longrightarrow 0.$$

We next choose the map β_1 so that it induces the composite $N_1 \longrightarrow C_1 \longrightarrow I_1$: again we may choose a complement Z'_1 to the cycles in Y to give a k-map $Y_1 \longrightarrow H_*(Y_1) = N_1$, and the resulting map clearly vanishes on cycles of X so that u_1 exists. Now $(\beta_1)_*$ has kernel M and cokernel ΣI_2. We therefore have a short exact sequence

$$0 \longrightarrow I_2 \longrightarrow N_2 \longrightarrow M \longrightarrow 0,$$

which necessarily splits, since I_2 is injective.

Finally, choose the map β_2 to kill the extraneous I_2, just as we chose β_1. The existence of u_2 is again automatic. □

We make a few brief remarks about the remaining cases.

Case 2: For the standard model the procedure is precisely similar. The basic injectives are suspensions of $(\mathbb{I}_H \longrightarrow 0)$, and suspensions of $(t^{\mathcal{F}}_* \longrightarrow t^{\mathcal{F}}_*)$ (see Section 5.5). From this all is automatic once we have observed that the basic injectives lie in the image of the inclusions $e : k\text{-mod} \longrightarrow \mathcal{A}$ and $f : \text{tors-}\mathcal{O}^f_{\mathcal{F}}\text{-mod} \longrightarrow \mathcal{A}$, which are left adjoint to the vertex and nub functors 5.4.3.

Case 3: For the torsion model the procedure is precisely similar. The basic injectives are suspensions of $(t^{\mathcal{F}}_* \otimes \text{Hom}(t^{\mathcal{F}}_*, \mathbb{I}_H) \longrightarrow \mathbb{I}_H)$, and suspensions of $(t^{\mathcal{F}}_* \longrightarrow 0)$ (see Section 6.3). From this all is automatic once we have observed that the basic injectives lie in the image of the inclusions $e_t : k\text{-mod} \longrightarrow \mathcal{A}$ and $f_t : \text{tors-}\mathcal{O}^f_{\mathcal{F}}\text{-mod} \longrightarrow \mathcal{A}$, which are left adjoint to the vertex and torsion functors 6.2.3.

We are frequently concerned with Quillen closed model categories of dg objects in abelian categories, and in functors between them which arise as functors between the associated abelian categories. If $\mathbb{A}$ and $\mathbb{B}$ are abelian categories and $F : \mathbb{A} \longrightarrow \mathbb{B}$ is a functor, we obtain an induced functor on dg categories, which we continue to denote F. We shall often have cause to discuss total right derived functors $RF : D\mathbb{A} \longrightarrow D\mathbb{B}$; the first observation is that it always exists.

LEMMA B.4. *Any functor $F : \mathbb{A} \longrightarrow \mathbb{B}$ induces a functor on dg categories which has a total right derived functor $RF : D\mathbb{A} \longrightarrow D\mathbb{B}$. Any fibrant approximation $X \longrightarrow \hat{X}$ induces an equivalence $RF(X) \simeq F(\hat{X})$. If F is exact then $RF(X) = F(X)$ for any X.*

Proof: Note that our categories have the property that weak equivalences between fibrant objects are homotopy equivalences. Since the functor F arises from the level of abelian categories, it preserves homotopies and hence homotopy equivalences; this is sufficient by [**22**, p I.4.4]. The final sentence is clear. □

However we only obtain really satisfactory behaviour (such as the preservation of composites) if the functor takes fibrations to fibrations. Note that when F is the right adjoint to an exact functor, it satisfies Conditions (i) and (ii) in the following.

LEMMA B.5. *Suppose $F : \mathbb{A} \longrightarrow \mathbb{B}$ is a functor so that*
(i) F preserves injectives,
(ii) F preserves limits.
Then with closed model structures defined by homology isomorphisms and basic injectives as above, F preserves fibrations.

Proof: Note first that the functor on dg categories preserves mapping fibres. It follows that $F(P(I)) = P(F(I))$, and hence by (i), that F preserves fibrations $P(I) \longrightarrow K(I)$ with I injective. Since any fibration is a retract of one formed from these by a finite number of pullbacks, it follows from (ii) that F preserves fibrations. □

We remark that it then follows that, if $F' : \mathbb{B} \longrightarrow \mathbb{C}$ is a second functor satisfying Conditions (i) and (ii), then $R(F'F)$ exists and $R(F' \circ F) = RF' \circ RF$.

COROLLARY B.6. *If $F : \mathbb{A} \longrightarrow \mathbb{B}$ is right adjoint to the exact functor G, RF is right adjoint to G.*

Proof: The exactness of G ensures that it preserves cofibrations and weak equivalences, and hence that $LG = G$. Now the fact that RF is right adjoint to G follows from B.5 by [**22**, I.4.5]. □

APPENDIX C

Conventions.

C.1. Conventions for spaces and spectra.

Quite generally, when we work in several categories, we should have notation for all functors if all ambiguity is to be avoided. However, in a natural attempt to aid readability and understanding, it is common to omit notation for certain functors, particularly forgetful functors. Once this is done, notation must be chosen in the two categories so that it respects this convention. Unfortunately, this has not been done for the category of G-spectra: broadly speaking there are two notational conventions. Both give rise to formal ambiguity, although in practice authors take trouble to ensure the correct interpretation is used at all places. Each has advantages and disadvantages, but it seems worth making them explicit, together with means for resolving these difficulties. We also explain the particular convention we have found convenient to use here.

The Adams convention is based on the idea that a spectrum is a generalized space, so that if Z is a space, we should continue to write Z for its suspension spectrum, leaving context to determine which is intended. This convention is very convenient when working on the homotopy level, and is basically the one adopted here. However, it leads to difficulties if the same symbol θ is adopted for functors on spaces and spectra unless they agree under passage to suspension spectra. We need the relation

$$\Sigma^{\infty}\theta(Z) = \theta(\Sigma^{\infty} Z)$$

if the notation $\theta(Z)$ is to be unambiguous. In particular, under the Adams convention, it is inconsistent to use $(\cdot)^H$ for both the fixed point space functor and for the Lewis-May fixed point functor. Because of its historical precedence we use Z^H for the fixed point space functor, and $\Psi^H X$ for the Lewis-May fixed point functor. To avoid confusion we use the Lewis-May notation $\Phi^H X$ for the geometric fixed point functor, rather than X^H as would be allowed by the Adams convention.

The May convention is based on the idea that notation for the suspension spectrum functor should not be omitted. If fanatically implemented this convention leads to total clarity. However there are few fanatical enough to write $\Sigma^{\infty} S^n$ for the n-sphere spectrum. The symbol $(S^1)^G$ is thus formally ambiguous, although the probable meaning is clear enough.

There is a second choice of convention which has to be made. In this case both conventions are consistent, but the reader must be aware of which convention is in force.

The first convention is that functors forgetting equivariance should be omitted. Thus, if X is a G-spectrum, and H is a subgroup of G then we should continue to write X for the corresponding H-spectrum. This is in accordance with usual practice for spaces, and consistent with the suspension spectrum functor. Thus, for a G-spectrum X, we may write

$$X^G_* = \pi^G_*(X) = [S^0, X]^G_*,$$

and

$$X^H_* = \pi^H_*(X) = [G/H_+, X]^G_* = [S^0, X]^H_*.$$

However the notation does conflict with certain accepted usages.

The second convention is to display the group of equivariance as a subscript at all times. Thus we would write X_G for a G-spectrum, and we might write X_H for an H-spectrum without any implication that X_H was X_G regarded as an H-spectrum. This has the advantage of being explicit, but the above displays become

$$(X_G)^G_* = \pi^G_*(X_G) = [(S^0)_G, X_G]^G_*,$$

and

$$(X_G)^H_* = \pi^H_*(X_G) = [(G/H_+)_G, X_G]^G_* = [(S^0)_H, res^G_H(X_G)]^H_*.$$

For ordinary purposes we find these too unwieldy, but there are occasions when it is important to be careful. We note also that this convention leads to difficulties in the case of equivariant K-theory with reality in the sense of Atiyah.

Finally, we suppress mention of universes wherever possible, on the grounds that there is usually a natural choice of complete universe for the ambient group of equivariance. This has implications for certain notation for change of groups. Thus, if N is a normal subgroup of a group G, there is a natural way of regarding a G/N-spectrum as a G-spectrum. With universes explicit, we suppose given a complete G-universe $\mathcal{U}$, and that our G/N-spectra are indexed on $\mathcal{U}^N$; we let $q : G \longrightarrow G/N$ denote the quotient and $j : \mathcal{U}^N \longrightarrow \mathcal{U}$ denote the inclusion. Properly speaking, we have a functor q^*, regarding a G/N-spectrum Y indexed on $\mathcal{U}^N$ as a G-spectrum q^*Y indexed on $\mathcal{U}^N$, and a change of universe functor j_* taking a G-spectrum Z indexed on $\mathcal{U}^N$ and building in stability under the representations not fixed by N to obtain a G-spectrum j_*Z indexed on $\mathcal{U}$. For work on the homotopy level, it is the composite functor $Y \longmapsto j_*q^*Y$ that is the important one. This was originally denoted $q^\# Y$ in [**19**], but it is now more common to abbreviate it to j_*Y, often without explaining what j is. Since we want to concentrate on the groups involved, and let universes take care of themselves, we write

$$\mathrm{inf}^G_{G/N} : G/N - \text{spectra} \longrightarrow G - \text{spectra}$$

for this composite, and refer to it as the inflation functor. However, since the inflation of the suspension spectrum of a G/N-space Y is the suspension spectrum of the same space Y regarded as a G-space under pullback, we prefer to omit notation for the inflation functor whenever possible, in accordance with the Adams convention. This is a dangerous convention, and should only be used with care. In particular, it must not be used when it conflicts with the convention omitting the forgetful functor. For example, the inflation of the non-equivariant spectrum representing complex K-theory is different from the equivariant K-theory spectrum. In any case, we always suppress inflation in an expression of form $\tilde{E}\mathcal{F} \wedge \Phi^{\mathbb{T}} X$; not only is the meaning clear, but also the non-fixed representations are essentially built in by the smash product, since $S^V \wedge \tilde{E}\mathcal{F} \simeq \tilde{E}\mathcal{F}$ at the space level when $V^{\mathbb{T}} = 0$.

C.2. Standing conventions.

The principal convention is that all spectra and modules are rationalized without comment. We also work at the homotopy level with topological objects unless indicated otherwise.

We regard it as natural to work equivariantly at all times, and hence we often omit the prefix '$\mathbb{T}$-' unless required for special emphasis: thus 'homotopy group' unmodified means 'equivariant homotopy group', 'equivalence' unmodified means 'equivariant equivalence' and so forth. When we wish to refer to non-equivariant notions we shall say so explicitly. Similarly, the term 'subgroup' refers exclusively to closed subgroups. The identity element of a group is usually written e, and a multiplicative group with one element is written 1.

We have also used homology grading throughout: in particular the cohomology of $B\mathbb{T}$ (arising as $[E\mathbb{T}_+, E\mathbb{T}_+]^{\mathbb{T}}_*$ and $\pi^{\mathbb{T}}_*(DE\mathbb{T}_+)$) is in negative degrees, and Chern classes are in degree -2. Of course, we also use the homology suspension: $(\Sigma^t M)_n = M_{n-t}$

There are several cases where a notational decision had to be made about suspensions. The problem is that notational simplicity in the topological and algebraic contexts suggest different decisions. The standard injective $\mathbb{Q}[c_H]$-module $\mathbb{I}_H$ has its lowest non-zero group in degree 0. We were directed by mnemonic appeal, together with the fact that $\mathbb{I}_H = \mathrm{Hom}(\mathcal{O}_{\mathcal{F}}, \mathbb{Q}_H)$ and that it is the injective envelope of $\mathbb{Q}$. This convention then extends to other standard injectives. However this means one must be careful about suspensions in certain places. Thus we have a short exact sequence $0 \longrightarrow \mathbb{Q}[c_H] \longrightarrow \mathbb{Q}[c_H, c_H^{-1}] \longrightarrow \Sigma^2\mathbb{I}_H \longrightarrow 0$, and we have $\pi^{\mathbb{T}}_*(E\mathcal{F}_+) = \Sigma\mathbb{I}$.

A thornier issue was whether we should always write $t^{\mathcal{F}}_* \otimes V \longrightarrow \Sigma T$ for an object of the torsion model, so that the torsion part in the standard model, would become prominent. The author decided against this on the grounds of brevity, and also because the the convention is easily forgotten, and is thereby liable to lead to confusion. Thus we write $t^{\mathcal{F}}_* \otimes V \longrightarrow T$ for an object of the torsion model, and the torsion part of the corresponding object of the standard model is $\Sigma^{-1}T$.

A Mackey functor M is a functor on the stable orbit category $h\mathcal{SO}$ with objects the natural cells $\mathbb{T}/H_+$. We write $M(H)$ for the value of this functor on $\mathbb{T}/H_+$. However M determines a functor on the category of retracts of objects of $h\mathcal{SO}$, and we let M^e denote the restriction of this to the category $h\mathcal{SB}$ of basic cells σ^0_H. We write $M^e(H)$ for the value of this functor on σ^0_H.

We abbreviate 'differential graded' to 'dg' with few exceptions. Given an abelian category $\mathbb{A}$ we have referred to the category of fractions in which the homology isomorphisms of dg$\mathbb{A}$ are inverted as the derived category $D\mathbb{A}$ of $\mathbb{A}$, rather than as the homotopy category of dg$\mathbb{A}$. This is to avoid confusion with the naive meaning of homotopy and to emphasize its algebraic nature: it should lead to no confusion since it would be perverse to form the category of chain complexes of $\mathbb{A}$ if $\mathbb{A}$ is graded. However, after initial work with sets of naive homotopy classes, we follow the convention common in topology of using the symbol $[X, Y]$ to mean morphisms in the derived category (i.e. homotopy classes of maps from a cofibrant approximation of X to a fibrant approximation of Y).

APPENDIX D

Indices.

D.1. Index of definitions and terminology.

A

Accessible	7.3	Constructible without indeterminacy
Adams spectral sequence		Spectral sequence recovering a triangulated category from a functor to an abelian category.

B

Based map	5.4	A map of based modules with an implicit map of vertices compatible with basing maps.
Based module	5.4	A module N with an implicit vertex and basing map.
Basic injectives	4.1	A collection of injectives so that an arbitrary module embeds in a product of suspensions of basic injectives.
Basic sphere	2.1	The principal part $\sigma^n_H := e_H\mathbb{T}/H_+ \wedge S^n$ of a natural sphere.
Basing map	5.4	A map $N \longrightarrow t^{\mathcal{F}}_* \otimes V$ which becomes an isomorphism when $\mathcal{E}$ is inverted.
Brown-Comenetz	12.4	Cohomology defined by $hI^n_G(X) = \mathrm{Hom}(\underline{\pi}^{\mathbb{T}}_n(X), I)$ for an injective Mackey functor I; the representing spectrum hI.
Burnside ring		Ring of stable self maps of S^0; rationally, the ring $C(\Phi G, \mathbb{Q})$ where ΦG is the space of conjugacy classes of subgroups with finite index in their normalizer.

C

Co-Eilenberg-MacLane		Spectrum JN representing ordinary homology with coefficients in the coMackey functor N for some N (see 12.3).
Cofibrant	B	Defined by lifting property, but all objects in our algebraic categories are cofibrant.
Cofibration	B	Defined by lifting property, but all dimensionwise injective maps in our algebraic categories are cofibrations.
CoMackey	12.3	A covariant additive functor on the stable orbit category.
Complete G-universe		A countable dimensional orthogonal representation of G containing all finite dimensional representations of G.

Completion	5.10	(1) (of an $\mathcal{O}_{\mathcal{F}}$-module) the map $M \longrightarrow \prod_H e_H M$ is the $\mathcal{F}$-completion. (2) I-adic completion for some ideal I.
Core	A	For a Mackey functor M: the kernel of the total restriction map $M(\mathbb{T}) \longrightarrow \prod_H M(H)$.
Cyclotomic	15.1	Spectrum equivalent to its geometric K-fixed points for all finite subgroups K.

D

Deflation	9.1	Pullback along $q^* : \overline{\mathcal{O}}_{\overline{\mathcal{F}}} \longrightarrow \mathcal{O}_{\mathcal{F}}$; left and right adjoint to inflation.
d-invariant		The homomorphism induced by a map in the homology theory under discussion.
Derived category	4.1	Category of dg objects with homology isomorphisms inverted.

E

Eilenberg-MacLane		Spectrum HM representing ordinary cohomology with coefficients in the Mackey functor M for some M.
e-invariant		The extension class defined by the cofibre of a map with zero d-invariant in the homology theory under discussion.
e-restriction	A	A structure map $M^e(\mathbb{T}) \longrightarrow M^e(H)$ for a Mackey functor M.
Euler classes		(1) The pullback of the identity along a map $e(V) : S^0 \longrightarrow S^V$ shifted into integer degrees.
	4.6	(2) An element of the form $c_\phi = e_\phi c + (1 - e_\phi)$ for a finite set ϕ of finite subgroups, or a finite product of such elements.
Euler-torsion		Module annihilated by inverting all Euler classes.

F

Family		A collection of subgroups closed under passage to subgroups and conjugates.
$\mathcal{F}$-contractible		H-contractible for all finite H; equivalent to an object of form $\tilde{E}\mathcal{F} \wedge Z$.
$\mathcal{F}$-finite		$\mathcal{O}_{\mathcal{F}}$-module M so that $M = \oplus_H e_H M$.
$\mathcal{F}$-free		Having isotropy groups in $\mathcal{F}$
Fibrant	B	Defined by lifting properties, but in our algebraic categories, constructed from injectives with zero differential.
Fibration	B	Defined by lifting properties, but in our algebraic categories, constructed by coattaching injectives with zero differential.
Fixed		A space on which all group elements act as the identity.
$\mathcal{F}$-space		Having isotropy groups in $\mathcal{F}$

$\mathcal{F}$-injective	A	An injective Mackey functor with $V(\mathbb{T}) \cong \prod_H V(H)$.
$\tilde{\mathcal{F}}$-injective	A	An injective Mackey functor concentrated at $\mathbb{T}$.
Formal	6.2	An object of the torsion model determined up to homotopy equivalence by its homology.
Free		Having 1 as the only isotropy group.
Fundamental class		The generator of the highest class of the nub of a wide sphere.

G

Geometric fixed points		The functor Φ^K extending the K-fixed point functor on G-spaces.

H

Hausdorff	5.10	An $\mathcal{O}_{\mathcal{F}}$-module M which embeds in $\prod_H M(H)$
Homotopy fixed points		(of a based G-space X) the space X^{hG} of G-maps G-map(EG_+, X); also the spectrum-level counterpart.
Homotopy quotient		(of a based G-space X) the space $X_{hG} = EG_+ \wedge_G X$; also the spectrum-level counterpart.
$\mathcal{H}$-semifree		Having isotropy in $\mathcal{H} \cup \{\mathbb{T}\}$.

I

Inflation functor	9.1	A functor from a category associated to a quotient group G/N to that for the group G.
Internal Hom	22.1	Right adjoint to tensor product.
Isotropy group		A group H for which $\Phi^H X$ is non-equivariantly essential.

J

K

L

Left derived functor		(of a functor) The left-universal counterpart of the functor at the derived category level (Appendix B).
Lewis-May fixed points		Right adjoint Ψ^K to the inflation functor.

M

Mackey functor	A	Contravariant additive functor on the stable orbit category.

Moore spectrum		Bounded below spectrum with Burnside-homology in exactly one dimension.

N

Natural sphere	2.1	Sphere $G/H_+ \wedge S^n$ used to build CW-objects.
Nub	5.4	The $\mathcal{O}_{\mathcal{F}}$-module in an object $N \xrightarrow{\beta} t_*^{\mathcal{F}} \otimes V$ of the standard model.

O

Object-accessible	7.3	(of a functor on categories of spectra) all objects may be constructed from primitive objects so that the models of objects and functions are known precisely throughout.
Over H		(of a $\mathbb{T}$-spectrum X) Having $\Phi^K X$ non-equivariantly contractible if $K \neq H$.

P

Pure parity		A dg module whose homology is only non-zero in one parity; a spectrum whose standard model has this property.

Q

R

Right derived functor		(of a functor) The right-universal counterpart of the functor at the derived category level (Appendix B).

S

Semifree		Having isotropy groups 1 and $\mathbb{T}$ only.
Sphere		See 'basic sphere', 'natural sphere' and 'wide sphere'.
Spike length	19.2	Measure of twisting in a wide sphere.
Split	4.2	(1) Split one dimensional: a one dimensional abelian category in which objects split into even and odd parts.
	4.2	(2) Split linear: a triangulated category with a suitable linearization functor to a split one dimensional abelian category.
Standard model	5.4	Category with objects $\mathcal{O}_{\mathcal{F}}$-maps $N \longrightarrow t_*^{\mathcal{F}} \otimes V$ which become isomorphic when Euler classes are inverted.

T

Tate cohomology	14.2	The cohomology represented by $t(k) = F(EG_+, k) \wedge \tilde{E}G$
Torsion		(1) A module annihilated by the inversion of all Euler classes.
	5.4	(2) An object of the form $f(T) = (T \longrightarrow 0)$ for a torsion module T.
	6.2	(3) An object of the form $f_t(T) = (t^{\mathcal{F}}_* \otimes \mathrm{Hom}(t^{\mathcal{F}}_*, T) \longrightarrow T)$ for a torsion module T.
Torsion free	5.4	(1) A module of the form $e(V) = (t^{\mathcal{F}}_* \otimes V \longrightarrow t^{\mathcal{F}}_* \otimes V)$.
	6.2	(2) A module of the form $e_t(V) = (t^{\mathcal{F}}_* \otimes V \longrightarrow 0)$.
Torsion model	6.2	The algebraic counterpart of viewing $\mathbb{T}$-spectra as fibres of $q_X : \tilde{E}\mathcal{F} \wedge X \longrightarrow \Sigma E\mathcal{F}_+ \wedge X$; objects $(t^{\mathcal{F}}_* \otimes V \longrightarrow T)$ with T a torsion module.
Type	19.1	The shape of the flag associated to a small monomorphic object in the semifree standard model.

U

Universe		Countable dimensional orthogonal G-space containing its finite dimensional subspaces infinitely often. See 'complete universe'.

V

Vertex	5.4	The graded vector space V in an object $N \stackrel{\beta}{\longrightarrow} t^{\mathcal{F}}_* \otimes V$ of the standard model.

W

Weak equivalence		(1) A map inducing an isomorphism of $[K, \cdot]^G$ for all finite complexes K.
	B	(2) Element of structural class for a Quillen closed model category structure.
Wide spheres	19.2	Special small semifree object.
	23.3	Special small object in standard model.
Width	19.2	Measure of size of a wide sphere.

X

Y,Z

D.2. Index of notation.

c_H	2.4	The part of the Euler class of $V(H)$ lying over H: $c_H = e_H c = e_H \chi(V(H))$.
c_ϕ	4.6	The Euler class $e_\phi c + (1 - e_\phi)$ associated to a finite subset $\phi \subseteq \mathcal{F}$.
c^v	4.6	The non-homogeneous element of $t^{\mathcal{F}}_*$ which is $c_H^{v(H)}$ at H.
c_0	15.1	The degree -2 generator of the ring $\mathbb{Q}[c_0]$ of cyclotomic operations.
$C(M)$	A	The core of a Mackey functor over $\mathbb{T}$: the kernel of the restriction map $M(\mathbb{T}) \longrightarrow \prod_H M(H)$.
C_M	A	The cokernel of the embedding $M \longrightarrow I_M$ of a Mackey functor in the canonical resolution.
$C\mathcal{F}[U]$	12.4	The injective Mackey functor constant at U for all finite subgroups, and $U^{\mathcal{F}}$ at $\mathbb{T}$.
$\chi(V)$	4.6	The integer graded Euler class of the representation V
$C(\mathcal{K}, \mathbb{Q})$		Ring of continuous $\mathbb{Q}$-valued functions on the (usually discrete) space $\mathcal{K}$.
cof	B	The class of cofibrations in a Quillen closed model category.
cof	6.5	The cofibre functor $dg\mathcal{A} \longrightarrow dgH\mathcal{A}_t$ from the dg standard model to the dg homology torsion model.
cof'	6.5	The cofibre functor $dgH\mathcal{A} \longrightarrow dgH\mathcal{A}_t$ from the dg homology standard model to the homology torsion model.

D

d	2.1	Same as c^{-1} in $R[c, c^{-1}]/R[c]$.
$d(f)$		The d-invariant (ie induced map) of f.
$\mathbb{D}$		Generic notation for a category.
δ_H	13.2	$(= pr^H_{\mathbb{T}} tr^{\mathbb{T}}_H)$ The standard generator of $[S^0, S^0]^{\mathbb{T}}_1$.
$D(\cdot)$	4.1	Operation of forming derived category.
def	9.1	The deflation functor $\mathcal{O}_{\mathcal{F}} \otimes_{\overline{\mathcal{O}}_{\mathcal{F}}} : \mathcal{O}_{\mathcal{F}}$-mod $\longrightarrow \overline{\mathcal{O}}_{\mathcal{F}}$-mod.
dg		Differential graded.
$\Delta^K X$	10.5	The free injective summand of the Lewis-May fixed points of X.
dgHom		Differential graded homomorphisms.
$dg\mathbb{C}$		Differential graded objects of $\mathbb{C}$.
$dgH\mathcal{A}$	6.5	Category of dg maps $N \longrightarrow t^{\mathcal{F}}_* \otimes V$, whose homology is an isomorphism when $\mathcal{E}$ is inverted.
$dgH\mathcal{A}_t$	6.5	Category of dg maps $t^{\mathcal{F}}_* \otimes V \longrightarrow T$, where the homology of T is Euler-torsion.
$\mathbf{D}_n$	19.3	The diagram corepresenting the nub of the semifree torsion functor.
$\mathbf{D}_{tv}$	19.3	The diagram corepresenting the extended vertex of the semifree torsion functor.
$\mathbf{D}_v$	19.3	The diagram corepresenting the vertex of the semifree torsion functor.
$\Delta(X)$	4.1	The fibrant approximation to X.
Δ	15.1	The functor $\mathcal{C}_t \longrightarrow \mathcal{A}_t$ corresponding to the inclusion of cyclotomic spectra in all spectra.

Δ_n	19.5	The object representing the nub part of the semifree torsion functor.
Δ_{tv}	19.5	The object representing the extended vertex part of the semifree torsion functor.

E

e		(1) generic notation for an idempotent
	8.4	(2) the idempotent supported at subgroups containing a fixed group K.
	5.4	(3) the functor $e(V) = (t_*^{\mathcal{F}} \otimes V \longrightarrow t_*^{\mathcal{F}} \otimes V)$ from graded vector spaces to the standard model.
e_t		The functor $e_t(V) = (t_*^{\mathcal{F}} \otimes V \longrightarrow 0)$ from the category of graded vector spaces to the torsion model.
e_t'		The functor $e_t'(V) = (V \longrightarrow 0)$ from the category of graded vector spaces to the category $\mathcal{A}_t'$.
$e^{\mathcal{C}}$	15.1	The functor $e^{\mathcal{C}}(V) = (\Sigma^2 \otimes \mathbb{I}_0 \longrightarrow \Sigma^2 \otimes \mathbb{I}_0)$ from the category of graded vector spaces to the cyclotomic standard model.
$e_t^{\mathcal{C}}$	15.1	The functor $e_t^{\mathcal{C}}(V) = (\Sigma^2 \otimes \mathbb{I}_0 \longrightarrow 0)$ from the category of graded vector spaces to the cyclotomic torsion model.
e_H		The primitive idempotent with support H in $\mathcal{O}_{\mathcal{F}}$.
e_ϕ	4.6	The idempotent with support $\phi \subseteq \mathcal{F}$.
$e(V)$	4.6	The inclusion $S^0 \longrightarrow S^V$.
$e(f)$		The e-invariant (ie extension class) of f.
$e\pi_n^H(\cdot)$		Abbreviation of $e_H\pi_n^H(\cdot) = [\sigma_H^n, \cdot]^{\mathbb{T}}$.
$E\mathbb{T}$		The universal free $\mathbb{T}$-space; a subscript '+' denotes a disjoint basepoint.
$E\mathcal{F}$		The universal $\mathcal{F}$-space; a subscript '+' denotes a disjoint basepoint.
$\tilde{E}\mathcal{F}$		The unreduced suspension of $E\mathcal{F}$.
$E\langle H\rangle$	2.3	The part $e_H E\mathcal{F}_+$ of $E\mathcal{F}_+$ over H; mapping cone of $E[\supset H]_+ \longrightarrow E[\supseteq H]_+$.
$\tilde{E}\langle H\rangle$	2.3	The mapping cone of the natural map $E\langle H\rangle \longrightarrow S^0$
$\hat{E}$		The object $(0 \longrightarrow t_*^{\mathcal{F}})$ (*not* in $\mathcal{A}$) corepresenting the vertex functor.
$\mathcal{E}$	4.6	The set $\mathcal{E} = \{c^v \mid v : \mathcal{F} \longrightarrow \mathbb{Z}_{\geq 0} \text{ with finite support}\}$ of Euler classes.

F

$\mathcal{F}$		The family of finite subgroups of $\mathbb{T}$.
$\overline{\mathcal{F}}$		The family of finite subgroups of $\overline{\mathbb{T}} = \mathbb{T}/K$.
$F(f)$		The mapping fibre of f.
$F_H[G, \cdot)$		The coinduction functor from H-spectra to G-spectra; right adjoint to the forgetful map.
f	5.4	The functor $f(T) = (T \longrightarrow 0)$ from the category of torsion $\mathcal{O}_{\mathcal{F}}$-modules to the standard model.

f_t	6.2	The functor $f_t(T) = (t_*^{\mathcal{F}} \otimes \mathrm{Hom}(t_*^{\mathcal{F}}, T) \longrightarrow T)$ from the category of torsion $\mathcal{O}_{\mathcal{F}}$-modules to the torsion model.
f_t'	6.2	The functor $f_t'(T) = (\mathrm{Hom}(t_*^{\mathcal{F}}, T) \longrightarrow \mathrm{Hom}(t_*^{\mathcal{F}}, T))$ from the category of torsion $\mathcal{O}_{\mathcal{F}}$-modules to the category $\mathcal{A}_t'$.
$f^{\mathcal{C}}$	15.1	The functor $f^{\mathcal{C}}(T) = (T \longrightarrow 0)$ from the category of torsion $\mathbb{Q}[c_0]$-modules to the cyclotomic standard model.
$f_t^{\mathcal{C}}$	15.1	The functor $f_t^{\mathcal{C}}(T) = (\Sigma^2\mathbb{I}_0 \otimes \mathrm{Hom}(\Sigma^2\mathbb{I}_0, T) \longrightarrow T)$ from the category of torsion $\mathbb{Q}[c_0]$-modules to the cyclotomic torsion model.
F_L^K	15.4	The inclusion map $\Psi^L X \longrightarrow \Psi^K X$, which corresponds to the Frobenius map in algebraic K-theory.
F_s	15.4	The Frobenius map F_L^K when L/K is of order s.
fib	B	The class of fibrations in a Quillen closed model category.
fib	6.5	The fibre functor $dg\mathcal{A}_t \longrightarrow dg\mathcal{A}$ from the dg standard model to the dg torsion model.
fib'	6.5	The fibre functor $dgH\mathcal{A}_t \longrightarrow dgH\mathcal{A}$ from the dg homology torsion model to the dg homology standard model.

G

G		A compact Lie group.
$\hat{\Gamma}$	22.1	The torsion functor: right adjoint to the inclusion $i : \mathbb{A} \longrightarrow \hat{\mathbb{A}}$.
Γ_c		The c-power torsion functor.
$\Gamma_{\mathcal{E}}$	17.3	The $\mathcal{E}$-torsion functor.
$\hat{\Gamma}'$	19.3	Corepresented description of the semifree torsion functor.
$\hat{\Gamma}''$	19.5	Represented description of the semifree torsion functor.

H

H		Generic notation for a finite subgroup of $\mathbb{T}$.
$\mathcal{H}$		A set of finite subgroups (not necessarily a family).
HM		The Eilenberg-MacLane spectrum with coefficient Mackey functor M.
hM	12.4	The Brown-Comenetz spectrum.
$h\mathcal{SB}$	A	The category of stable maps of basic spheres.
$h\mathcal{SB}_*$	13.2	The category of graded stable maps of basic spheres.
$h\mathcal{SVB}$	A	The category of finite wedges of basic spheres.
$h\mathcal{SO}$	A	The Burnside category of stable maps of natural spheres.
$h\mathcal{SVO}$	A	The category of finite wedges of natural spheres.
$\hat{\mathbf{H}}\mathbf{om}$	22.1	The Hom object in $\hat{\mathbb{A}}$.
$\hat{\mathbf{H}}\mathbf{om}_{\overline{\mathcal{O}}_{\mathcal{F}}}$	24.4	The relative Hom object in $\hat{\mathbb{A}}$.

I

I		Generic notation for injective or image.

$\mathcal{I}$	4.1	The collection of fibrant objects used to construct derived categories of our algebraic categories.
$\mathbb{I}$		(1) The standard injective $\mathcal{F}$-finite $\mathcal{O}_{\mathcal{F}}$-module $\Sigma^{-2}t_*^{\mathcal{F}}/\mathcal{O}_{\mathcal{F}} = \bigoplus_H \mathbb{I}_H$. (2) The standard injective $k[c]$-module $\Sigma^{-2}k[c,c^{-1}]/k[c]$.
$\mathbb{I}(H)$	2.4	The standard injective $\mathbb{Q}[c_H]$-module $\Sigma^{-2}\mathbb{Q}[c_H, c_H^{-1}]/\mathbb{Q}[c_H]$.
$\mathbb{I}_H$		Variant of $\mathbb{I}(H)$.
$\mathbb{I}_0$	15.1	The standard cyclotomic injective $\Sigma^{-2}\mathbb{Q}[c_0, c_0^{-1}]/\mathbb{Q}[c_0]$.
$I(Y)$	3.4	The first stage in an injective Adams resolution $Y \longrightarrow I(Y) \longrightarrow J(Y)$
I_V	A	The canonical injective in which a Mackey functor V embeds.
$I_{\mathcal{F}}$	A	The quotient of the Mackey functor I by the $\tilde{\mathcal{F}}$-injective $L[C(I)]$ on its core.
I^*	12.4	The Mackey functor $\underline{\pi}^{\mathbb{T}}_{-1}(hI)$.
IntHom	22.1	The internal Hom functor.
$\mathbf{IntHom}_{\overline{\mathcal{O}}_{\mathcal{F}}}$		The relative internal Hom functor (see 24.4).
$\inf_{\mathbb{T}/K}^{\mathbb{T}}$	10.1	(1) The inflation functor taking a $\mathbb{T}/K$-spectrum, regarding it as a $\mathbb{T}$-spectrum indexed on a K-fixed universe and then building in the other representations.
	9.1	(2) The algebraic counterpart of the topological inflation functor.

J

J		Generic notation for an injective.
$J(Y)$	3.4	The second stage in an injective Adams resolution $Y \longrightarrow I(Y) \longrightarrow J(Y)$
JN	12.3	The coEilenberg-MacLane spectrum with coefficient coMackey functor N.

K

k		(1) Generic notation for a field.
	17.3	(2) The inclusion $\Gamma\phi M \longrightarrow \phi M$.
K		(1) Generic notation for a finite subgroup of $\mathbb{T}$. (2) Generic notation for the kernel. (3) Generic notation for a cohomology theory. (4) Representing spectrum for complex K-theory.
$K\mathbb{Z}$		The *integral* representing spectrum for complex K-theory
$K(I)$	B	The Eilenberg-MacLane object associated to an injective I.
K_V	A	The kernel of the map $P_V \longrightarrow V$ in the canonical projective resolution of a Mackey functor V.
$\underline{K}_G^q$		The Mackey functor $G/H_+ \longrightarrow K_G^q(G/H_+) = K_H^q$.

L

l	17.3	The inclusion $\phi M \longrightarrow M$.
L		(1) Generic notation for a finite subgroup of $\mathbb{T}$. (2) Generic notation for a module.
$L(\cdot)$	B	Total left derived functor operation.
$L[U]$	A	The $\mathcal{F}$-contractible injective Mackey functor with value U at $\mathbb{T}$.
$\lambda(V)$	13.4	The degree zero K-theory Euler class of the representation V.
ΛX	15.2	The space of free loops on a space X.
L_H	5.9	The algebraic counterpart of the basic cell σ^0_H.
$\Lambda^K X$	10.5	A summand of the Lewis-May fixed points of X.

M

M		(1) Generic notation for a module. (2) Generic notation for a Mackey functor regarded as an $h\mathcal{SO}$-module $G/H_+ \longmapsto M(H)$.
$M(H)$		(1) The summand $e_H M$ of the $\mathcal{O}_{\mathcal{F}}$-module M. (2) The value of the Mackey functor M at H.
M^e	A	The $h\mathcal{SB}$-module associated to the Mackey functor M.
$M^e(H)$	A	The value of the Mackey functor M on the basic cell σ^0_H: $M^e(H) = e_H M(H)$.
$M(\mathcal{F})$		(for a Mackey functor M) the limit $\lim_{\rightarrow H} M(H)$; equal to $\oplus_H M^e(H)$.
$M^\wedge_{\mathcal{F}}$	5.10	The completion $\prod_H M(H)$ of a $\mathcal{O}_{\mathcal{F}}$-module M.
$\mathfrak{M}_H$	8.1	The category of H-Mackey functors.
$\mathfrak{M}^{triv}_H$	8.1	The category of H-Mackey functors M whose associated representations $e_K M(K)$ are trivial.

N

N		(1) Generic notation for a module.
	5.4	(2) Generic notation for the nub of an object in the standard model.
	12.3	(3) Generic notation for a coMackey functor.

O

$\mathcal{O}_{\mathcal{F}}$	4.4	The ring $C(\mathcal{F}, \mathbb{Q})[c]$ of operations for $\mathcal{F}$-spectra.
$\mathcal{O}_{\mathcal{F}}(v)$	5.8	The v-twisted ring of operations.
$\mathcal{O}_{\mathcal{H}}$		The ring $C(\mathcal{H}, \mathbb{Q})[c]$ of operations for $\mathcal{H}$-free spectra.
$\mathcal{O}_H$	2.4	The ring $\mathbb{Q}[c_H] = [E\langle H\rangle, E\langle H\rangle]^{\mathbb{T}}_*$ of operations for $\mathbb{T}$-spectra over H.

P

P		Generic notation for a projective.
P_H	A	The indecomposable projective Mackey functor $e_H\mathbb{A}_H = [\cdot, \sigma_H^0]^{\mathbb{T}}$ concentrated over H.
P_V	A	The canonical projective mapping onto the Mackey functor V.
$P(c^v)$	20.1	The subset $\Sigma^v\mathrm{Hom}(S^v, C) \subseteq P$. Related to $\gamma^{-1}(c^v \otimes W)$.
p	5.6	The equivalence $\mathbb{T}$-Spec $\longrightarrow D\mathcal{A}$, or one of the other equivalences from a topological to an algebraic derived category.
$P(I)$	B	The path object associated to an injective I.
$pr_{\mathbb{T}}^H$	13.2	The canonical map $\sigma_H^0 \longrightarrow \sigma_{\mathbb{T}}^0$.
ϕ	5.4	(1) Generic notation for map of vertices.
	17.3	(2) The functor $\phi M = \oplus_H e_H M$.
Φ^K	8.4	The geometric K-fixed point functor.
Φ_n		The nth cyclotomic polynomial.
$\pi_*^{\mathbb{T}}(X)$		The stable homotopy of X.
$\underline{\pi}_*^{\mathbb{T}}(X)$		The Mackey functor $G/H_+ \longmapsto \pi_*^H(X)$.
$\pi_*^{\mathcal{A}}(X)$	5.6	The $\mathcal{A}$-valued homotopy groups $(\pi_*^{\mathbb{T}}(X \wedge DE\mathcal{F}_+) \longrightarrow \pi_*^{\mathbb{T}}(X \wedge DE\mathcal{F}_+ \wedge \tilde{E}\mathcal{F}))$ of X.
$\pi_*^{\mathcal{A}_t}(X)$	6.6	The $\mathcal{A}_t$-valued homotopy groups $\pi_*^{\mathcal{A}_t}(X) = \left(\pi_*^{\mathbb{T}}(X \wedge DE\mathcal{F}_+ \wedge \tilde{E}\mathcal{F}) \longrightarrow \pi_*^{\mathbb{T}}(X \wedge \Sigma E\mathcal{F}_+)\right)$ of X.
ψ_s		The sth power map $\mathbb{T} \longrightarrow \mathbb{T}$.
Ψ^K	10.1	(1) The Lewis-May K-fixed point functor.
	9.2	(2) The algebraic counterpart of the Lewis-May K-fixed point functor.
Ψ_c^K	10.5	The crude algebraic counterpart of the Lewis-May K-fixed point functor on the standard model.

Q

Q		Generic notation for a projective.
$\mathbb{Q}$		The rational numbers: usually suppressed.
$\mathbb{Q}(H)$		The rationals viewed as a $\mathcal{O}_{\mathcal{F}}$-module via the H-augmentation $\mathcal{O}_{\mathcal{F}} \longrightarrow e_H\mathcal{O}_{\mathcal{F}} \longrightarrow \mathbb{Q}$.
$\mathbb{Q}_H$		Variant of $\mathbb{Q}(H)$.
Q_H	12.3	Standard projective coMackey functor.
q		The quotient homomorphism $\mathbb{T} \longrightarrow \mathbb{T}/K$.
q_*		The function $\mathcal{F} \longrightarrow \overline{\mathcal{F}}$ induced by q.
q^*		The map $\overline{\mathcal{O}}_{\overline{\mathcal{F}}} \longrightarrow \mathcal{O}_{\mathcal{F}}$ induced by the quotient homomorphism q.
q_L	5.9	The map $e(V_L) \longrightarrow f(\Sigma T_L)$.
q_X	5.9	The map $\tilde{E}\mathcal{F} \wedge X \longrightarrow \Sigma E\mathcal{F}_+ \wedge X$.
$\hat{q}_X$	6.1	The map $t_*^{\mathcal{F}} \otimes \pi_*(\Phi^{\mathbb{T}}X) \longrightarrow \pi_*^{\mathbb{T}}(\Sigma E\mathcal{F}_+ \wedge X)$ in the derived category which classifies q_X.

R

R_H	A	The Mackey functor $\underline{\pi}_1^{\mathbb{T}}(\sigma_H^0)$; the non-split extension of P_H over $L[\mathbb{Q}]$.
$R(\cdot)$	B	Total right derived functor operation. In most cases arising here, $RF(X) = F(\hat{X})$ where $X \longrightarrow \hat{X}$ is a fibrant approximation.
ρ_K	15.1	The $\lvert K\rvert$th root isomorphism $\mathbb{T} \stackrel{\cong}{\longrightarrow} \mathbb{T}/K$.
$\rho_K^!$	15.1	The functor from $\mathbb{T}/K$-spectra indexed on $\mathcal{U}^K$ to $\mathbb{T}$-spectra indexed on $\mathcal{U}$ arising from a cyclotomic universe.
r_K	15.1	(1) The cyclotomic structure map $\rho_K^! \Phi^K X \stackrel{\simeq}{\longrightarrow} X$.
	A	(2) The rank function on projective Mackey functors associated to the subgroup K.
R_L^K	15.4	The 'restriction map' $\rho_{K/L}^! \Psi^K X \longrightarrow \Psi^L X$ for a cyclotomic spectrum X.
R_s	15.4	The restriction map R_L^K when L/K is of order s.
$R'\mathrm{Hom}$	23.4	Right adjoint of left derived tensor product; not always the right derived Hom functor.

S

S	4.1	(1) The class of morphisms to be inverted in forming the derived category. (2) The multiplicative set generated by the K-theory Euler classes $1 - z^n$ for $n \geq 1$.
$S^{-1}\mathbb{C}$	4.1	The category of fractions formed from $\mathbb{C}$ by inverting morphisms in the class S.
s	4.1	The fibrant approximation map $X \longrightarrow \Delta X$.
S^V		The one point compactification of the representation V.
$S^{kV(\mathcal{F})}$		The direct limit of $S^{kV(H)}$, equal to $S^0 * S(kV(\mathcal{F}))$.
$S(V)$		The unit sphere in the orthogonal representation V.
$S(kV(\mathcal{F}))$	1.4	The direct limit of $S(kV(H))$.
SS	15.4	The solenoid $\underleftarrow{\lim}(\mathbb{T}, \psi^s)$.
σ_H^n	2.1	The basic n-sphere $e_H(\mathbb{T}/H_+ \wedge S^n)$ over H.
$\sigma^{kV(H)}$	2.3	The part of $S^{kV(H)}$ which is H-semifree.
Σ		Suspension functor.
$S_{n(2)}^{n(1)}(\mathbf{a})$	19.2	Wide sphere in the semifree model.
$S_{\mathbf{c^v} \otimes \mathbf{e}}^{g}(\mathbf{a})$	23.3	Wide sphere in the standard model.

T

T		(1) Generic notation for a torsion module. (2) Multiplicative set generated by the cyclotomic polynomials.
$\mathbb{T}$		The circle group.
$\overline{\mathbb{T}}$		The quotient group $\mathbb{T}/K$.

$t_*^{\mathcal{F}}$	5.2	The twisting module arising as the $\mathcal{F}$-Tate cohomology of S^0 and also as $\mathcal{E}^{-1}\mathcal{O}_{\mathcal{F}}$.
$TC(X)$	15.4	Topological cyclic homology spectrum of a cyclotomic spectrum X.
$THH(F)$	15.2	Topological Hochschild homology of a functor F with smash products.
$TR(X)$	15.4	Topological restriction homology spectrum of a cyclotomic spectrum X.
$\mathbb{T} \ltimes_H (\cdot)$	8.1	The induction functor from H-spectra to $\mathbb{T}$-spectra; left adjoint to the forgetful functor.
$\mathbb{T}/\!\!/\mathcal{F}_+$	1.4	The 'cell with isotropy $\mathcal{F}$'.
$\mathbb{T}$-Spec		The category of rational $\mathbb{T}$-spectra.
$\mathbb{T}$-Spec$_{sf}$		The category of semifree rational $\mathbb{T}$-spectra.
$\mathbb{T}$-Spec$/H$		The category of rational $\mathbb{T}$-spectra over H.
$\mathbb{T}$-Spec$/\mathcal{F}$		The category of rational $\mathbb{T}$-spectra over $\mathcal{F}$.
tors-$\mathcal{O}_{\mathcal{F}}^f$-mod		The category of $\mathcal{F}$-finite torsion $\mathcal{O}_{\mathcal{F}}$-modules.
$tr_H^{\mathbb{T}}$	13.2	The transfer map $\sigma_{\mathbb{T}}^1 \longrightarrow \sigma_H^0$ dual to the projection $pr_{\mathbb{T}}^H$.
τ_H	13.2	$(= tr_H^{\mathbb{T}} pr_{\mathbb{T}}^H)$ Standard generator of $[\sigma_H^0, \sigma_H^0]_1^{\mathbb{T}}$
θ	5.4	Generic notation for a map of nubs.
$T_{n(2)}^{n(1)}(\mathbf{a})$	19.4	Dual wide sphere in the semifree model.

U

U		Generic notation for a graded vector space.
$\mathcal{U}$		Generic notation for a universe, usually complete.

V

v	4.6	A twisting function $\mathcal{F} \longrightarrow \mathbb{Z}$ with finite support.
V		(1) Generic notation for a graded vector space.
	5.4	(2) Generic notation for the vertex of an object in the standard model.
$V(H)$		A one dimensional representation of $\mathbb{T}$ with kernel H.

W

W		Generic notation for a graded vector space.
we	B	The class of weak equivalences in a Quillen closed model category.

X

X		Generic notation for space or spectrum.
$X(\mathbb{T})$		The part of a $\mathbb{T}$-spectrum over $\mathbb{T}$: $\tilde{E}\mathcal{F} \wedge X \simeq \tilde{E}\mathcal{F} \wedge \Phi^{\mathbb{T}} X$.

$X(H)$		The part of a $\mathbb{T}$-spectrum over H: $E\langle H\rangle \wedge X$.
$X(\mathcal{F})$		The part of a $\mathbb{T}$-spectrum over $\mathcal{F}$: $E\mathcal{F}_+ \wedge X$.
$\hat{X}$		A fibrant approximation to X.
χ		The Euler class $1 - z$.
$\chi(V)$	4.6	The Euler class of V.

Y

Y		Generic notation for space or spectrum.

Z

z		The standard representation of $\mathbb{T}$.
zL		The subobject of cycles in a dg object K.
$\mathbb{Z}$		The integers.
$\mathbb{Z}((\chi))$		The ring $\mathbb{Z}[[\chi]][\chi^{-1}]$.
Z		Generic notation for space or spectrum.

Symbols

1		(1) The subgroup of $\mathbb{T}$ with one element. (2) The identity morphism.
$\ltimes$		(1) For differential graded objects, $A \ltimes B$ denotes an object with quotient A and kernel B; the differential is given by those of A and B together with $A \longrightarrow \Sigma B$. (2) Used in the induction functor $\mathbb{T} \ltimes_H X$.
$\circledast$	23.4	The left derived tensor product or right derived suspended torsion product.
$[\subseteq H]$		The family of subgroups of H.
$[\subset H]$		The family of proper subgroups of H.
$\tilde{\cdot}$		The unreduced suspension $S^0 * (\cdot)$.

APPENDIX E

Summary of models.

E.1. The standard model.

Refer to Chapter 5 for more details.

Ingredients: The following are fundamental

- the set $\mathcal{F}$ of finite subgroups of $\mathbb{T}$,
- the ring $\mathcal{O}_{\mathcal{F}} = C(\mathcal{F}, \mathbb{Q})[c]$ with c of degree -2 of operations (see Section 4.5)
- the set $\mathcal{E} = \{c^v \mid v : \mathcal{F} \longrightarrow \mathbb{Z}_{\geq 0}$ with finite support$\}$ of Euler classes (see 4.6.1)
- the ring $t_*^{\mathcal{F}} = \mathcal{E}^{-1}\mathcal{O}_{\mathcal{F}}$ (see 5.2.1)

Objects: Maps $(N \xrightarrow{\beta} t_*^{\mathcal{F}} \otimes V)$ of $\mathcal{O}_{\mathcal{F}}$-modules which become isomorphisms when $\mathcal{E}$ is inverted.

Morphisms: Squares

$$\begin{array}{ccc} M & \xrightarrow{\theta} & N \\ \beta\downarrow & & \downarrow\gamma \\ t_*^{\mathcal{F}} \otimes U & \xrightarrow{1\otimes\phi} & t_*^{\mathcal{F}} \otimes V. \end{array}$$

Nomenclature: In an object as above we call

- N the nub
- V the vertex
- β the basing map

Special objects: (See Section 5.4).

- $e(V) = (t_*^{\mathcal{F}} \otimes V \longrightarrow t_*^{\mathcal{F}} \otimes V)$ for a graded vector space V
- $f(T) = (T \longrightarrow 0)$ for an Euler torsion $\mathcal{O}_{\mathcal{F}}$-module T

Injectives: (See Section 5.5.) The objects $e(V)$ for a graded rational vector space V and $f(I)$ for an injective Euler torsion module I give enough injectives. The injective dimension of the category is 1.

Relationship to topology: (See Section 5.6).

- Given any $\mathbb{T}$-spectrum X,

$$\pi_*^{\mathcal{A}}(X) = \Big(\pi_*^{\mathbb{T}}(X \wedge DE\mathcal{F}_+) \longrightarrow \pi_*^{\mathbb{T}}(X \wedge DE\mathcal{F}_+ \wedge \tilde{E}\mathcal{F}) = t_*^{\mathcal{F}} \otimes \pi_*(\Phi^{\mathbb{T}}X)\Big)$$

  gives an object of $\mathcal{A}$ which is a complete invariant of equivariant homotopy type.
- There is a short exact sequence

$$0 \longrightarrow \mathrm{Ext}_{\mathcal{A}}(\pi_*^{\mathcal{A}}(\Sigma X), \pi_*^{\mathcal{A}}(\Sigma Y)) \longrightarrow [X, Y]_*^{\mathbb{T}} \longrightarrow \mathrm{Hom}_{\mathcal{A}}(\pi_*^{\mathcal{A}}(X), \pi_*^{\mathcal{A}}(Y)) \longrightarrow 0.$$

- There is an equivalence of triangulated categories

$$\mathbb{T}\text{-Spec} \simeq D\mathcal{A}.$$

Torsion functor: (See Chapter 20)

- $\hat{\mathcal{A}}$ is the category of all $\mathcal{O}_{\mathcal{F}}$-maps $N \longrightarrow t_*^{\mathcal{F}} \otimes V$.
- The right adjoint $\hat{\Gamma}$ to the inclusion $i : \mathcal{A} \longrightarrow \hat{\mathcal{A}}$ is defined (see 20.2.2) on $C = (P \longrightarrow t_*^{\mathcal{F}} \otimes V)$ by
  - $\hat{\Gamma}C = (P' \xrightarrow{\gamma'} t_*^{\mathcal{F}} \otimes V)$, where
  - $V' := \mathcal{E}^{-1}P(c^0) = \varinjlim_v (\mathrm{Hom}(S^v, C), c^*)$
  - There is a pullback square

$$\begin{array}{ccc} P' & \longrightarrow & P \\ {\scriptstyle \gamma'}\downarrow & & \downarrow \\ t_*^{\mathcal{F}} \otimes \mathcal{E}^{-1}P(c^0) & \xrightarrow{\tilde{\nu}} & \mathcal{E}^{-1}P. \end{array}$$

Tensor product: Given by tensoring nub and vertex over $\mathcal{O}_{\mathcal{F}}$:

$$(M \longrightarrow t_*^{\mathcal{F}} \otimes U) \otimes (N \longrightarrow t_*^{\mathcal{F}} \otimes V) = (M \otimes_{\mathcal{O}_{\mathcal{F}}} N \longrightarrow t_*^{\mathcal{F}} \otimes U \otimes V).$$

Hom functor: (See 22.6.1)

- With $B = (N \xrightarrow{\beta} t_*^{\mathcal{F}} \otimes V)$ and $C = (P \xrightarrow{\gamma} t_*^{\mathcal{F}} \otimes W)$ we take
  - $\hat{\mathbf{Hom}}(B, C) = (Q \xrightarrow{\delta} t_*^{\mathcal{F}} \otimes H)$ with
  - $H = \mathrm{Hom}(V, W)$
  - There is a pullback square

$$\begin{array}{ccccc} Q & & \longrightarrow & & \mathrm{Hom}(N, P) \\ {\scriptstyle \delta}\downarrow & & & & \downarrow{\scriptstyle \gamma_*} \\ t_*^{\mathcal{F}} \otimes \mathrm{Hom}(V, W) & \xrightarrow{\Delta} & \mathrm{Hom}(t_*^{\mathcal{F}} \otimes V, t_*^{\mathcal{F}} \otimes W) & \xrightarrow{\beta^*} & \mathrm{Hom}(N, t_*^{\mathcal{F}} \otimes W). \end{array}$$

- $\mathbf{IntHom}(B, C) = \hat{\Gamma}\hat{\mathbf{Hom}}(B, C)$.

E.2. The torsion model.

Refer to Chapter 6 for more details.

Ingredients: The following are fundamental

- the set $\mathcal{F}$ of finite subgroups of $\mathbb{T}$,
- the ring $\mathcal{O}_{\mathcal{F}} = C(\mathcal{F}, \mathbb{Q})[c]$ with c of degree -2 of operations (see 4.5)
- the set $\mathcal{E} = \{c^v \mid v : \mathcal{F} \longrightarrow \mathbb{Z}_{\geq 0}$ with finite support$\}$ of Euler classes (see 4.6.1)
- the ring $t_*^{\mathcal{F}} = \mathcal{E}^{-1}\mathcal{O}_{\mathcal{F}}$ (see 5.2.1)

Objects: Maps $(t_*^{\mathcal{F}} \otimes V \xrightarrow{q} T)$ of $\mathcal{O}_{\mathcal{F}}$-modules where T is an Euler torsion module.

Morphisms: Squares

$$\begin{array}{ccc} t_*^{\mathcal{F}} \otimes U & \xrightarrow{1\otimes\phi} & t_*^{\mathcal{F}} \otimes V \\ p\downarrow & & \downarrow q \\ S & \xrightarrow{\psi} & T. \end{array}$$

Nomenclature: In an object as above we call

- V the vertex
- T the torsion part

Special objects: (See Section 6.2)

- $e_t(V) = (t_*^{\mathcal{F}} \otimes V \longrightarrow 0)$ for a graded vector space V
- $f_t(T) = (t_*^{\mathcal{F}} \otimes \mathrm{Hom}(t_*^{\mathcal{F}}, T) \longrightarrow T)$ for an Euler torsion $\mathcal{O}_{\mathcal{F}}$-module T

Injectives: (See Section 6.3). The objects $e_t(V)$ for a rational vector space V and $f_t(I)$ for an injective Euler torsion module I give enough injectives. The injective dimension of the category is 2.

Relationship to topology: (See Section 6.6).

- Any $\mathbb{T}$-spectrum X is modelled by a dg $\mathcal{A}_t$-object $T(X)$, but $T(X)$ is defined indirectly via the standard model.
- $H_*(T(X)) = \pi_*^{\mathcal{A}_t}(X)$ by 6.6.1 where

$$\pi_*^{\mathcal{A}_t}(X) = \left(t_*^{\mathcal{F}} \otimes \pi_*(\Phi^{\mathbb{T}}X) = \pi_*^{\mathbb{T}}(X \wedge DE\mathcal{F}_+ \wedge \tilde{E}\mathcal{F}) \longrightarrow \pi_*^{\mathbb{T}}(X \wedge \Sigma E\mathcal{F}_+)\right)$$

- $\pi_*^{\mathcal{A}_t}(X)$ does not determine X up to homotopy in general; if it does we say X is *formal*. If $\pi_*^{\mathbb{T}}(X \wedge E\mathcal{F}_+)$ is injective then X is formal.
- There is an Adams spectral sequence 6.6.2
$$E_2^{*,*} = \mathrm{Ext}_{\mathcal{A}_t}^{*,*}(\pi_*^{\mathcal{A}_t}(X), \pi_*^{\mathcal{A}_t}(Y)) \Longrightarrow [X, Y]_*^{\mathbb{T}}$$
which collapses at E_3.
- There is an equivalence (6.5.1) of triangulated categories
$$\mathbb{T}\text{-Spec} \simeq D\mathcal{A}_t.$$

Bibliography

[1] J.F.Adams "On the groups J(X) – IV" Topology **5** (1966) 21-71

[2] M.Bokstedt, W.C.Hsiang and I.Madsen "The cyclotomic trace and algebraic K-theory of spaces." Inventiones Math. **111** (1993) 865-940

[3] J.-L. Brylinski "Cyclic homology and equivariant theories." Ann. Inst. Fourier, Grenoble **37** (1987) 15-28

[4] G.Carlsson and R.Cohen "The cyclic groups and the free loop space." Comm. Math. Helv. **62** (1987) 423-449

[5] M.C.Crabb "Invariants of fixed-point-free circle actions" London Math. Soc. Lecture Notes **139** (ed. S.M.Salamon, B.Steer and W.A.Sutherland) Cambridge U. P. (1989) 21-47

[6] V.Ginzburg, M. Kapranov and E. Vasserot "Elliptic algebras and equivariant elliptic cohomology." Preprint.

[7] T. Goodwillie "Notes on the cyclotomic trace." Draft mimeographed notes (1991)

[8] J.P.C.Greenlees "The geometric equivariant Segal conjecture for toral groups." J. London Math. Soc. **48** (1993) 348-364

[9] J.P.C.Greenlees "Tate cohomology in commutative algebra." J. Pure and Applied Algebra **94** (1994) 59-83

[10] J.P.C.Greenlees "Rational Mackey functors for compact Lie groups I." Proc. London Math. Soc. (to appear) 34pp.

[11] J.P.C.Greenlees "A rational splitting theorem for the universal space for almost free actions." Bull. London Math. Soc. **28** (1996) 183-189.

[12] J.P.C.Greenlees "Rational $O(2)$-equivariant cohomology theories." Fields Institute Communication Series, Stable Homotopy (ed S.O.Kochman and P.Selick) American Math. Soc. (to appear) 8pp.

[13] J.P.C.Greenlees and J.P.May "Derived functors of I-adic completion and local homology" Journal of Algebra **149** (1992) 438-453.

[14] J.P.C.Greenlees and J.P.May "Generalized Tate cohomology." Mem. American Math. Soc. **543** (1995).

[15] J.P.C.Greenlees and Hal Sadofsky "The Tate spectrum of v_n-periodic complex oriented theories." Math. Z. **222** (1996) 391-405

[16] I. Grojnowski "Delocalized equivariant elliptic cohomology." Preprint (1994)

[17] J.-P. Haeberly "For $G = S^1$ there is no Chern character." Contemporary Math. **36** (1985) 113-118

[18] L.Hesselholt and I.Madsen "On the K-theory of finite algebras over Witt vectors of a perfect field." Topology **36** (1997) 29-101.

[19] L.G.Lewis, J.P.May and M.Steinberger (with contributions by J.E.McClure) "Equivariant stable homotopy theory." Lecture notes in maths. **1213** Springer-Verlag (1986).

[20] J.E.McClure "Restriction maps in equivariant K-theory." Topology **25** (1986) 399-409.

[21] I.Madsen "Algebraic K-theory and traces." Aarhus Preprint **26** (1995).

[22] D.G.Quillen "Homotopical algebra." Lecture notes in Maths. **43** Springer-Verlag (1967)

[23] D.G.Quillen "Rational homotopy theory." Annals of Maths. **90** (1969) 205-295

Editorial Information

To be published in the *Memoirs*, a paper must be correct, new, nontrivial, and significant. Further, it must be well written and of interest to a substantial number of mathematicians. Piecemeal results, such as an inconclusive step toward an unproved major theorem or a minor variation on a known result, are in general not acceptable for publication. *Transactions* Editors shall solicit and encourage publication of worthy papers. Papers appearing in *Memoirs* are generally longer than those appearing in *Transactions* with which it shares an editorial committee.

As of December 31, 1998, the backlog for this journal was approximately 5 volumes. This estimate is the result of dividing the number of manuscripts for this journal in the Providence office that have not yet gone to the printer on the above date by the average number of monographs per volume over the previous twelve months, reduced by the number of issues published in four months (the time necessary for preparing an issue for the printer). (There are 6 volumes per year, each containing at least 4 numbers.)

A Copyright Transfer Agreement is required before a paper will be published in this journal. By submitting a paper to this journal, authors certify that the manuscript has not been submitted to nor is it under consideration for publication by another journal, conference proceedings, or similar publication.

Information for Authors and Editors

Memoirs are printed by photo-offset from camera copy fully prepared by the author. This means that the finished book will look exactly like the copy submitted.

The paper must contain a *descriptive title* and an *abstract* that summarizes the article in language suitable for workers in the general field (algebra, analysis, etc.). The *descriptive title* should be short, but informative; useless or vague phrases such as "some remarks about" or "concerning" should be avoided. The *abstract* should be at least one complete sentence, and at most 300 words. Included with the footnotes to the paper, there should be the 1991 *Mathematics Subject Classification* representing the primary and secondary subjects of the article. This may be followed by a list of *key words and phrases* describing the subject matter of the article and taken from it. A list of the numbers may be found in the annual index of *Mathematical Reviews*, published with the December issue starting in 1990, as well as from the electronic service e-MATH [**telnet e-MATH.ams.org** (or **telnet 130.44.1.100**). Login and password are **e-math**]. For journal abbreviations used in bibliographies, see the list of serials in the latest *Mathematical Reviews* annual index. When the manuscript is submitted, authors should supply the editor with electronic addresses if available. These will be printed after the postal address at the end of each article.

Electronically prepared papers. The AMS encourages submission of electronically prepared papers in $\mathcal{A}_{\mathcal{M}}\mathcal{S}$-TeX or $\mathcal{A}_{\mathcal{M}}\mathcal{S}$-LaTeX. The Society has prepared author packages for each AMS publication. Author packages include instructions for preparing electronic papers, the *AMS Author Handbook*, samples, and a style file that generates the particular design specifications of that publication series for both $\mathcal{A}_{\mathcal{M}}\mathcal{S}$-TeX and $\mathcal{A}_{\mathcal{M}}\mathcal{S}$-LaTeX.

Authors with FTP access may retrieve an author package from the Society's Internet node `e-MATH.ams.org` (130.44.1.100). For those without FTP access, the author package can be obtained free of charge by sending e-mail to pub@ams.org (Internet) or from the Publication Division, American Mathematical Society, P.O. Box 6248, Providence, RI 02940-6248. When requesting an author package, please specify AMS-TEX or AMS-LATEX, Macintosh or IBM (3.5) format, and the publication in which your paper will appear. Please be sure to include your complete mailing address.

Submission of electronic files. At the time of submission, the source file(s) should be sent to the Providence office (this includes any TEX source file, any graphics files, and the DVI or PostScript file).

Before sending the source file, be sure you have proofread your paper carefully. The files you send must be the EXACT files used to generate the proof copy that was accepted for publication. For all publications, authors are required to send a printed copy of their paper, which exactly matches the copy approved for publication, along with any graphics that will appear in the paper.

TEX files may be submitted by email, FTP, or on diskette. The DVI file(s) and PostScript files should be submitted only by FTP or on diskette unless they are encoded properly to submit through e-mail. (DVI files are binary and PostScript files tend to be very large.)

Files sent by electronic mail should be addressed to the Internet address pub-submit@ams.org. The subject line of the message should include the publication code to identify it as a Memoir. TEX source files, DVI files, and PostScript files can be transferred over the Internet by FTP to the Internet node e-math.ams.org (130.44.1.100).

Electronic graphics. Figures may be submitted to the AMS in an electronic format. The AMS recommends that graphics created electronically be saved in Encapsulated PostScript (EPS) format. This includes graphics originated via a graphics application as well as scanned photographs or other computer-generated images.

If the graphics package used does not support EPS output, the graphics file should be saved in one of the standard graphics formats—such as TIFF, PICT, GIF, etc.—rather than in an application-dependent format. Graphics files submitted in an application-dependent format are not likely to be used. No matter what method was used to produce the graphic, it is necessary to provide a paper copy to the AMS.

Authors using graphics packages for the creation of electronic art should also avoid the use of any lines thinner than 0.5 points in width. Many graphics packages allow the user to specify a "hairline" for a very thin line. Hairlines often look acceptable when proofed on a typical laser printer. However, when produced on a high-resolution laser imagesetter, hairlines become nearly invisible and will be lost entirely in the final printing process.

Screens should be set to values between 15% and 85%. Screens which fall outside of this range are too light or too dark to print correctly.

Any inquiries concerning a paper that has been accepted for publication should be sent directly to the Editorial Department, American Mathematical Society, P. O. Box 6248, Providence, RI 02940-6248.

Editors